A BOOK OF WAVES

The Lewis Henry Morgan Lectures
Robert J. Foster and Daniel R. Reichman,
Co-directors

A BOOK OF WAVES

Stefan Helmreich

DUKE UNIVERSITY PRESS
Durham and London
2023

publication supported by a grant from
The Community Foundation for Greater New Haven
as part of the **Urban Haven Project**

Printed in the United States of America on acid-free paper ∞
Project Editor: Ihsan Taylor
Designed by A. Mattson Gallagher
Typeset in Source Serif 4 and IBM Plex Sans
by Westchester Publishing Services

Library of Congress Cataloging-in-Publication Data
Names: Helmreich, Stefan, [date] author.
Title: A book of waves / Stefan Helmreich.
Other titles: Lewis Henry Morgan lectures.
Description: Durham : Duke University Press, 2023. | Series: The Lewis Henry Morgan lectures | Includes bibliographical references and index.
Identifiers: LCCN 2022044728 (print)
LCCN 2022044729 (ebook)
ISBN 9781478020417 (paperback)
ISBN 9781478019947 (hardcover)
ISBN 9781478024538 (ebook)
Subjects: LCSH: Human ecology. | Ocean and civilization. | Ocean waves—Climatic factors. | Sea level—Social aspects. | Ocean—Philosophy. | BISAC: SOCIAL SCIENCE / Anthropology / Cultural & Social
Classification: LCC GF65 .H45 2023 (print) | LCC GF65 (ebook) | DDC 304.209162—dc23/eng20230407
LC record available at https://lccn.loc.gov/2022044728
LC ebook record available at https://lccn.loc.gov/2022044729

Cover art: Clifford Ross, *Digital Wave* (still, detail), 2017.
Courtesy of the artist.

CONTENTS

FOREWORD

The Lewis Henry Morgan Lectures at the University of Rochester were inaugurated in 1962 and have been presented annually ever since, with the exception of 2020, when they were postponed due to the COVID-19 pandemic. The lectures commemorate the contributions of Morgan (1818–81) to the University of Rochester, his support for the founding of a women's college, and his legacy in anthropology, as reflected in the topics of the first three lectures, which focused on kinship (Meyer Fortes, 1963), Native North Americans (Fred Eggan, 1964), and the origins of the state (Robert M. Adams, 1965).

As the oldest and longest-running anthropology lecture series in North America, the Morgan Lectures have produced some of the most influential texts in modern anthropology. To name but a few: Victor Turner's *The Ritual Process: Structure and Anti-Structure* (1966), Marilyn Strathern's *After Nature: Kinship in the Late Twentieth Century* (1992), and Marisol de la Cadena's *Earth Beings: Ecologies of Practice across Andean Worlds* (2015). A view of the lectures after sixty years illustrates the ways that anthropologists have moved well beyond Morgan's Enlightenment roots and how they have expanded on and reconsidered the topics that preoccupied him: kinship and social organization, political economy, Indigenous peoples, and cross-cultural comparison.

Stefan Helmreich presented the Morgan Lectures over three days in October 2014. His public lecture was followed by a daylong workshop during which members of the Department of Anthropology, along with invited scholars, discussed several draft chapters of the manuscript that became this book. Formal discussants included Michael Fortun (University of California, Irvine), Anand Pandian (Johns Hopkins University), Daniel Reichman (University of Rochester), Nicole Starosielski (New York University), and Holly Watkins (Eastman School of Music).

Helmreich's lectures track the figure of the wave across various sociocultural domains, including oceanography, climate modeling, maritime infrastructure, statistics, music, sports, and art. His work brilliantly queries the relationship between waves' material dimensions and the myriad formalizations meant to capture them, which, he argues, express as well as format different understandings of time, nature, and culture. Waves can be rhythmic, regular processes that point toward a knowable future or unruly, destructive forces that portend an unknowable, unpredictable set of threats. For populations facing the imminent dangers of rising seas due to climate change, the material, representational, predictable, and unpredictable meanings of waves all interact to orient how people think about what is to come on a planet in precarity. In this regard, the wave serves as a key symbol to think about human responses to climate change.

The reality of climate change has led to a paradigm shift in anthropology, transforming our understandings of temporality, nature, and culture in the Anthropocene. Lewis Henry Morgan lived through another paradigm shift in anthropological notions of time, during which the biblical account of human origins was challenged by recognition of the Earth's long history (Trautmann 1992). Morgan's sometime interest in the intelligence of nonhuman species, particularly the American beavers he studied in Michigan's Upper Peninsula, was thus shaped by a scientific and cultural reassessment of the place of humanity in the natural world, and it gave a Darwinian charge to his questions: What separated humans from other animals? Did nonhuman animals possess intelligence, sociality, and culture? In the beavers Morgan found nonhumans that transformed their landscape to adapt to changing water levels. It was their architectural skill—their ability to construct sophisticated canals, dams, channels, and embankments—that convinced Morgan of their humanlike intelligence. In 1868, he even described an event in which a beaver dam was cut in half by newly laid railroad track, lowering the water to its original level, after which the beavers "immediately repaired the breach" (Morgan 1868, 102), restoring the water to a suitable level for them—which, of course, was unsuitable for the railroad engineers (Cheng 2006; Feeley-Harnik 2021, 21). Although Morgan famously referred to the beavers as "mutes," we might also think of them as nonhuman hydroengineers.

Morgan worked as an attorney for a railroad company that developed Michigan's Upper Peninsula for iron mining, and he personally profited from the endeavor. The railroads and mines were built on land that had been unjustly ceded by Chippewa bands only a decade before the railroad work began. And Rochester, where Morgan resided, then, as now, sits on the unjustly occupied land of the sovereign Onöndowa'ga:' (Seneca) Nation of the Haudenosaunee (Iroquois) Confederacy, vital to acknowledge since Morgan's anthropological career depended on a mix of advocacy for and racializing and assimilationist views of Indigenous peoples in what settlers, in the ongoing wake of colonialism, call North America. The very processes of industrialization and extractivism in which Morgan participated, and that he heralded as "progress," have had long legacies, contributing, now, to the climate disaster. The coastal engineers whom Helmreich describes in segments in this work are seeking to undo—or live amid—the damage. How the waves with which they, as well as their oceanographer colleagues, reckon may themselves act or be enlisted as agents of hydroengineering—amplifying, damping, or rescripting sea level rise on a climate-changed planet—remains to be seen.

DANIEL R. REICHMAN
ROBERT J. FOSTER
Department of Anthropology
University of Rochester
May 2022

PRELIMINARY

FORWARD AND BACK

The coronavirus arrived in waves. Rolling graphs of case and death rates from COVID-19 came chronicled and projected in uneven lulls and surges. I finished writing this book during the pandemic, and as I revised my stories of ocean wave science—accounts I had gathered throughout the 2010s—they seemed to slip into another era, even as I knew that one of the tales I wanted to tell, of sea level rise and climate calamity, was still being written, inexorably, in and by the waves. Like many sequestered at home, I read. The edition of Albert Camus's *The Plague* that I obtained was fronted by a photo, taken as if from a sea cliff, showing a span of spray-crested waves, at once repeating and sporadic (plate 1). What were we meant to see? Epidemic waves? The dissolution of individual fortune in the stir of aggregation? A vaguely sinister postcard image of the kind of watery pathway that brought French colonialism to *The Plague*'s North African setting, the waves legible as a medium of politics? I took the image (which turned out to be from Australia's Bondi Beach, famous for surfing and known to its Aboriginal traditional owners, the Gadigal people of the Eora Nation, as an early site of British imperial incursion) to be about the riddle of reading itself, that puzzle in pattern recognition, in discerning meaning, that preoccupies me and the wave scientists and publics who are the subjects of this book. The question of how to interpret the roil of the world—moving nowadays through a surge of syndemic crisis, ascendant authoritarianism, and renewed nuclear threat—requires, like wave forecasting, looking backward to look ahead, orienting to how the wakes of history tell on the present. But it also demands attention to the unforeseen; to that which, like suddenly breaking waves, may overwhelm expectation. Waves, this book proposes, may be oracular forms and forces through which to apprehend today's climatological and political sea changes, as oceanic and social processes increasingly churn into one another, running forward and back.

PREFACE
WAVE CLUTTER

Imagine a rippling sea. Imagine, next, the invisible waviness of a radio transmission skittering along the surface. As radio waves propagate over ocean waves, where seawater and atmosphere meet, oscillations of multiple sorts materialize. Often, the waves, both watery and electromagnetic, interfere with one another, appearing on ship radar readouts as a mess of squiggles.

One day in 2017, finding my feet on a research vessel operated out of the Scripps Institution of Oceanography, in La Jolla, California, I was directed by a sea scientist to a sensor display of this scribbly disorder. I already knew what this chaos was called: wave clutter. I had first heard the phrase a few years back from the poet Paul Muldoon on a seagoing study abroad program on which he was leading writing classes and I was teaching cultural anthropology (on leave from my usual post, at the Massachusetts Institute of Technology). Muldoon had just returned from a tour of the bridge, and, taking me aside, asked, in a conspiratorial voice, "Have you heard of this? Wave clutter?" Muldoon, a Pulitzer Prize–winning poet, had a way of making ordinary words sound like dispatches from a secret history of the English language. *Wave clutter*. I turned the phrase over in my head: *wave* suggesting a rising and diving flow; *clutter*, a crosscutting confusion. The term came back to me as I swayed slowly before the Scripps ship's radar screen, which was visualizing waves in the waters around us, thirty-five miles off Malibu.

Waves of many sorts had at that time been scrolling steadily through news headlines, and this book, *A Book of Waves*, arose in the wake of the multifarious, dramatic waves that had been capturing wide, sometimes world, attention since around the turn of the millennium. A partial list:

The first officially measured instance of a "rogue wave," an eighty-four-foot wave that hit an oil platform in the North Sea in 1995 and established the real, not mythical, status of such "freak waves," now defined as twice the size of their immediate neighbors

The 2003 foundation of Wavestar, a Danish company dedicated to generating fossil-free electricity from the energy of ocean waves arriving at the shores of the Jutland Peninsula

The 2004 Indian Ocean tsunami, which killed some 227,000 people

The 2005 storm surges of Hurricane Katrina, which devastated New Orleans and claimed the lives of more than 1,800 souls

The 2008 and 2014 waves that flooded the Marshall Islands, leading to declarations of a state of emergency and ongoing anticipation of a sea level rise that could inundate the archipelagic republic

The 2009 Green Wave postelection protests in Iran, followed by what the anthropologist David Graeber (2013, 64) called "a wave of resistance sweeping the planet" that he named as beginning in Tunisia in January 2011, moving from Egypt to Libya to Yemen to manifest as the Arab Spring and traveling on to the Occupy protests that materialized later that year across the United States. In 2016, the election of Donald Trump to the US presidency ushered in a counterwave, a courier of a worldwide tide of populism and authoritarianism, whose full effect is yet unknown

The Tōhoku tsunami of 2011, which hit the east coast of Japan, killing some twenty thousand people and causing a level 7 meltdown at the Fukushima Daiichi Nuclear Power Plant

The 2011 surfing of a record-breaking 23.8 meter wave in Nazaré, Portugal

The 2012 approval by the US Food and Drug Administration of a Wi-Fi enabled implantable cardioverter defibrillator, which administers shocks to the human heart to keep its ventricular waves on pace, but which can also be jammed by ambient electromagnetic waves

The 2012 waves of Hurricane Sandy, which hit the Caribbean and then left New York and other East Coast cities in the dark, an Atlantic calamity superseded by 2017's Hurricane Maria, which killed thousands in Puerto Rico

The gravitational waves detected by cosmologists in 2016, a burst consequent upon the collision of two black holes in a distant galaxy

The 2017 opening of the surfer Kelly Slater's artificial wave pool in inland California

Finally, to flip the script into a future I could not in 2017 see: the waves of coronavirus infection that began in 2020

Such a list may seem unaccountably cluttered—joining different orders of things; fusing the disastrous, the mundane, and the recreational; collapsing diverse modes of description; breaking across various planes of analysis, and including not just the oceanographic, but also the social, biological, and cosmological. All of these waves, however, manifest as energies that in some way gather up or substantiate anxieties, terrors, and, sometimes, optimisms about the shape of history and of the future.

In the years I have spent thinking about waves—motivated, initially (full disclosure) by enthusiasm for the sport of bodysurfing, later by an interest in ocean environmentalism, and still later by an anthropological commitment to examining the cultural and political making of natural-science knowledge—I have been fascinated time and again by how waves have been considered both abundantly material and abstractly formal and by how they have been described as phenomena that can unfurl in diverse media (electromagnetic, sonic, oceanic), even as water waves remain the reference example. As I researched this book, in the context of a world in growing turbulence and uncertainty, I came to see something more. I came to understand waves as objects through which people seek to apprehend time, to foretell futures: ecological, scientific, political, local, planetary.

Waves are forms, patterns, and material carriers of change, sometimes regular and periodic, sometimes abrupt and irreversible. The futures waves bring may unfold with slow inexorability (sea level rise) or disastrous speed (storm surges, tsunamis), in synchrony with infrastructural transformation (the shoring up or falling apart of coastal defenses), and in uneven resonance with the churn of collective social process (including intensified currents, these days, of protest, migration, and pandemic relay). If the coming decades on our superheating planet will see increased ice melt, urban flooding, redrawn coast lines, and accompanying courses of social rearrangement, waves will be heralds of those dynamics, symbols and forces of future sea changes.[1]

This book centers an anthropological attention on wave science—and primarily, though not exclusively, ocean wave science, a science that has been organized around projects of maritime, civil, military, recreational, and infrastructural planning. Other books offer more traditional institutional, disciplinarily internal histories of the science, as well as more technical primers on how waves are known in oceanography, hydrodynamics, and mathematics (see, e.g., Parker 2012; Pretor-Pinney 2010; Zirker 2013). My account is based on ethnographic research I conducted among oceanographers, coastal engineers, programmers, surfers, and others, mostly in Europe and North America, though also with scientists in Australia, Japan, and Bangladesh. I followed them as they worked to understand the Earth's wavescape through fieldwork at sea, wave modeling in the lab, and, increasingly, computational simulation. I examine how scientific portraits of waves have transformed over the past century or so in calibration with shifting methods of technoscientific representation, as well as with changing sociopolitical concerns and demands, and I argue that waves should be approached not only as natural phenomena but also as culturally significant entities.

A Book of Waves understands waves to be not only material processes of energy propagation across space or of vibration at some frequency. Waves are also made manifest through abstractions crafted by scientists who decide what will count as wave activity, whether in a passive medium (as with water waves, sound waves), an excitable medium (as with waves in biological tissue), or a vacuum. Waves are empirical and conceptual phenomena both. Empirical: there they are. Conceptual: discerning that *there they are* requires abstraction, whether accomplished by an electronic transducer visualizing radar returns at sea, a gravitational wave detector listening for sounds of the early cosmos, an electrocardiogram mapping heartbeats, a watercolor painter seeking to capture the evanescence of a cresting wave, a network of coastal buoys detecting incoming storm surges, a surfer judging potential rides, or a social theorist seeking to make a claim about epochal political change.

In her review of Paul Muldoon's *To Ireland, I* (1998), Clair Wills (2000, 6) suggests that, for Muldoon, "the political dimension of literature . . . is not to be found in public position-taking, nor purely in a willful disruption of linguistic codes, but in a kind of interference of wavebands." The chapters here, some growing from Lewis Henry Morgan Lectures I delivered in the

Department of Anthropology at the University of Rochester in 2014, are very much tuned to the interference of wavebands, to worldly clutter, though the reader will also find sharp moments of position-taking and willful disruption in the service of understanding wave science, like any science, as fully immersed in the disputations of human life and politics. I treat wave clutter not as confusion fully to be cleared up or as always a generator of unwelcome artifacts but, rather, as indicative of the inextricability of the world and its representation.

A Book of Waves presses a bit against the coherence of the "book," against the singular form of the monograph. Ethnographic chapters on ocean wave science move from the Netherlands to San Diego, then to Oregon and Japan, to Washington, DC, and to the Bay of Bengal (by ricocheting way of Australia), and are interleaved with shorter, interstitial pieces I have gathered into *sets*. A *set*—a term of art from surfing—is a group of waves of like speed that travel together, entrained by a common wavelength. In this book, each set contains three essays that reapproach, from varied angles, themes in the ethnographic chapters they follow. They reflect on waves in folklore, surf culture, music, movies, painting, forensics, environmental disaster, and more. Some (typically the final entry in a set) consider waves other than ocean waves—gravitational, cardiac, social—examining the rippling exchange of analogy among natural, cultural, and social domains.[2] These essay sets offer thematic refractions—or, better, diffractions, interferences of wavebands, constructive clutter.[3]

As I set to writing this book, I had already spent a lot of time with ocean scientists in work toward *Alien Ocean: Anthropological Voyages in Microbial Seas* (Helmreich 2009). In that book, I examined how research on marine microbes in the early 2000s was reshaping how marine biologists viewed the ocean, coming to see the sea as not just populated by microorganisms but as, to some extent, made of them—made, that is, of life evolutionarily enduring, but also of life damaged and at risk. The global state of the sea, I came to see, may be illuminated through tracking the making of knowledge about its composing parts. Just so, the waves of wave science—whitecaps, swells, breakers, surges, rogues—may have much to reveal about the social, cultural, and political state of this beleaguered ocean planet and its future.

Introduction

Significant Waves

THE SUN HAD NOT YET SET. Overlooking the Pacific Ocean from the chalky cliffs of La Jolla, California, I was talking with the ninety-seven-year-old oceanographer Walter Munk at his house, a mid-century modern perched just north of the Scripps Institution of Oceanography. Munk, born in Vienna in 1917, arrived at Scripps in 1939—the same year he became an American citizen; the same year the Nazis annexed his native Austria—and was soon recruited into projects of ocean wave prediction for World War II. Working with colleagues from California to Cornwall, he forecast the forms and forces of waves breaking on the beaches of Normandy, providing critical information for the amphibious landing of Allied troops in northern France on June 6, 1944—D-Day. When the Cold War heated up, he was tapped by the US Navy to undertake measurements of wave action in the western Pacific, calculating the terrible wake of experimental detonations of nuclear bombs in Micronesia. In the 1960s, he led a team tracing the storm-born Southern Hemispheric origin of the swells that hit California, knowledge still relied on by weather forecasters and the surfing faithful. A few months before our conversation, Munk had joined in a public tête-à-tête with the Dalai Lama, the Tibetan Buddhist monk and spiritual leader, on a matter of common concern: climate change and sea level rise. And the day before we spoke, in August 2015, the *New York Times* published a biographical profile, calling Munk the "Einstein of the Oceans" (Galbraith 2015), a designation that, in an age when the sea's ecological health is under siege from anthropogenic insult, summoned up the significance of the ocean as a problem space for physics, with the Einstein comparison suggesting pioneering genius, a historical implication in deadly technological enterprise, and, perhaps, a weathered arrival at worldly wisdom (see also Munk and Wunsch 2019).

Munk's house was called "the Seiche" (Swiss French for "undulating wave"), a name proffered by his late first wife's mother in 1954, designating a form of wave—one that oscillates back and forth in a bounded space—that stands for a contained dynamism. I had found Munk in a contemplative mood. As he adjusted his hearing aid to tune me in, I scouted for a way to ask about the relation, for him, among waves; science; and the politics of war, peace, and everyday life. I started with a technical question, asking him about the difference between studying waves in the ocean and in the controlled environment of the laboratory, or even, these days, on a computer. He declared that he had always been "on the field side," then said something remarkable:

> I've often asked this question: if we met somebody from another planet who had never seen waves, could he dream about what it's like when a wave becomes unstable in shallow water? About what it would do? I don't think so. It's a complicated problem. If you asked the best mathematicians who have worked with waves—who have never been in a lake or in an ocean—what happens to waves when they come to the coast, I doubt they could calculate it.[1]

No mathematical abstraction alone, Munk urged, could properly predict a wave's breaking. Thinking about wave dynamics requires conceptual work, to be sure, but there is something empirical, irreducible, about waves in the wild. Putting on his glasses to look out over the dusking Pacific, Munk directed my attention to bands of waves arriving from the horizon, heading toward us in neat lines, and continued:

> I've never gotten tired of watching waves. And that's the amazing thing. There's sufficient variability. I've been in certain bathhouses where they make waves by plungers. The ones that I saw were single-frequency—terribly dull, one wave like the other. The interesting thing is waves

that are not either so stochastic that they have no predictability associated with them nor so regular that one is like the other. In between, you get the most interesting things. Some degree of predictability, but certainly no certainty.

"What appeals to you about that?" I asked. Munk leaned forward. "That's a question for *you*, the anthropologist."

Munk's flipping of the interview script set me reorienting. I looked down at the La Jolla waves, recalling my young adult years as a bodysurfer trying to read them, seeing in them, from my vantage as an East Coast transplant, some Pacific promise of an open, unfurling future, unaware then of their historical and ongoing roles as watery supports for colonial adventure and war, as elements in calculations around coastline development, and as media in the sea changes of climate transformation—all matters that now, as an anthropologist studying the culture and politics of science, shape my seeing of the sea. Munk's rejoinder, pointing to the matter of how wave scientists assign cultural value to the boundaries between predictability and unpredictability, unlocked for me a set of social questions. At stake in ocean wave science and its dedication to forecasting are orientations to time and to the future—futures to do with coastal infrastructure, beach and sea recreation, the logistics and ongoing projection of military power (US and otherwise), disaster preparedness, atmospheric transmutation, the seaborne trajectories of plastic and other pollutants, human and multispecies pathways of shipping and migration, marine insurance bureaucracies, the harnessing of ocean energy, and much more.[2] Scientific accounts of waves not only offer modes of analyzing oceanic process; they are also conditioned by the frames of anticipation and value—national, regional, hemispheric—within which such reading takes place.

Such frames, for wave science, build on a scaffold put together, in one canonical outline, by European, Enlightenment, and colonial preoccupations. Roll back to the Italian polymath Leonardo da Vinci, who, writing in his journal at the turn of the sixteenth century about "the numberless waves of the sea" (see Baskins 2010), delivered a cosmologically expansive vision of the sea as an immeasurable force, one of which he was famously terrified, inking toward the end of his life image after image of the world ending in swirling deluge (figure I.1).

Western mariners and scientists later strove to bring this realm into the sphere of the accountable, though they were hardly the first, or alone.[3] From 1405 to 1433, the Ming dynasty mariner Zheng He created sailing charts

FIGURE I.1 Detail of Leonardo da Vinci, *A Deluge*, ca. 1517–18, pen and black ink with wash on paper. Royal Collection Trust, London.

that outlined ocean winds between South India and East Africa (Pereira 2012). The fifteenth-century Arab navigator and cartographer Ahmad ibn Mājid wrote works on the currents, tides, and winds of the Indian Ocean, naming breaking waves (*ouqod al-ma*) "sea signs" (*isharat*) of close-by coasts and describing a phenomenon later translated as the "wave of the Cross (Southern Cross)," which enabled trade across the monsoon ocean ([1490] 1971, see also Aleem 1967; Al Hosani 2005). Sixteenth-century fishers on the West African coast of what is now Ghana designed dugout surf canoes (*ali lele* in the Fanti language) to ride waves safely in to shore (Dawson 2018). And wayfinders and surfers in the Pacific developed techniques of navigation and surf riding (*he'e nalu* in Hawaiian) pitched to a range of wave configurations (see Genz 2016; Walker 2011).

Well before the rise of oceanography, back in Europe, researchers in fluid mechanics had been describing waves as moving patterns of crests and troughs—patterns that could be characterized by their wavelength (horizontal distance from crest to crest), height (vertical measure from crest to trough), and period (the time it took for a crest to advance one wavelength).[4] A roll call of European mathematicians including Isaac Newton, Leonhard Euler, Joseph-Louis Lagrange, Baron Augustin-Louis

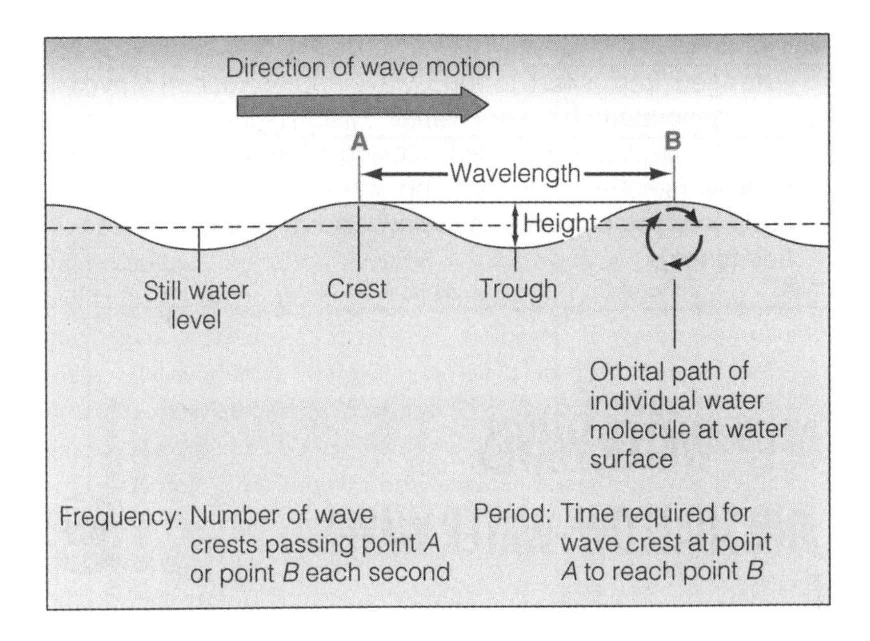

FIGURE I.2 "The Anatomy of a Progressive Wave." From T. Garrison 2005, 230, fig. 10.2.

Cauchy, Heinrich Martin Weber, and Lord Kelvin made formal models of wave action—from field observations, in glass wave tanks, and using calculus (with infinitesimals a tool for bounding Leonardo's "numberless waves" [Craik 2004]). Water in waves, they emphasized, did not hurtle itself forward but, rather, relayed a kinetic form. Water particles looped up and down, back and forth, beneath a traveling, progressive oscillation. With the rise of wave theories of light and sound, ocean waves came to be understood as a specific genre of energy transmission, working through processes of compression and rarefaction and prone to reflection, refraction, and diffraction.[5] But while sound and light waves moved every which way, ocean waves were bound by gravity and atmospheric pressure, organized around that wriggling, horizontally oriented boundary that divides the realm above (air, mostly) from the world below (water).

A now standard water wave diagram is a side-view line drawing of a full cycle of an individual wave (figure I.2) in deep water, in forward motion, not yet breaking (see figure 4.2). The diagram emphasizes that it is the pattern, not the water, that is the wave. That pattern unfurls a sine wave: a description that results from treating the line traced out by wave motion as graphing the travel of a point around a circle—an abstraction that renders the wave a kind of nonhuman inscription, making water the author of an automatic nature writing.[6] But the canonical wave formalism also bears traces of cultural technique. Beneath the wave, the coiling motion of water under curvy crests in circular motions is called an "orbit," a concept derived from work in celestial mechanics by the nineteenth-century astronomer George Biddell Airy. Pointing to origins for wave idioms inspired by clockwork mechanics, the physicist J. B. Zirker (2013, 16) offers, "Under the surface, a traveling water wave looks like the inside of a fine clock, filled with carefully synchronized 'gears.'" (The analogy is inexact. An orbiting water particle never returns *exactly* to a previous point but slowly edges forward in a wave dynamic called "Stokes drift," a process described in 1847 by the University of Cambridge mathematician George Stokes.) And the correlation between the wavelength of the wave and the period it takes for a crest to travel that distance was worked out using pendulums as a model (where the length of a cord supporting a pendulum weight was analogous to wavelength). Wave science is stocked with comparisons to mechanistic process, posing waves as kinds of mechanisms, even machines. The individuality of the "wave," meanwhile, has been an important conceptual fiction, a way to keep comprehension uncluttered. It has also been a building block for an account of waves as traveling in trains and groups and as belonging to populations, underwriting a statistical framing elaborated in the mid-twentieth century and central for wave science since.

Here is a quick primer on the maritime histories that prepare the wave science elaborated by people such as Munk: in the eighteenth century, the British Parliament sought methods for finding longitude at sea. In recognizing John Harrison's marine chronometer as the solution, it came to envisage the gridded globe as a kind of clock face (Sobel 1995). In 1835, the Cambridge natural philosopher William Whewell—who coined the word *scientist*—led a "great tide experiment," organizing seven hundred "associate laborers, including dockyard officials, harbormasters, expert calculators, tide table makers, and professional military men associated with the trigonometric and coastal surveys of the British Navy" in a project that "measured, tabulated, graphed, and charted the tides around the world at exactly the same time." It was a project that enabled the extension of British colonial enterprise, calibrating ocean temporality—time and tide (with tide a kind of giant, Earth-hugging wave)—to imperial objectives (Reidy 2008, 8–9). The United States in the 1840s saw the naval officer Matthew Fontaine Maury (1860, 343) leading projects to map the oceans' currents and winds, seeking to create "mile-posts . . . set up on the waves" to make ocean currents and winds into infrastructure for extending national commerce—and, notably, for Maury himself, who defected to the Confederacy in 1861, the institution of slavery, which he advocated exporting to South America (Hardy and Rozwadowski 2020; Hearn 2002). What the geographer Philip Steinberg (2001) has described as a European model of the ocean as a "great void" between nations acquired ever sharper surface textures. Shipbuilders had been paying close attention. "From at least the sixteenth century onward," writes Christina Sharpe (2016, 40), "a major part of the ocean engineering of ships has been to minimize the bow wave and therefore to minimize the wake," meaning that wave science emerged in calibration with merchant, military, and colonial desires for seafaring speed, particularly in the Atlantic. In the 1860s, knowledge of waves became crucial for the laying of the submarine transatlantic telegraph cable (see Starosielski 2015). By the time of the publication of Sir Horace Lamb's *Hydrodynamics* in 1879, the study of water waves had become an established science, and by 1934, with Vaughn Cornish's *Ocean Waves and Kindred Geophysical Phenomena*, oceanography had its still more specific enunciation. And just as longitude and oceanic wind tracks became thinkable through techniques of their time (clocks, railroads), so would ocean waves become readable to scientists using conceptual schemes made available by twentieth-century technologies and media (photography, film, computing).

In response to Munk's extraterrestrial thought experiment—his provocation on mathematically minded aliens—I told him I sought to understand how wave scientists, from the mid-twentieth century to now, navigated the fit between abstraction and the world, the relation of formal knowledge to the phenomenology of ocean experience. I wanted to know how scientists had ushered waves, as material forces and as icons, indexes, and symbols of watery process, into technical representation. What historical, political, and cultural conditions had been bound up with rendering ocean waves systematically knowable, countable, legible—where, for whom, and to what ends? How, in turn, did the waves of wave science, mathematically and diagrammatically apprehended, curl back into different communities' intuitions about and encounters with ocean worlds, politics, and futures?

My questions, I continued, were about the *significance* of waves—which I fully intended to resonate, I told Munk, with a term of art he had early introduced to wave science: *significant wave height*. Munk's coinage of that phrase, in 1944, emerged from his work with Marine Corps steersmen, whose eyeballed (and, he thought, exaggerated) estimates of wave height he hoped to square with scientific measure. Munk told me about his conversations with Marine coxswains training off the coast of Oceanside, California, as they prepared for their invasion of Normandy:

> We'd ride in their boats, and we would ask them, "What do you call the wave height today?" Now, statisticians talk about the *root mean square* of . . . but it was clear that a marine wouldn't give you a number that was *higher* than the root mean square of. I made the policy decision that it would be easier to accept their statistics than to try and teach them mathematics. So I invented using *significant wave height* to accommodate the community that works with waves. And for some reason that has taken tremendous hold.[7]

Significance, of course, is about meaningfulness, and I am concerned in this book with how, for whom, and at what scales, from personal to planetary, waves become meaningful. I am interested in how scientific measurement, monitoring, and modeling have informed that meaningfulness, and I am curious about how scientific research into waves has been shaped by and, in turn, has shaped culturally contoured accounts of waves as unceasing, wild, ephemeral, meditative, useful, inexorable, sublime, terrifying.

I argue this: oceanographers' and coastal engineers' apprehensions of waves encode ideas about—orientations toward—time, nature, and culture. These might bespeak a vision of the past as a horizon of eternal and recurring cycles, which may be used as a template for prediction. They might pose waves as disorderly entities to be tamed by shore and maritime structures crafted in coordination with judgments of wave height probabilities and risks over year-, decade-, or century-long spans of time. One generation of scientists might see waves as belonging fully to the order of nature, a ready-made matrix for seafaring activity, while a next comes to see them as a medium carrying the historical, material freight of anthropogenic ocean harm (radiation, pollution). Engineers may treat waves as amenable to abstractions that can be scale modeled and time stepped in the lab, or *in silico*, only to find waves in real time unfold faster, slower, or more nonlinearly than anticipated. The dynamics of wave-land interactions in one part of the world—due to sea level rise or planetary warming—may be construed as heralds of what is coming elsewhere. Waves broker negotiations between the foreseen and unforeseen, with their breaking the symbol of that uncertain arbitration, which may carry the heaviness of storms, hazards to human construction, and, these days, a future of climate transmutation.

Waves may also carry memory. I take lessons from the Black studies scholar Christina Sharpe's *In the Wake*, which presses readers to remember "the transverse waves of the wake" (2016, 57) of Atlantic slave ships. For Sharpe, following the Saint Lucian poet Derek Walcott, *the sea is history*—an archive of imperial violence, an unmarked grave—and waves, particularly ship-made waves, may be read as its haunting inscriptions.[8] Living "in the wake" is to recall the tracks of oceanic and human damage and to bear witness, to mourn. Waves, in repetition, may keep recollection, even trauma, alive. They may also mix with unexpected futures. The Barbadian poet Kamau Brathwaite (1999) captures this doubleness with his notion of "tidalectics," which refuses the resolution of the dialectic (thesis, antithesis, synthesis), turning instead to the back-and-forths of watery flux, pasts and futures stirring into one another (see also Hessler 2018).[9]

In remembrance, in hope, in fear, in anticipation, waves are significant.

Such significance, to adapt Munk's phrasing, takes a particular form in the practices of "the community that works with waves"—and there exist these days, as ever there have been, many such communities. Meteorologists, fishers, shippers, surfers, coastal engineers, climate activists, maritime human rights advocates, and many others take waves (their shapes, frequencies, powers, material effects) as matters of concern and, in so doing, read waves through measures and abstractions developed by scientists.

Let me introduce today's wave science community, then, through an account of a conference where I first met many of its members and from which I set out for anthropological fieldwork in the Netherlands, the United States, Australia, and, through videoconferencing, with scientists in Japan and Bangladesh. I follow this orientation with a first-pass account of how wave scientists think about the effects of human action on the ocean, a topic that leads me to ask what the uneven changes associated with the lately proposed geological epoch of the Anthropocene (ancient Greek *anthropos*, "human," plus *cene*, "new") look like from the sea. I deliver, next, an explication of what it means (for scientists, for me) to *read waves*—to interpret something that is not written. Scientific readings unfold at sea, in labs, online—in watery and communications media, mobilizing abstractions that have real-world consequences and that sometimes come to be read into the nature of the wavy world itself. Those readings carry concepts of time—recurring, irreversible, momentary, planetary—as well as loyalties to particular framings and schedules (e.g., national, global, market-based, humanitarian), all of which implicate visions of the future as a time of coherence and continuity, though also, increasingly today, of rupture, breaking.[10]

Wave Science, an Orientation

Forecasting Dangerous Sea States, the Thirteenth International Workshop on Wave Hindcasting and Forecasting, was held in Alberta, British Columbia, in October 2013.[11] Oceanographers, mathematicians, coastal engineers, ship and platform designers, and meteorologists gathered in the château-esque halls of the Banff Springs Hotel, a UNESCO National Heritage Site, to present results on monitoring, modeling, and managing hazardous wave activity, particularly the sort on the rise with climate change, including storm surges consequent on hurricanes.

When I arrived, some eighty or so participants were circulating through the hallways; they had come mainly from Australia, Canada, France, Germany, Japan, the Netherlands, the United Kingdom, and the United States, with a handful from elsewhere (Italy, Malaysia, Mexico, Portugal, Russia). The organizers opened the workshop by observing that its first day fell on the one-year anniversary of Hurricane Sandy, the 2012 superstorm that brought so much devastation to the Caribbean and Atlantic United States. The meteorological past was invoked as a charter for conversation about the future, underscoring how *hindcasting*—retrospective, model-checking prediction using past data—prepares the way for forecasting. Possible futures become visible to scientists when they think backward, when they frame what they call *inverse problems,* scenarios about *what will have had to have been the case* for particular outcomes to materialize.

Over the next days, dozens of talks offered angles on capturing global wave climate, long-term trends in wave weather. Wind is the force that initiates waves, transferring energy from air to sea (with energy imparted to air, in the earlier instance, by the sun). A light wind will give rise to capillary waves, what sailors call "cat's-paws." A more intense wind will lead to larger, traveling (and lengthening) waves, impelled by both wind and the force of gravity.[12] Persistent wind across an area of water, known as a "fetch" of "windsea," generates waves with a predictable range of heights. Out from under the influence of wind, waves are called "swells," packets of energy that continue to travel on their own. As trains of swells of different wavelengths run into each other, variously reinforcing or canceling one other out (interfering, as they move in and out of phase), they gather into groups (*sets* to surfers). These groups then move at half the speed of their constituent waves. (Think of people moving briskly along an airport conveyer belt: people are waves; the belt is the group.) When waves arrive to shoal and shore (defined as when water is shallower than half a wavelength), they begin to "feel" the bottom, which slows their bases, causing their crests to steepen. When the ratio of wave height to wavelength hits about 1:7, waves begin to break—as spillers (which sloppily collapse on themselves), plungers (which pitch over themselves to create the "barrels" beloved of surfers), and surgers (which slosh onto shore without generating whitewater), all of which next create "swash," a turbulent layer of moving water that marks the end of a wave. The story of waves is a tale about energy moving across space.

To track such motion, waves these days are largely measured by floating buoys and orbiting satellites, owned and operated by a collage of governments, companies, and other agencies. Buoys have their own internet protocol addresses, used to transmit information to computers that work the data into forecasts for meteorological organizations, shipping companies, coastal planners, fishers, boaters, and surfers. Wave prediction is all about infrastructure, and it starts with buoys.

Buoys were a big workshop topic. Val Swail, manager of the Climate Data and Analysis Section of Environment Canada, offered an overflowing inventory of wave-measuring buoys in a talk entitled, "Are Wave Measurements Actually Ground Truth?" (Jensen et al. 2013). Showing us photos of buoys hewing to a bevy of standards, he asked: "How to 'ground truth' the 'ground truth'?" (figure I.3). With good-humored world-weariness, Swail advised us: "You need to define what a wave *is* before you can measure it. Is your device measuring whitecaps, foam, green water, blue water? And in which *direction* are you measuring? Given that a typical buoy costs around $60,000, it's worth some thought." Swail's question—a mixed-sea-and-land metaphor—made clear what practitioners were after: a level reference plane in the ocean, bringing into this technology histories of trying to fix, in practical, world-spanning abstraction, the notion of *sea level* (see Hardenberg 2020; Sammler 2019).[13] Models of wave dynamics start with a stationary sea, and a lot of work goes into factoring out the pitch, yaw, and roll of buoys so wave data can be presented as oscillation against a fixed baseline.[14]

Knowledge about waves, Swail was acknowledging, is not just an *empirical* matter but also an *epistemological* one—to do with how wave scientists warrant what they know. The mathematical physicist Elzbieta Bitner-Gregersen, working for Extreme Seas, a European university consortium dedicated to ship safety, underscored this point. In her talk, she named two kinds of uncertainty in wave measurements: *physical* and *epistemic*. *Physical uncertainty* includes the randomness of waves in the world, whereas *epistemic uncertainty* points to uncertainties in data sets, statistical analyses,

FIGURE I.3 Various wave-measuring buoys. From Jensen et al. 2013.

and models of waves (Bitner-Gregersen and Magnusson 2013). Waves as scientific objects flicker between reality and representation. Waves are mash-ups, amalgams of watery events, instrumented captures of those events, and mathematical portraits of those happenings, often described statistically rather than singularly.

The statistical view came into focus in the mid-twentieth century, when oceanographers found that water waves, like light or radio stations, might be sorted by magnitude (frequencies) into a *spectrum*.[15] Instead of using oscillating lines to visualize the simple rise and fall of water, oceanographers turned to a representation that broke down a large wave into the smaller, diversely sized waves of which it might be composed. (Math aficionados will know this as Fourier analysis.) Waves came to be known not as individuals but as collections of superimposed waves, little and big, with different generating origins and histories. In the "wave spectrum" model, waves were rendered as collisions of bell curves; these might index the many time scales and processes that come together in any "wave" (figure I.4).

A "wave" might be made up of energy generated a month earlier on some faraway shore, by a hurricane a week earlier, and by fresh energy

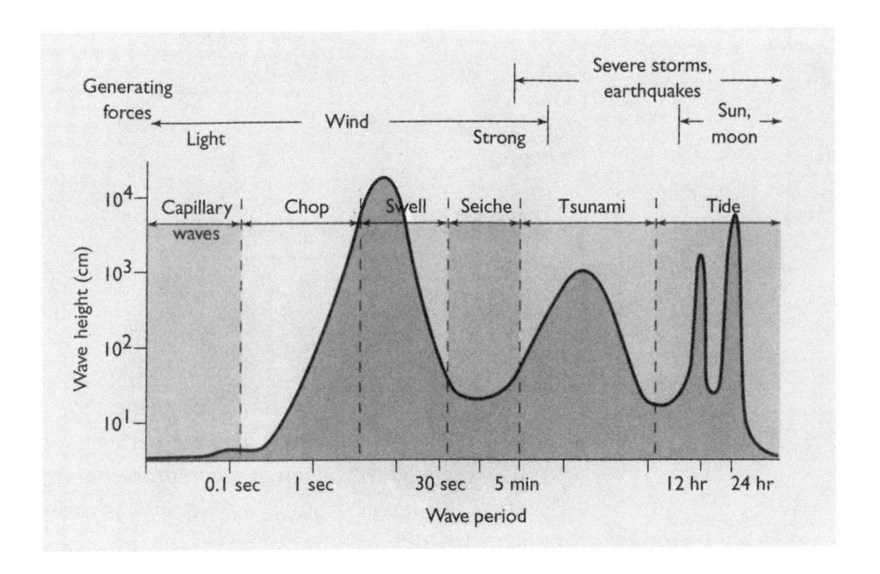

FIGURE I.4 "Idealized Wave Spectrum." From Pinet 2009, 232, fig. 7.1b.

from wind-swept ripples (figure I.5). As the historian of oceanography David Irvine (2002, 382) writes about wave spectra, however, "'Observed' spectra are not really observed; they are the finished products of a sophisticated mathematical analysis. . . . Spectra [stand] midway between raw observation and fundamental theory."[16] Swail's talk made that clear—and the days are long gone when such epistemic hybrids made scientists balk. Munk once wrote, "Inasmuch as these terms—'fetches,' 'finite durations'—are really great idealizations of the wind field over the sea, to try and write spectra for given fetches and finite durations is to endow these meteorological notions with more claim to reality than they deserve" (quoted in Irvine 2002, 380). The reality of waves nowadays, however, is known through precisely such idealizations, through instruments of mechanical objectivity (Daston and Galison 2007), a representational idiom putatively free of human judgment.

Where are buoy data sent? The computer wave modeling framework WAVEWATCH is employed in more than eighty countries and operated by the US National Weather Service, under the National Oceanic and Atmospheric Administration (NOAA). The designer of the system, Hendrik Tolman, who began the model as a master's thesis project in the 1980s at the Delft University of Technology, in his native Netherlands, gave a passionate update. The WAVEWATCH framework is built on millions of lines

FIGURE I.5 US Coast Guard, GAPA 05-05-71, *Rough Sea Swells in Mid-ocean, Far from Shore: The Texture of a Great Wave*, Willard Bascom Papers, SMC 62, Special Collections and Archives, University of California, San Diego, Library, box 12, folder 6.

of the FORTRAN computer programming language and, he announced, would eventually be integrated into the General Circulation Models on which the Intergovernmental Panel on Climate Change (IPCC) depends. This simulation framework offers a digital double for the ocean, crafted for anticipatory governance (see Lehman 2016; see also Gabrys 2016). When I caught up with Tolman at the conference banquet, he told me that he had lately been traveling around the world leading WAVEWATCH workshops, engaging in a kind of scientific missionary work. He also told me that if I wanted to learn the deep history of wave modeling I would need to do fieldwork in the Netherlands, where his teachers framed some of the earliest models in wave science (see chapter 1). He invited me to learn more about WAVEWATCH by visiting the offices of NOAA (chapter 4).

Wave models run on an armada of supercomputers that require huge amounts of air conditioning, contributing, scientists at the conference wryly remarked, to the very weather transformations they track (see Gonzalez Monserrate 2022). As computationally intensive as they are, wave models often demand *more* data points than there are buoys. To fill the gaps, models conjure proxy data, creating virtual buoys by interpolating between known points. Such phantom buoys speed up prediction. Waves can also be simulated in real water—either through using theoretical models to generate oscillations in wave flumes and basins or through reincarnating real, historical waves in scale-model form. Such work can be useful for ship and port design, the results of which may be delivered to logistics planners (Gray 2020; see also chapter 3 in this volume).

The centrality of prediction to wave science means that waves are vehicles—media—for scientists to orient to the shape of things to come, whether imminent, medium range, or emergent from the epochal climate-changing modulations of human activity. Wave futures are usually imagined against the grid of a uniform, metered, secular, homogeneous time—the tick-tock clock time of modernity (and of state and market bureaucracies [see Bear 2014; Munn 1992])—which invites researchers to pose cycle and succession in a stream of time whose material correlate is something like the unspooling rolls of graph paper scientists once used to gather wave records (see figure 2.6). To employ a term from the philosopher Mikhail Bakhtin (1982), wave science has operated within a *chronotope*—a representation of time in space—in which time and space are abstract.[17] If the figure of the wave offers what the literary theorist Nathalie Roelens (2019) describes as "revolving movement as the unveiling of a latency," this is always "implicit against Euclidian geometry." The futures believed to be prefigured by waves express *orientations*—"anticipation, expectation, speculation, potentiality, hope, and destiny" (Bryant and Knight 2019, 2)—to possible worlds. Think, then, of wave knowledge, drawing on Donna Haraway's (1991b) notion of "situated knowledges," as *oriented knowledge*, knowledge for and toward specific aims. The forward aims of oriented wave knowledge depend on the steady work of hindcasting, of past reference—though waves, as Munk reminded me, also *break*, a capacity that can upend their predictability, that may clutter, disorient, fracture, or overtop the neat models or infrastructures prepared to meet them.

It was during the Banff workshop that I first began to accumulate a sense of waves as all at once physical, epistemic, computational, and virtual. And political. Waves, captured by buoys and computer programs, may exist inside jurisdictions. In the United States, waves have been officially considered part of the weather since 1973, when the United Nations ratified the Convention for the Safety of Life at Sea, which tasked national meteorological rather than oceanographic agencies with wave measurement. If you want to get US data about waves, consult the National Weather Service, not the National Ocean Service—though neither will do you much good if there's a government shutdown, as there was just before Banff, which meant US waves were . . . closed.

Waves have lives as commodities, too. For European polities, wave data are stewarded by the European Center for Medium-Range Weather Forecasts (ECMWF), an intergovernmental organization that hosts the world's largest store of numerical weather-prediction information.[18] Data are available to "national meteorological and hydrological services and research institutions," and countries that are members of the ECMWF have custom subscription arrangements.[19] Such instrumented measures of wave phenomenology become purchasable products.[20] And WAVEWATCH is a model that, while open source, may not, because of US trade embargoes, be distributed to countries such as North Korea and Iran. Waves, then, can be read as political-economic texts, formatted by national, military, and corporate infrastructures. They also have legal lives; one speaker in Banff reported that, for the World Meteorological Organization, "only windsea and two swells are regulated in ship reports"—that is, only part of the wave spectrum. Waves, then, are *phenomenological-technical-mathematical-political-legal objects*. Sometimes their ghosts are carried into legal gray areas, as when pirates pillage buoys for parts. One conferee reported that, while reading wave data from a buoy off the coast of Somalia, he realized it was on land, perhaps being ferried in the back of someone's truck.

When I signed up for Forecasting Dangerous Sea States, I imagined the event would be mostly math and modeling. But the gathering was full of stories about people. There was an anthropology inside wave science.

Wakes, Breaks, and Churns in the Chronic Ocean

The figure of humanity—individual, corporate, regional, global—loomed over the gathering. That became visceral as news streamed into the conference of Saint Jude, an extratropical cyclone that arrived in northern Europe on October 27, 2013, and killed seventeen people. Just hours into the workshop, a presenter showed the day's headline from *The Mail*: England was being "lashed by 25 f[oo]t waves." Participants from the ECMWF were gratified that their models seemed to be predicting the path, though they were upset about people heading into the waves anyway. "We can't save everyone," one participant exclaimed at a picture of a kite surfer. Talk of probability becomes talk of *risk* when human elements are introduced, one speaker remarked.

Such attention to individual humans contrasted with talks about populations of coastal dwellers. Conversation turned to storm surges, what one speaker called "the stupid big brothers of waves." Here, differences among nation-states differently positioned geopolitically—in the Global North and Global South—emerged as a topic. The UK scientist Matt Lewis spoke about the World Meteorological Organization's Coastal Inundation Forecasting Demonstration Project, aimed at improving predictions for such places as Bangladesh, a low-lying country with a history of monsoon-inflected, cyclone-activated storm surges that have killed hundreds of thousands of people (see chapter 5). He commented that it is hard to "get data into a model" for such settings because they are "data poor"—that is, not dotted with measuring instruments. More, surge models imported from elsewhere do not always work; the mangrove forests and coastal villages of Bangladesh are not written into off-the-shelf models. Facts on the ground vary and change.

Open-ocean and hemispheric dynamics do not stay still, either. I encountered a dramatic, global picture on a conference poster. Lured in by lustrously colored maps of world seas, I found the Portuguese oceanographer Alvaro Semedo explaining future wave climate. His team had taken global data about significant wave heights from the present, defined as 1979–2009, and then, assuming a steady increase of carbon dioxide into the atmosphere—the Intergovernmental Panel on Climate Change's anthropogenic scenario—projected significant wave heights into the future, 2070–2100 (see plate 2; see also Semedo et al. 2013). His prognosis? Climate change will generate ever larger significant wave heights in southern oceans, which will correlate with more extreme weather—storms, droughts—and an accelerated breaking up, by wave action, of Antarctic ice.[21] This was a claim I would hear again and again, suggesting that thinking about climate change from Southern Hemisphere oceans might unsettle dominant perspectives from the North (see chapter 5). As I stood in front of Semedo's map—also a kind of graph, using saturated colors to underscore its message—an Australian wave scientist joined the conversation around the poster. He jumped up and down. *Homo sapiens* are driving the planet toward disaster: heat waves and fires in Australia, floods in South Asia, collapsing agricultural infrastructure in Africa, and waves, he said, of climate refugees.

The ocean and atmospheric scientists around me spoke animatedly about how humans were agents and destinations for thinking about global climate. As a wave scientist from Oregon State University put it to me:

> Changing climate means changing the wave climate—and that's a relation with *people*. We're producing too much carbon dioxide. So part of the destruction of our houses is because we're doing something wrong with the environment. So we should be interested in waves not only because we have fun in the waves or we have to protect ourselves from the waves, but also because what happens now with the waves is partly *because of us*. The loop closes that way with wind waves.

Ocean waves have become legible as materializations of social process. Against a vision of their forms as pertaining merely to the order of an endlessly cycling nature, enacting a steady beat of physical time, they have been brought into history.

Back in 2013, the term *Anthropocene,* though coined at the turn of the millennium, was just beginning to gain public traction. The atmospheric chemist Paul Crutzen and the ecologist Eugene Stoermer (2000) proposed the word to designate the contemporary geological epoch, during which human activity began to have planetary effects—effects now layered into a geological record marked with evidence of coal extraction, atomic testing, plastic and other pollutants, and accompanying species extinction.[22] Semedo's poster mapped a world in the wake of those processes, an Anthropocene sea.

The analytic of the Anthropocene has gathered copious commentary pointing out the historically specific human practices that initiated it, which come not from humans in general, but from economic, political, and military forms animated by large-scale resource extraction under colonialism and industrial enterprise, mostly inaugurated in the Global North. A raft of possible renamings has surfaced, prominently the *Capitalocene,* since so many recent geophysical transformations have followed from capital-intensive generation of fossil fuels, and the *Plantationocene,* pointing to how the transoceanic slave trade and the plantation complex shifted world ecologies toward a tipping point (Haraway 2015; Moore 2016; Jobson 2021; McKittrick 2013).[23]

What about the oceans? "What would it mean to take the Anthropocene out to sea?" asks the ocean humanities scholar Stacy Alaimo (2017, 153).[24] For one thing, it would usher the analytic away from thoroughly geological time, mixing temporalities deep and shallow (think ocean circulation, upwelling, freshwater flux) in different cadences from what happens in the layering realm of the stratigraphic (neither itself always linear [see Roosth 2022]).[25] Such "a possible oceanic turn," offers the literary theorist Steve Mentz (2015, xviii, xix), might suggest the frame of the *Thalassocene* (*thalasso*, "sea," plus *cene*, "new"), naming "a recurrent material counterforce, a pressure the inhuman ocean exerts on all histories."[26] If such recurrence also folds in the *old*, call that the *Thalassochronic: thalasso*, "sea," plus *chronic*, "old," "long lasting"—or perhaps, simply, the *chronic ocean*. The repetition and churn of such a sea is unevenly distributed, socially intensified, and felt. It is choppy, subject, like waves, to cluttered, broken motion.[27] It conjoins, unsteadies—and, sometimes, breaks—scales of analysis and experience. Sea level rise, the continued growth of trash vortexes, acidification-driven extinction and coral bleaching, the legacy of nuclear weapons detonations—these pressures churn together natural, cultural, colonial, imperial, postcolonial, and climate histories.[28] The historian Dipesh Chakrabarty (2021, 26) writes, "Anthropogenic explanations of climate change spell the collapse of the age-old humanist distinction between . . . natural history and human history." For wave scientists at Semedo's poster, humans operated as a kind of doubled figure—as an embodied (and unequally positioned and impacted) collectivity and as a scaled-up actor with inhuman capacities, activating the powers of a geophysical force.[29]

Emblematizing conference conversations on futurity, the problem of predictability, and the constant interruption of clutter and chop was the figure of the *rogue wave*, the statistically unexpected wave, twice the significant wave height of its surroundings.[30] Bigger storms in contemporary oceans might lead, some wave scientists held, to increases in numbers of rogue waves (see Bitner-Gregersen and Toffoli 2013; Rosenthal and Lehner 2008). Conferees were divided on that claim, though all believed such waves existed. Although rogue waves were, once upon a time, considered mythical, conjured by credulous mariners, they are now accepted as real.[31] Such waves may emerge from the superimposition of waves in "crossing seas," from wave-current interactions (Africa's Cape of Good Hope is notorious), or from resonance events in which one wave sucks energy from another.[32] One speaker said, "We can expect that in some ocean areas, where the wind severity increases and we get more 'crossing seas,' we will see more rogue waves."[33] The phrase *rogue waves* not only echoes terms such as *rogue elephant* and *rogue shark* (used by naturalists to refer to anomalous, wild, individual animals that depart from the dominant socialities of their species); it also resonates with the "rogue state" idiom forwarded by political scientists in the 1990s, entities that disturb geopolitical business as usual (Hoyt 2000). Rogue waves appear as newly cast characters in an Anthropocene drama, materializations of the inhuman human—manifestations of, to lean into the workshop's title phrase, "Dangerous Sea States." Storm surges, tsunamis, and rogue waves have become consequential forms in the twenty-first century, avatars of an ocean encroaching in an age of climate change, rearranging maritime globalization, and breaking down infrastructure at world coasts. The tidalectic ocean, bearing as well as breaking Anthropocene analysis, carries waves that exist in the stream of history and at the edge of a future difficult for institutions to keep calibrated to their routines and rhythms. This chronic ocean is an ocean made of churning time and, these days, too often—think chronic illness—of recurring conditions of compromised health.

This is the ocean of *the wake*, that "region of disturbed flow" in which, as Sharpe (2016, 3, 9) writes, "the past that is not past appears, always, to rupture the present." It is also the ocean of *the break*, "a broken or disturbed portion on the surface of water," according to the *Oxford English Dictionary* (*OED*), that brokers "an interruption of continuity." Recall Munk's assertion that wave breaking is best understood empirically, not theoretically (it is always, to some extent, a complex surprise) and torque this into dialogue with Sharpe's *wake*, drawing on the poet Fred Moten's *In the Break* (2003, 99), an analysis of the aesthetics of interruption (particularly in the Black radical tradition, from the jazz break to the breaking of language in poetry) in which he argues that a *break* is a "temporal-spatial discontinuity" that requires of those who experience it a "fundamental reorientation." Contemporary wave science, I suggest, operates, especially in a time of climate uncertainty and its unequal social effects, *in the wake* and *in the break*.

I left Forecasting Dangerous Sea States, then, with a set of questions, as well as a sense of where to travel to learn more. I conducted anthropological fieldwork in the Netherlands in 2016, a country below sea level where many wave modeling practices originated and from which many models have been globally exported. I worked at the Scripps Institution of Oceanography over several summers, from 2015 to 2019, learning—through archival work, interviews, and a trip to sea on the singular "FLIP ship"—about the outsize influence of field research on waves undertaken at this site, largely in calibration with US Navy priorities. I spent a month in 2015 at the world's largest water wave modeling basin, at Oregon State University, permitting me to contrast Scripps's field approach with wave science in the lab—and to learn about scale-model studies of tsunamis, which had intensified at this laboratory since Japan's disaster in 2011, as well as fresh news about the possibility of a similar event, soon, in the US Pacific Northwest. In 2017, I enrolled in a summer school on WAVEWATCH, led by scientists and engineers at NOAA. I visited Australia to understand how wave science looked from the Southern Hemisphere, moving from there to think from the Indian Ocean about the shape of waves to come, forms that may look, worldwide, more and more like those strong waves spinning off from southern seas (waves that cannot be understood as made of water alone but must be understood as also composed of silt and sand, demanding a rethinking of land-and-sea distinctions). In 2020, I conducted online fieldwork with wave scientists in Bangladesh, interviewing and attending events by videoconference after the COVID-19 pandemic made travel impossible. All of this work required examining the many ways wave scientists read waves and how I might, in turn, read these readings in anticipation of writing this, *A Book of Waves*.

Reading Waves

In Italo Calvino's short story "Reading a Wave," an eccentric amateur philosopher called Mr. Palomar stands on a beach, seeking to follow with his eye the arrival, passage, and decoalescence of a single ocean wave. Hoping to achieve a calming reverie, Palomar endeavors to "carry out an inventory of all the wave movements that are repeated with varying frequency within a given time interval" (Calvino [1983] 1985, 6). He finds that any single "wave" is crosscut by others, interrupted by "many dynasties of oblique waves" (7). But if Calvino, as he later explained, meant "Reading a Wave" as an inquiry into how to "read something that is not written" (quoted in Lucente 1985, 248), Palomar already comes to the sea with highly tutored reading habits, parsing what he sees in terms of "frequencies" and "intervals." These measures make sense to him because of the mathematical grammar that has formatted wave research, a grid of interpretation that patterns water waves as abstract arcs of energy to be regarded with a passionate dispassion, as texts to be read in the stillness of scientific meditation.[34]

Contrast Palomar's reading with early attempts by one of Munk's colleagues, Willard Bascom, to take wave readings near shore. One question before Bascom, who began work in the 1940s, was, "Did large waves in the ocean do the same things as those in a model tank?" In his memoir, *The Crest of the Wave*, he writes about an encounter with the sea in Humboldt Bay: "The surf roared at us, which is to say that the wide spectrum of frequencies created by all the waves crashing, colliding, swashing, and releasing bubbles produced a high volume of white noise—a hiss of astonishing proportions broken only by the occasional crack of a single breaker" (Bascom 1988, 7, 5). He recalls trying to bring waves into measure from an amphibious truck (repurposed versions of which are now known to American tourists as "duck boats" [figure I.6]) about to be swamped by a wave: "While balancing under this incipient waterfall, I would estimate the height of the wave that was about to come crashing down, add one third of that . . . to the trough depth, call the answer into the microphone, and duck. Then the reaching crest of the plunging wave would collapse on us, not quite capsizing the DUKW (Bascom 1988, 9)."[35]

A narrative entanglement of observation, audition, olfaction, and theory was a signature of Bascom's rhetoric. (Think of his shouting out wave heights as reading waves *aloud*.) He writes in *Waves and Beaches: The Dynamics of the Ocean Surface* (1964, 40–41):

> Now, full of confidence that we understand waves both in theory and by actual test, we fling open the laboratory door, stride to the edge of the cliff and look to the sea. Good grief! The real waves look and act nothing like the neat ones . . . that march across the blackboard in orderly equations. These waves are disheveled, irregular, and moving in many directions. Should we slink back inside to our reliable equations and brood over the inconsistencies of nature? Never! Instead, we must become outdoor wave researchers. It means being wet, salty, cold—and confused.[36]

This is an expression of scientific affect at home simultaneously at the blackboard and outdoors. Bascom's biography can be situated in a history that has seen American natural science researchers, mostly white men, conducting work in lands and seascapes imagined as rugged zones of uncontaminated nature, a vision institutionalized in such famous US work-vacation sites as Woods Hole, Cold Spring Harbor, Los Alamos, and Monterey Bay (see Kohler 2002; Pauly 2000), all places, too, with significant and too often effaced Indigenous histories and presents. This early data point on scientific readings of waves is at the same time a reminder of an embodied, sensory method that nowadays has been fractionated and mostly delegated to machines (which do not, in the same way, listen, smell, taste, swim, or gasp for air).[37]

Wave scientists now read waves through fieldwork at sea, research in laboratories, and computational modeling, as well as through mathemati-

FIGURE I.6 *Willard Bascom Standing on Duck Boat with Wave Ruler, 1940s*. Willard Bascom Papers, SMC 62, Special Collections and Archives, University of California, San Diego, Library, box 11.

cal formalisms, techniques of visual recording and replay, and networked information technologies. *A Book of Waves* asks how wave researchers craft their knowledge within historically shifting technical, social, and political settings—*contexts* within which waves as *texts* come to significance. I take *reading*, here, to be not only about interpreting formal properties—think of the morphologies of languages, images, music—but also about the sensory impressions on which such activity relies (Gandorfer 2016). As Karin Littau (2006, 3) writes, reading is "not only about sense-making but also about sensation." Therefore, particularly in the sciences, it is also about distributed anatomies of perception, nowadays often a hybrid of human and machine-meditated interpretation (see also Gabrys and Pritchard 2018).

The reading tools wave scientists have to hand include eyes and ears, measuring devices such as buoys (consider Swail's question, "How to 'ground truth' the 'ground truth'?," as cautioning against what literary scholars would call a "surface reading," taking wave readouts from buoys at face value [see Best and Marcus 2009]), underwater pressure sensors, beachside and floating labs, radar, water basins for scale model wave reenactment, video cameras, mathematical approximations, and computer simulations (with WAVEWATCH permitting the kind of "distant reading" advocated by practitioners of digital humanities, scanning swaths of big-data results, discerning large-scale patterns of significance in a corpus of texts [see Jänicke et al. 2015]). The range of practices bundled under this ecumenical definition of reading—taking readings, enumerating, calculating, watching, even, sometimes, machine-aided hydrophonic listening—underscores an additional claim I make, which is that waves emerge in these reading relations as *media*, or material forms that convey meaning (oceanic, climatic, anthropogenic) to wave researchers and their publics. This may not be a surprise, since the ocean, especially its farther reaches, for humans often requires access through the mediation of technology.

Media technologies (e.g., radio, photography, buoys) have enabled and guided how scientists understand ocean waves. More, waves have, in the process, become understood and experienced as media themselves.[38] The literary theorists John Durham Peters (2015) and Melody Jue (2020) approach the elemental properties of water, atmosphere, and weather as media, both in the environmental sense (think of seawater as that medium in which dolphins swim, methane hydrate bubbles, and waves uncoil) and in the sense of communication infrastructure (think of *sonar* as that medium that channels information through the propagation of echoing underwater acoustic signals).[39] Waves *as media* carry signals about where they come from, what they are made of, and where they are going.

In the early twentieth century, the Swiss linguist Ferdinand de Saussure used a wave analogy to describe how humans make meaning through language. He illustrated the relation between thinking and speaking with a diagram to be read from left to right as a wavy flow of time (figure I.7). For Saussure, (A) "the indefinite plane of jumbled ideas" and (B) "the equally vague plane of sounds" mutually inform one another, creating units—divided, in the diagram, by dotted lines—which he called *signs*, composed of *signifieds* (ideas) and *signifiers* (sounds, or, more broadly, articulations). Extending the analogy with a meteorologically inspired reading of his illustration, Saussure ([1915] 1959, 112–13) offered this guid-

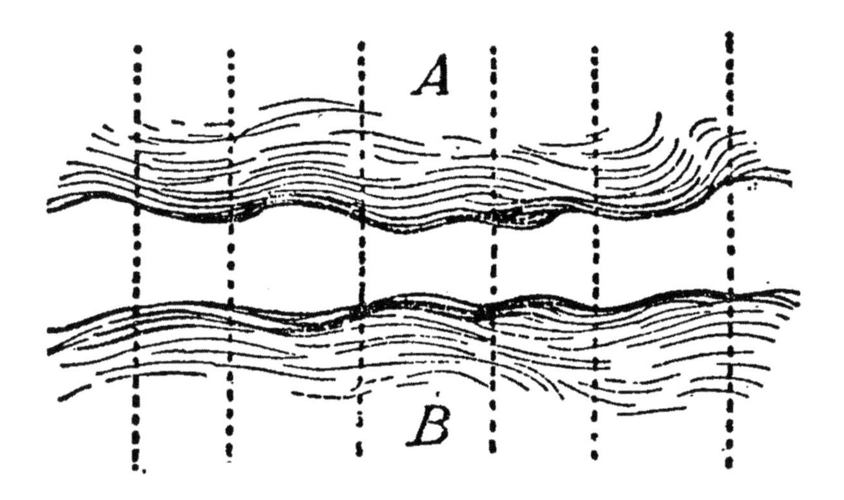

FIGURE I.7 Diagram of language as a wave-like meeting of signifieds (A) with signifiers (B). From Saussure [1915] 1959, 112.

ance: "Visualize the air in contact with a sheet of water; if the atmospheric pressure changes, the surface of the water will be broken up into a series of divisions, waves; the waves resemble the union or coupling of thought with phonic substance." In the "borderland where the elements of sound and thought combine," Saussure argued, "their combination produces a form, not a substance."[40] Such forms may be multiple, overlapping, and, if one speaks sign language, of many textures, and they may generate interference, clutter. Saussure, who thought signs emerged as socially conventional (or, as he put it, arbitrary) links between ideas and sounds, was here using the wave as an analogy for the sign. Over in the realm of the literal, oceanographers who study waves do indeed study them just so: as forms that emerge at the interface of atmosphere and water. If they treat these as signs less arbitrary than Saussure's—waves carry quasi-indexical impressions of recent pressure changes, of distant and coming storms, of oceanic displacements—scientists do sometimes, even simultaneously, understand them through stacks of abstractions that have more to do with conventions of mathematics than with one-to-one mappings onto the world. Waves may be reformalized many times over (see esp. chapter 4) as they are represented on paper, in equations, in wave tanks, and in computer simulation. Think of these hybrid mimetic and nonfigurative representations, forgive the pun, as *sign waves*.

Waves, then, are material forces that are apprehensible, simultaneously, only through abstract formalisms (whether these are curvilinear simplifications or nonfigurative mathematical descriptions). In *Phenomenon of Life*, the philosopher Hans Jonas (1966, 77) wrote that a "'wave' . . . has its own distinct unity, its own history, and its own laws, and these can become independent objects of mathematical analysis, in abstraction from the more immediate identities of the substratum." In the essay *Précisions sur les vagues* (*Waves in Detail*), Marie Darrieussecq (2008, 36) writes, "The wave is by its essence a formal modification of matter, creating froth, vapor, sprays, breaks and shocks" (my translation). Waves enter signification—in prose, philosophy, music, film, oceanography—as forces of generation, as sorts of quasi-agents that, while empty of intentionality, are full of animate action, lapping, running, roaring, their forms, as I occasionally discuss in this book, sometimes compared to those of horses, dolphins, wolves, lovers, monsters.[41]

Waves are therefore also useful entities to think with and against recent demands in cultural theory to attend to *materiality*, to a physical world beyond or before signification. The political theorist Jane Bennett (2010, 349) calls for recognizing the vital agencies inherent in materials such as trash, metal, and oil, naming "materiality as a protean flow of matter-energy." Scholars in the "blue humanities"—literary and media theory concerned with oceans—have also turned to materialities, or what the geographers Phil Steinberg and Kimberly Peters (2015) call a "wet ontology." They argue for conceptualizing oceans as "open, immanent, and ever-becoming" and, more, as elemental volumes (elemental: watery and molecular; volumes: not merely horizontal surfaces or vertical depths) in turbulent motion.[42] Such a view facilitates tracking the multiplicitous entities that assemble ocean space, matter, and time.[43] Equally important to keep in mind, however, is the role of concept work in formatting what scientists take to be material in the first instance (a demand with a reflexive charge; scholars in the critical humanities should also think about the historical origins of their own theories of materiality, which often come from classical physics). Waves must be read as things material *and* formal, concrete *and* conceptual.[44]

A Book of Waves

If the scientifically described ocean is a kind of text, and its waves are like pages in a modern, published book—reproduced in many identical, standardized copies, bound by law to represent the agency of a consistent, knowable author (here, physical nature [see Johns 1998])—my hope is to survey some of the ways this book has been, and might yet be, read, even as scientists have also rendered it into a flip-book movie, transduced its sounds into audio data, and, of course, in the age of the e-book, reformalized it in computer models. I argue against the political theorist Carl Schmitt, who, writing about the sea as a blank space between nations, dismissed waves this way: "The sea has no *character*, in the original sense of the word, which comes from the Greek *charassein*, meaning to engrave, to scratch, to imprint. . . . On the waves there is nothing but waves" (Schmitt [1950] 2003, 42–43).[45] This is a modern, disenchanted reading of the ocean as an empty chaos (Corbin [1988] 1995). It construes ocean expanse as timeless horizon (see Rozwadowski 2019), a vision reinforced, argues the historian of oceanography Eric Mills (2009, 82, 43), by the way nineteenth-century physical sciences, working at "making the ocean mathematical," approached it "ahistorically and atemporally" (in contrast to nineteenth-century geology and fossil-grounded evolutionary biology, which depended on deep, layered time).[46] In Schmitt's picture, waves become guardians of the ocean's secrets, orderly scriveners ever overwriting the matter of seawater, erasers unmaking their own inscriptions. This is only one way to see the sea. In this book, I track how waves may be and may become texts, agents, heralds, and, indeed, *characters*.

The first chapter, "From the Waterwolf to the Sand Motor: Domesticating Waves in the Netherlands," offers an ethnographic itinerary around the Netherlands, a country shaped by centuries-old endeavors to hold waves back from a land below sea level. Waves, long interpreted as forces of a wild, enemy nature, have come to be read as entities that might be rewritten, domesticated, allies in sculpting resilient environmental infrastructure. The chapter introduces, in miniature, themes that recur in this book—to do with the historical matrices out of which wave science arrives, with folklore, record keeping, physical scale models, computer simulations, remote sensing instruments, field measurement campaigns, and visions of waves as natural objects yoked to projects of cultural control.

Essays in my "Set" sections, similarly to the interludes in Virginia Woolf's novel *The Waves* (1931), break away from the book's throughline story, offering reflections, refractions, and diffractions on topics in the chapters that they follow. Chapter 1's theme of zoomorphic or anthropomorphic waves finds an alternative angle in "Set One, First Wave: The Genders of Waves." "Set One, Second Wave: Venice Hologram," elaborates on visions of waves as infrastructure, revisiting a narrative Walter and Judith Munk wrote about their stay in Venice in the 1970s, during which they imagined unorthodox ways to save the city from drowning. "Set One, Third Wave: Wave Navigation, Sea of Islands," sets out from an account a Dutch scientist relayed to me about a trip he took to Micronesia, where he sought to place computational wave models in dialogue with Marshallese wave navigation. I read his story in diffraction with postcolonial and decolonial discussions of Indigenous wave piloting in Oceania.

"Flipping the Ship: Oriented Knowledge, Media, and Waves in the Field, Scripps Institution of Oceanography," the second ethnographic chapter, recounts fieldwork at the Scripps Institution of Oceanography, in La Jolla, California. I am concerned with science *in the field*—at shore and on the sea—and organize my account around my stay on one of oceanography's most storied vessels: the FLoating Instrument Platform (FLIP), a craft able to "flip" itself vertically to become a live-aboard buoy that stays stationary amid rolling wave fields. Scientists, seeking fixed positions for observation, try to hold still frames within which they read waves, flipping between orientations to science as objective and as a source of wonder and between visions of science as pure or applied, civil or military. I offer a history of Scripps scientists reading waves through analogies to media (sonic, filmic, infrared, biotic), a history that also tracks generational shifts among physical oceanographers from military motivations to concerns with climate and contamination.

"Set Two, First Wave: Being the Wave," the first essay in Set Two, dives into the sea off San Diego, documenting my entry into the Thirty-sixth Annual World Bodysurfing Championship. It argues that surfing techniques, contoured by gender, race, and class, shape what surfers take waves to be. "Set Two, Second Wave: Radio Ocean" picks up on wave sound, in poetry and music. And "Set Two, Third Wave: Gravitational Waves, Sounded," listens to how the detection in 2016 of gravitational waves from the distant collision of two black holes 1.3 billion years ago was made audible through media that, similar to oceanographic formalisms, modeled the profiles of waves in advance of their arrival.

Staying with the theme of models, the third ethnographic chapter, "Waves to Order and Disorder: Making and Breaking Scale Models inside and outside the Lab, from Oregon to Japan," recounts fieldwork at the O. H. Hinsdale Wave Research Laboratory at Oregon State University, home of the world's largest tsunami simulation basin. I am concerned with what wave science looks like *in the lab*, as scientists make scaled-down replicas of real-world waves. Wave tanks, using water as a modeling medium, turn textbook waves into liquid movies, offering *theory, animated*. The chapter also reports on how Oregon-based scientists, living on the Cascadia fault line, grapple with Japan's 2011 tsunami as a premonition of their own possible Pacific future. Wave scientists' experiences in Japan itself, meanwhile, lead them to speculate on what happens when the very notion of *scale fails*, when lab time cannot prophecy real time.

"Set Three, First Wave: Massive Movie Waves" examines how waves in cinema deliver their effects. "Set Three, Second Wave: Hokusai Now" contemplates the world's most iconic representation of waves, the Japanese woodcut "Under the Wave off Kanagawa" (1829), examining how it has lately been used to speak about sociogenic ocean damage. The theme of health opens into "Set Three, Third Wave: Blood, Waves," which examines the electrocardiogram—the formalism that treats the heartbeat as a wave that can be managed by devices implanted into heart patients and that can be monitored remotely, like wave buoys, for signs of future danger.

The fourth ethnographic chapter, "World Wide Waves, *In Silico*: Computer Memory, Ocean Memory, and Version Control in the Global Data Stack," draws on fieldwork at NOAA, where, since the 1980s, the WAVEWATCH computer model has organized national wave prediction in coordination with a global infrastructure of buoys and satellites. Interested in what wave science looks like in *computer models*, I enrolled in a summer school on WAVEWATCH, meeting an international collection of wave scientists from countries that included Bangladesh, Brazil, China, Iraq, Korea, Mexico, and Turkey. Culminating in an account of how we learned to model 2005's Hurricane Katrina, the chapter argues that, as computer models of waves work with the time of waves at sea, the retrodictive time required to generate predictions, the speedy time of simulation, the reshuffling "version" time of computing in the data stack, and the staggered time of global wave science, they both depend on and create idiosyncratic, biographical, and political memories of all the waves yet to be included.

The first essay in Set Four, "First Wave: Middle Passages," examines the work of organizations repurposing surveillance data to reconstruct human rights violations in today's Mediterranean, which has seen thousands of migrants drown as they escape war in the Middle East and Africa, a tragedy that Sharpe (2016) has read alongside and through the Middle Passage. Drawing next, in "Set Four, Second Wave: Wave Power," on a field visit I made to Denmark to learn about experiments in wave energy, I consider utopian dreams of human sovereignty over waves. "Set Four, Third Wave: Wave Theory ~ Social Theory" then considers how waves have become figures in describing and predicting social change, from waves of opinion, immigration, and protest to waves of fascism and pandemic.

The final ethnographic chapter, "Wave Theory, Southern Theory: Disorienting Planetary Oceanic Futures, Indian Ocean," reports on wave science in the Southern Hemisphere, drawing from a conference I attended in Australia, as well as from Zoom fieldwork with scientists in Bangladesh, on the Bay of Bengal, where legacies of colonialism mix with future-facing projects to refashion land, sea, and siltscapes to meet rising seas. Thinking from the Indian Ocean may re- and de-orient knowledges about how to read oceans and their waves.

The postface, "The Ends of Waves," turns to where I live, Massachusetts, to glance at preparations for sea level rise in Boston, reading those preparations against Indigenous calls to remember histories of Nipmuc relocation and death on the Harbor Islands, modes of reckoning with pasts and futures of waves and the stories they carry.

A Book of Waves offers readings of wave science—readings of scientists reading waves—that spell out histories and present states of this cultural knowledge. We must think of waves, following Munk, not from the point of view of an alien, but from an Earth Ocean vantage. The historian Paul Gilroy (2018) argues that our time of planetary crisis calls for "sea level theory." Sea level, of course, is always oscillating and locally uneven, as June Pattullo, one of Munk's early graduate students, argued in foundational research she conducted in the 1950s (Pattullo et al. 1955; and see Baker-Yeboah et al. 2009 for recent work on sea surface height variability). Nowadays, sea level is also rising, overtopping any steady-state vision of fixed future horizon.

This book offers an anthropology of waves and wave science. It participates in what scholars in the humanities and social sciences have called an "oceanic turn"—where that turn is not just toward marine topics, but also toward anticipations, expectations, and anxieties about the shape of maritime worlds to come. Mindful of the heterochronic time of the ocean, however, it may be better to call this reorientation an *oceanic churn*, where to churn, according to the OED, is "to agitate, stir, and intermix any liquid, or mixture of liquid and solid matter; to produce (froth, etc.) by this process." The waves discussed in this book are forces of agitation and intermixture, of churn. They are media whose significance stirs together the wakes and breaks of ocean history and futurity.

Chapter One

From the Waterwolf to the Sand Motor

Domesticating Waves in the Netherlands

Station 1: The Waterway Forest

THE WOODS WERE FILLED with a fall mist. I was walking into a scrubby forest in the Netherlands with two Dutch hydraulic engineers. "We are now at the bottom of the sea," Jan reported. "Are you nervous?"

I laughed apprehensively and said no, affirming that I trusted the distant Dutch dikes that protected us, that kept dry this nation that sits, one-third of it, below sea level. We were near the eastern edge of the IJsselmeer, a freshwater lake that decades prior was a saltwater bay known as the Zuiderzee, or Southern Sea (see map 1.1). In 1932, this sea was dammed up, partly drained, and transmogrified into the IJsselmeer to shield once seaside settlements from the inundating forces of saltwater waves—waves that in 1916, one hundred years earlier, had submerged several towns. Jan, Gerbrant, and I stood on a polder—the Northeast Polder, land reclaimed from this sea-become-lake in 1942 by diesel-driven pumps. The land here, despite its recent vintage, was host to some singular archaeological relics. Jan pointed to the ruins of metal machines sitting derelict in leaf-strewn ponds, all that now remain of mid-twentieth-century scale

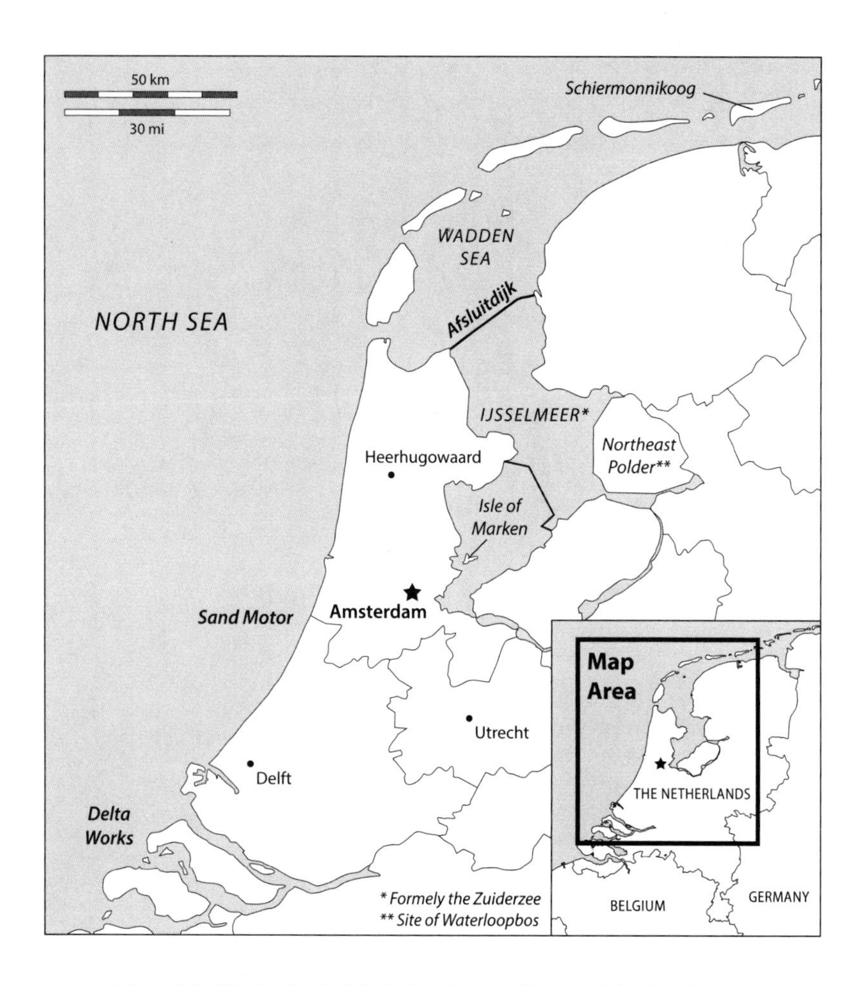

MAP 1.1 Map of the Netherlands labeled with sites discussed in the chapter. Created by Christine Riggio.

models of ocean ports. "We are arriving in Denmark," he instructed me. "Next, we'll head to Thailand, and then to Libya."

Where *were* we on that October 2016 day? In the *waterway forest*, or the "Waterloopbos," where hydraulic engineers working from the 1950s into the 1990s built scale models of ports in the Netherlands and elsewhere. The models' ruins, rusted and overgrown, are nowadays newly cared for as cultural heritage. The Waterloopbos is a national "nature monument," a site curated so visitors can look back on the accomplishments of Dutch hydrological engineering.[1] The afternoon took us through models representing the ports of Rotterdam, Copenhagen, Bangkok, and Marsa al Brega, Libya.

Though I stood markedly shorter in stature than my two Dutch guides, we three were like giants striding through a Lilliputian scene.

Scientists and engineers once worked all around this outdoor lab. Before I visited, one hydraulic engineer recounted to me that "it was unique. We didn't have a central building. We had no computers. We were all over the place in bungalows, little wooden shacks. You would work in the forest with one other colleague. The first few years I spent in splendid isolation."[2] Jan took out some old photos of young Dutch men on bicycles delivering coffee to scientists scattered throughout the woods.

The reach of Dutch hydraulic expertise, organized through Delft Hydraulics (a research organization founded in 1927) along with the Netherlands Development Corporation (a branch of the Dutch Ministry of Foreign Affairs), can be read as an echoing trace of the transoceanic enterprises captained by the Dutch East India Company during the Dutch Golden Age. Jan showed us a 1963 photo of Thai engineers who, seeking to model the port of Bangkok, found themselves faced by an unseasonably cold winter that made their model freeze over. With nothing for it, they were then invited by their colleagues to enjoy the typically Dutch pastime of skating—over the harbor models. The photo was strange, scripting these engineers as visitors from an emerging nation being tutored in European technique and culture.[3] It was a technique and culture much scaled down, diminished, from its once imperial range, though it was also a reach that still lingers in Dutch-led hydraulic engineering projects around the world, from Guyana to Indonesia and Bangladesh (see Dewan 2021; Ley 2021; Thompson 2018; Vaughn 2017).

The Waterloopkundig Laboratorium modeled the flow of water and, importantly, the action of waves. Hydraulic engineers could create waves of specific periods and heights, scale models meant to stage the taming of their real-world counterparts. The whole affair was embedded in the material affordances of land, wind, and water: the place itself is three meters below sea level, permitting water to flow downward from the diked-off IJsselmeer. The Waterloopkundig Laboratorium offered a means of generating and then reading waves very unlike those nonlinear mathematical and computational approaches that would arrive later to model, measure, and monitor wave dynamics.

Gerbrant van Vledder, a coastal engineer in his late fifties who divided his time between working at the Delft University of Technology and consulting projects, had become my guide into the world of Dutch wave science.[4] He had a fondness for the antique material culture of the field and a biography

FIGURE 1.1 Wave machine ruins at the Waterloopbos in Marknesse, Netherlands, situated in waters that last represented a port of Libya. Photograph by the author.

that had taken him from the days of the Waterloopkundig Laboratorium to today's computer era.[5] As we traipsed through the woods, he pointed out some hulking, rusty devices: mechanical wave-generation machines whose steampunk-like gears were once set to make waves of specific periods and heights (figure 1.1). These machines, made by IJsselmeer blacksmiths, materialized models of the internal structure of waves. Think of their bicycle-like gears as engineered to draw diagrams—in water—of waves, understood as oscillating forms made of orbiting particles of water.

These apparatuses once reliably generated scale-model water waves so that real-world counterparts might be predicted. These waves were uncomplicated, made of ups and downs, backs and forths. What did they look like? Perhaps like those recorded in a 1941 documentary about Delft Hydraulics, a film in which one scene shows an engineer with a notepad marking down characteristics of waves as they scroll by in an indoor wave tank, a modern, rational practice of holding parts of "nature" still so they might be measured (figure 1.2).

FIGURE 1.2 Still from Polygoon Hollands Neuws film on Waterloopkundig Laboratorium Delft, June 1941. Nederlands Instituut voor Beeld en Geluid/NOS.

There was something nostalgic, folkloric, about the corroded outdoor wave makers. They materialized dreams of doubling, shrinking, and taming the wildness of waves. For many Dutch hydraulic engineers—and publics—waves have come to be thought of as entities that can be domesticated by learning how properly to diagram, read, and even, through infrastructural interventions, rewrite their effects. The archaeologist Ian Hodder (1990, 45) notes that the notion of *domestication* draws on the word *domus* (household) and "obtains its dramatic force from the exclusion, control and domination of the wild, the outside" (see also Cassidy and Mullin 2007). *Domestication* might seem a weird word to describe the apprehension and management of watery waves, animate but not biological—entities that, though they are *generated*, do not come in *generations* in the sense of arising, like plants or animals, as individuals that belong to successive populations linked by lines of heredity that can be molded by artificial selection. In what follows, however, I demonstrate that Dutch wave scientists seek to draw waves into a give-and-take relation with cultural enterprise, excluding, attenuating, or harnessing them for purposes of human housekeeping, intervening in their propagation and arrival.[6] In so doing, they script waves into Dutch national narratives of progress and futurity.

Waves are regularly replicated in representation—in scale models, in simulations—with the aim of creating analogues to actual waves. Often, that work involves making diagrams of waves—where *diagram* describes an illustrative, often geometric, figure representing a process—some of which are treated as blueprints to be used in writing computer programs that model, mime, and make waves (e.g., in wave tanks). Diagrams are apparatuses for visual reasoning about relations, for tracing trajectories so they might be acted on. Waves in this way become phenomena amenable to *reading* and to *rewriting* (*diagram*, etymologically, is *through-writing*). That relay between reading and writing maps out a zigzag pedigree for wave models, created in the name of framing and taming water waves. Waves come, in recent Dutch science and national discourse, to be regarded as nonhuman entities with which to ally—to work *with* rather than against—in building protective sandy and grassy beach infrastructure, for example.

In what follows, I narrate travels I took in the Netherlands in fall 2016—travels on foot, by bike, on trains, by ferry. The chapter takes the form of an excursion around the country (see map 1.1), stopping at nine sites where waves have been met and modeled. I call these sites "stations" not only because I spent a lot of time at train and port stations, nor only because my travel had the feel of a secular pilgrimage to locations in Dutch wave history, but also because these were my (and sometimes my interlocutors') field stations. The Waterloopbos has been the first, archaeological, stop on this itinerary. Next is a station made of two historically significant sites, both to do with twentieth-century, high-modernist hydraulic engineering projects in the Netherlands and with keeping waves away from the shore.

Station 2: The Afsluitdijk and the Delta Works

There is a deep Dutch history with waves. Some sources tell the history of the Netherlands in "seven floods," with the most recent an event of 1953 that killed some 1,800 people in Zeeland, along with forty-seven thousand cows and many other animals, in the southern part of the country (van de Stadt 2013; see also Agostinho 2015; Hoeksema 2006; Hughes 2009; Leenaers 2013; van Veen 1962). A durable national narrative has it that "the Dutch" have long been "at war" with seawater and, over the centuries, have held it back or tamed it (or, sometimes, used it as a weapon of war, calculatedly

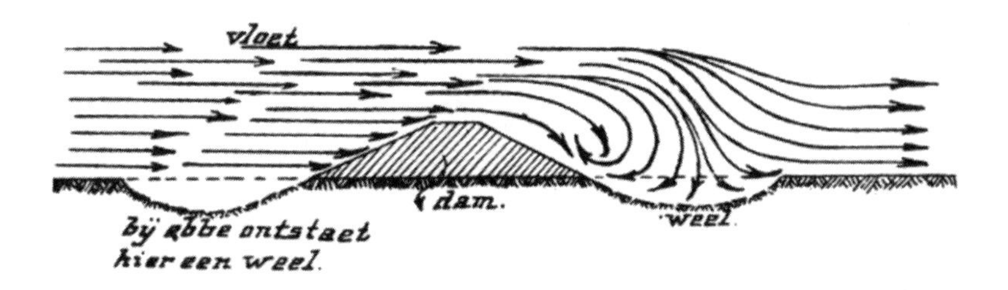

FIGURE 1.3 Diagram from Vierlingh (1579) 1920, plate VI.

flooding enemy attackers). The historian Simon Schama (1987) has suggested that the fight with water has become part of the "moral geography" of the country, part of a "hydrographic culture" that generates historical genealogies that rhetorically tether modern innovations in dike and dam building to millennium-old constructions of mounds (see also Wielenga [2012] 2015). As I read back into Dutch history, more than a few hydrologists referred me to a 1579 text on embanking written by the dike builder Andries Vierlingh, offering it as a 450-year-old example of Dutch interest in the flow of water. The text is full of diagrams (figure 1.3). A commonly repeated Dutch aphorism summarizes a triumphalist vision of the Netherlands as long holding back the waves: "God created the Earth, but the Dutch created the Netherlands" (see van der Cammen and de Klerk 2012).[7]

The thematic shift from wild to tame water can be tracked through Dutch seascapes, which in the 1600s featured lots of ships among storm waves and by the twentieth-century postcard age often pictured polder-calmed waters, sometimes with domesticated Holstein Friesian cows wading in mirrored shallows. The Dutch landscape tradition of picturing quotidian life employs a low horizon as a mark of the everyday, though it also offers, alongside this invitation to think, close-up, near to the nation, an expansive (imperial) navigational point of view.[8]

The Northeast Polder that hosts the Waterloopbos sits next to one of the key accomplishments in Dutch flood control history: the IJsselmeer, the freshwater lake made from the Zuiderzee. In 1916, a devastating flood inundated fishing towns on the circumference of this sea (figure 1.4), an event that put into motion plans in play since the 1890s to cordon this body of water off from the North Sea. The dike that was constructed—the Afsluitdijk, finished in 1932—has become an icon of Dutch hydrology, described by the historian Maarten Asscher as "thirty kilometers of naked dike that

FIGURE 1.4 Postcard marking the occasion of the Zuiderzee flood of 1916. Collection of the Zuiderzeemuseum, Enkhuizen, Netherlands.

severs the Wadden Sea from the former Zuiderzee like a heavenly blade" (quoted in Metz and van den Heuvel 2012, 40; see also Asscher 2009).

The 1953 flood, commonly known as "the Disaster" (as many people died as did in the aftermath of Hurricane Katrina in the United States), is also remembered for its waves. Some of the most poignant representations are in children's books, which pose the waves not as unthinking enemies, but as divine tests. In *Sijtje and the Island of Wooden Shoes*, villagers build a great raft of wooden clogs to keep themselves "safe from the waves, praying to God" (van Looy 1955).[9]

The aftermath of the 1953 flood saw the construction, over the next four decades, of the Delta Works (figure 1.5), a network of dikes, dams, and locks that control the flow of water in the estuaries of the Rhine, Meuse, and Scheldt rivers, south of Rotterdam, in Zeeland and South Holland. The hydraulic architectures of which the Delta Works consist are icons of rational control over nature and are visible from space. Their scale was such that they needed to be built with construction infrastructure that could survive the battering of waves, one impetus for Dutch work on wave prediction. (Another was the expansion of the Port of Rotterdam, deepened during the energy crisis of the 1970s, when the Suez Canal was closed.) Over the course of the Delta Works' construction, the local political context changed.

FIGURE 1.5 *Strong Waves Crashing on Oosterscheldekering Flood Barrier, Vrouwenpolder, Zeeland, Netherlands.* Photograph by Mischa Keijser.

In the 1970s, environmentalists protested against the effects of stoppering exchanges—biotic, ecological—between ocean and delta waters, with the result that later structures in the Delta Works incorporated sluices and movable barriers.

The emphasis on Dutch hydrological enterprise in the telling of the story of the Afsluitdijk and the Delta Works underscores a feature of Dutch tales about water I came across time and again. The Dutch science studies scholar Wiebe Bijker (2002, 575), enunciating an analytic view that is of a piece with national narratives, writes, "The Dutch have a long tradition of planning and actively shaping their environment. This applies not only to the physical landscape of the Netherlands but also to society; Dutch political culture displays a general belief in the malleability of society. Finally, the political culture of the Netherlands is distinctly consensual and oriented toward cooperation and compromise. . . . [I]n the end the Dutch need to cooperate with each other, under penalty of being flooded."[10] This is a story of a reflexive national character, one generative of democracy from necessity. Bijker, himself the son of a hydraulic engineer, told me that he played in the Waterloopkundig Laboratorium as a boy. It may be no wonder that an awareness of the cultural construction of landscape, of nature, comes easily to Dutch traditions of science and technology studies.

This vision of the Netherlands as amenable to reflexive intervention is accompanied by a persistent interest in questions of scale. The models of the Waterloopkundig Laboratorium are foundational. Bijker (2002, 581) points out that the Delta Works also required thinking through scale: "In physical models, dimensions are scaled down by factors of one hundred and four hundred, time is scaled up by a factor of forty, sand is scaled down by using finely ground Bakelite. . . . For detailed studies of dikes and constructions, wind and wave flumes were used. The organization of this model research was as difficult and crucial as interpreting the scaling principles." As if to underscore the point, almost everywhere one goes in the Netherlands there are museums that include scale models of towns and waterways; one outdoor museum, Madurodam, is a tourist attraction that features 1:25 scale-model replicas of Dutch landmarks. Interactive models made especially for children are common; they teach kids to take a God's-eye view of familiar landscapes, tutoring them in seeing themselves as bigger than the land they inhabit, something like the way that Cornelis Lely, the civil engineer who envisioned the draining of the Zuiderzee, was pictured in 1916 in a cartoon in which he was represented as an eminently practical Dutch washerwoman doing nation-state-scale work of domesticity (figure 1.6).[11]

Station 3: Marken and the Waterwolf

There are folkloric reference points, too. One arrives with the story of the "Waterwolf," a zoomorphic representation of the eroding power of waves (the Dutch word for wave, *golf*, rhymes with *wolf*). In one canonical seventeenth-century representation, the waterwolf is pinned down by the Dutch lion (figure 1.7). An accompanying poem exhorts the lion to "enclose this perilous beast with a dyke"; to "Let the Lord of the Winds with his fine mills / Empty the Water Wolf into the sea" so that land may be "won for the Lion by spinning gold from foam" (van den Vondel 1641). The last line is an evocation of Golden Age dreams of windmill-powered pumps transubstantiating watery waves into wealth. As the anthropologist Ghassan Hage (2017, 33) notes, "In the Western imaginary, the wolf is the ultimate representative of the threatening undomesticated other of nature." With the figure of the wolf haunting Christians' and medieval Europeans' wor-

FIGURE 1.6 Werkvrouw Lely cartoon. From *Groene Amsterdammer*, September 16, 1916.

ries about lines between corruptible bodies and redeemable souls (think werewolves), this animal sets boundaries around the figure of the northern European human (O'Neill 2018, 514).[12]

The waterwolf symbol has been revived in recent years.[13] Shortly after my visit to the Waterloopbos, I learned of an event called Waterwolf 2016, to be staged in the small Dutch municipality of Marken, sited on a segment of the IJsselmeer called the Markermeer. Back in 1916, this island town, situated on a large saltwater bay, was inundated by waves rolling in from the breach of nearby dikes—waves that left sixteen people drowned (see Vissering 1916).[14] Waterwolf 2016 was a one-hundredth anniversary flood-preparedness exercise. I emailed the organizers to ask whether I could participate, to see what imagining invisible waves from the vantage point of an individual in a crowd might be like. I was welcome to join, an organizer told me: "You can be a tourist who got stuck on the island. I think that's a nice bonus."

FIGURE 1.7 Detail from Willem van der Laegh and Joost van den Vondel, "Provisional Draft Plan and Proposal for the Diking of the Large Lakes," 1641.

When I arrived at the town center, I was provided with a name tag that read "Evacué: gezond. Bij nood roep 'NO-PLAY'! (Evacuee: healthy. If in real distress, say "NO PLAY!"). Like everyone else, I was given an identifying fleece hat, screen-printed with a logo designed for the event. Those dozens of people gathered to participate turned out mostly to be the town's retirees, primed for a change of routine. When I explained that I was an anthropologist, a self-described grandmother took me under her wing to make sure I was provided with a piece of typical Marken cake. Marken, I knew, had earlier been of interest to anthropologists, treated as a time capsule for antique Dutch customs. Johann Friedrich Blumenbach in 1828 and, later, Rudolf Virchow in 1877 even speculated that inhabitants of Marken were possessed of throwback body types, though it turned out that the characteristic sloping of Marken skulls that fascinated these naturalists was likely the result of Marken children wearing tight caps in their early years (Schild 2015). I imagined myself a different kind of anthropologist—although, with people excited to talk about Marken culture with me, I was quickly filled in on some ritual events of early 2016, when schoolchildren around Marken (now not an island but, with a causeway added in 1957, a peninsula) made papier-mâché sculptures of the waterwolf to burn as a sign of conquering the waves. Everyone had photos of these fiery figures on their phones.[15] As pleasant as everyone was, I half-worried I had walked into an old-time movie about a small town with a grim secret. Effigy images of the waterwolf bolstered a sense I had been developing that, for many in the Netherlands, waves could be imagined simultaneously as mischievous agents and as nervous-making presences—haunting figures from disasters past and, if one wasn't careful, a flooding future.

After a pre-evacuation briefing in the town center, the one hundred or so of us who had signed on for the event (all Dutch except for me and one Swedish woman) were escorted to a church to obtain life jackets. We were told to imagine water rising around us and referred to markers of the 1916 flood carved onto stilted houses nearby. A few people not participating walked by in amusement, including a couple of girls wearing hijabs and carrying grocery bags. Real members of the Dutch police, army, and Red Cross (who had been preparing for weeks) took charge, with some townspeople feigning confusion and injury, having, some of them, even covered themselves with fake blood.

Red Cross workers escorted us to Zodiac escape boats, which ferried us at speed across a segment of water to the town of Monnickendam, to "safety" (an endeavor that, at a time when migrants across the Mediterranean were drowning for want of life jackets [see "Set Four, First Wave: Middle Passages"], was more than a little surreal). We were meant to imagine waves chasing us, the waterwolf at our heels. Here, again, as with the Waterloopbos, was an exercise in simulation, with the scales of people and waves inverted: the Waterloopbos had people larger than waves; *here*, the imagined waves chasing us were outsizing us. I was put in mind of a work of flood-themed installation art: the "Waterlicht," a "virtual flood" rendered in wavy lines of LED-generated blue light projected against steamed-up air, an ephemeral work that blanketed Amsterdam to visualize what water levels would look like without Dutch water defenses in place (see Chin 2015).

Think back to the Waterloopbos and we have here two moments: one archaeological, at the Waterloopbos, and one ritual, in Marken—Dutch moments—about controlling and conquering waves; stories that pose waves as having bodies, lives, stories, anatomies, structures. The first is about waves as objective entities, well understood by a Dutch science ready to be deployed around the world (see Mrázek 2002). The second is about waves as historic, folkloric enemies. These stories appear in more recent, scientific registers, even as they also become more generalized, less nationally specific—registers that have waves as possessed of internal structure and even population dynamics, which scientists and engineers can read and write mathematically and computationally.

Station 4: Books of Waves

Let me get back, then, to Dutch wave science proper. Many founding figures were still active, if retired, in 2016 and happy to chat. My fieldwork often had me traveling by train and bike, setting out from my base in Utrecht, where I lived for four months, to visit these older men in their modest homes, visits that had them tutoring me in the history of wave science by guiding me through books in which wave knowledge was archived—part of what inspired the title of this volume. My initial point of introduction to this older generation was van Vledder, with whom I had taken the Waterloopbos tour and who, when I first met him, at a storied Utrecht restaurant housed in a building constructed in 1280, came armed with a satchel of texts to guide our conversation. Affirming that there was "a strong Dutch signal in wave science," van Vledder, who belongs to the generation just after the founding figures, had worked for Delft Hydraulics in the 1980s and was developing computer programs back then of the sort that would displace many physical scale models. He was keen to put me in touch with his teachers.

One of my first trips was a fifteen-kilometer bike ride from Utrecht to visit Gerbrand Komen, retired from the Royal Netherlands Meteorological Institute (in the nearby town of De Bilt), where he had worked since 1977 and where he had served as head of the Oceanographic Research Division. The Royal Netherlands Meteorological Institute, he told me, had been founded in 1854. One of its tasks was to collect knowledge about ocean navigation to and from what was then the colony of Indonesia. That impetus had long vanished, with the end of Dutch Empire.

Our conversation was punctuated by his taking books from his shelves to guide me through key moments in wave science history. He handed me a 1949 book titled *Zeegolven* (Ocean Waves), written by the Dutch scientist Pier Groen, who had visited Walter Munk at the Scripps Institution of Oceanography in La Jolla, California, and hoped, with this book, to ferry American findings into Dutch conversations.[16] Komen emphasized the significance of the 1953 flood, which became the impetus for putting wave science to work in the Netherlands, largely to help engineer the Delta Works and predict the long waves known as storm surges. His studies had been in how wind waves grow. (He had traveled to Malaysia in the 1980s to see whether he could adapt wave models to the South China Sea; the lack of measurements of the sea floor—bathymetry—had hampered the models, though it became clear that reefs were damping swells.) He had eventually been involved in the development of a computer model called WAM (for Wave Model) (see Komen et al. 1994; see also Woods 2006), which is now employed by the European Center for Medium-Range Weather Forecasts.

I took a train to see the civil engineer Jurjen Battjes, whom I visited at his canalside home in Delft—a city storied in maritime history not only because it had been the site of Delft Hydraulics, but also because it is the birthplace of Hugo Grotius, the Dutch jurist who wrote *Mare Liberum* (The Free Sea) in 1609, setting out the doctrine of the freedom of the seas, framed to bolster Dutch contestations with the Portuguese over access to trade routes.[17] Battjes had retired from a professorship in fluid mechanics at the Delft University of Technology in 2004. When I arrived, he was looking out his window at boats going down the canal.

Battjes has been thickly involved in research on wind waves. Grappling with the matter of wave propagation in coastal waters, particularly as waves arrive at sloping coastal structures, he devised the notion of "surf similarity," defining it in terms of a parameter proportional to the change in depth over one wavelength of the surf zone (Battjes 1974). This formulation, which predicts similar behavior for different combinations of wave height, period, and slope angle if the overall parameter holds, became a tool that could be transported across cases, permitting researchers to infer whether and how wave breaking and reflection would occur as waves of known height and period arrived at a slope of known angle.[18] The construction of the Delta Works, Battjes told me, required the calculation of wave loads on the movable gates of the Oosterschelde storm-surge barrier, for which Battjes modeled the reducing effects of short-crestedness in incident waves (see Battjes 1982).[19] Thinking through how to chart long-term (climatological) wave statistics, Battjes (1972) also developed an approach eventually called the "Battjes method" for predicting what might, over a fifty- to one hundred-year period, stand as the highest waves likely to hit a coastal structure in a specific place. His work offered ways to protect Dutch infrastructure that could be translated to cases beyond; he had lectured over the years in China, Japan, Vietnam, and elsewhere.

Our conversation wound down with Battjes telling me about his love of teaching. As I left, he took me outside to show me how he illustrated a key principle of wave science: the fact that waves have not only individual speed but also "group velocity." He asked me to throw a rock into the canal and observe how waves radiated out in a concentric band. The first thing to notice, he instructed, was that shorter and longer wavelengths are initially mingled together. As they propagate, however, a process of sorting called *dispersion* unfolds, with longer waves moving quickly to the front of the group to outrun shorter waves, which lag behind. By fixing one's eye on an individual crest, one can see that it travels forward through the group, appearing at the rear of the group, seemingly out of nowhere, and vanishing at the front. He said, "An individual wave is an ephemeral thing," and offered an analogy to an animal: "You have seen caterpillars that move by having compression waves move through their body. The caterpillar is the group, and the wiggles are the individual waves. And if the group velocity and the wiggle velocity were the same, the thing would never move." This appeal to animate life—and, more, to a worm as a diagram of itself—brought group velocity to life.

Another train ride took me to Leo Holthuijsen, who lived in a small house in Dordrecht, part of a compound of identical homes, many with model ships in the windows. The anthropologist Karen-Sue Taussig (2009) has observed that a dominant Dutch value is "to be ordinary," not to stick out. Here again was the unpretentious domicile of an older Dutch man, with materials for hobbies—and more bookshelves. Like Komen, Holthuijsen had a library filled with books of waves—records of formal wave knowledge, anatomy books perhaps (or, if one is a wave modeler, something like animal-breeding books, complete with ideal types), an archive of a life lived creating ways of writing down waves. I had been carrying around Holthuijsen's own contribution, a book from 2007 titled *Waves in Oceanic and Coastal Waters*. I wanted to refresh my memory of that key formalism in wave science: the wave spectrum.

Wave spectra might be understood as a largely theoretical concept, but some of Holthuijsen's colleagues became interested, in the late 1960s, in whether there might be an empirical case they could use to understand how wind generated wave spectra. That curiosity resulted in the Joint North Sea Wave Project, which gathered oceanographers to take measurements of actually existing wave fields, treating a portion of the North Sea cradled by the coasts of the Netherlands, Germany, and Denmark as "a lab" to figure out spectra that might be generalized to other places. (Holthuijsen had the idea of measuring wave height stereoscopically, from the sky, and was able to borrow helicopters from the Royal Netherlands Air Force to do this.) Holthuijsen showed me his copy of "Measurements of Wind-Wave Growth and Swell Decay during the Joint North Sea Wave Project (JONSWAP)," from 1973. Drawing on the work of Dutch, American, English, German, and Japanese scientists, it had created the "JONSWAP spectrum," to this day a kind of touchstone for understanding how waves grow everywhere (Hasselmann et al. 1973). Holthuijsen said, "It turned out to be universal. Even in a hurricane, you'll find the JONSWAP spectrum." (For different views on JONSWAP universality, see chapter 4.)

Before I arrived, Holthuijsen had prepared for me a diagram of wave science history, a left-to-right chronology with threads for theory, empirical work, institutional contexts, and technological transformations. Pointing to the phrase *computer modeling*, to tell me where his work took him next, he now took out another book, from 1985. By that year, a number of models of waves had been created, and it was time to compare them—in a project called the Sea Wave Modeling Project (SWAMP). The SWAMP book documented a pruning event, which left only WAM standing (SWAMP Group 1985). This became the standard model and the beginning of what would later be the framework used by the US National Weather Service, WAVEWATCH, created by Hendrik Tolman, a student of Leo Holthuijsen and Jurjen Battjes who eventually immigrated to the United States (see chapter 4).

My final trip on this itinerary took me to Germany to meet a scientist everyone told me had revolutionized wave theory. Klaus Hasselmann lived in Hamburg with his wife, Susanne Hasselmann, an applied mathematician who reduced many wave formalisms to computational tractability in early versions of WAM. Eighty-five at the time we spoke, Hasselmann was the retired director of the Max Planck Institute for Meteorology in Hamburg and emeritus professor of theoretical geophysics at the University of Hamburg. He is a towering figure in wave science, having early on transported his thinking from physics into oceanography. He was interested in how waves interact; in how, within a spectrum, they might exchange energy (Hasselmann 1962). The formalism that emerged from this question was the "four-wave" model, which models a system of four waves in resonance. Representing the interwave interaction that results—the nonlinear transfer of energy—requires the use of some otherworldly mathematics, mathematics that thinks with so many integrals that the problem ends up being six-dimensional. I had looked at a (two-dimensional) diagram of the problem in Holthuijsen's book (figure 1.8) and learned that, while there was a geographic mode of understanding the problem, there was also one that worked in "wave number" space. We were now far from thinking of wave anatomies in three-dimensional (3D), bodied terms, conjuring, rather, with abstract space that made me think of theoretical biologists' geometric diagrams of "morphological space," the space of possible organic forms.

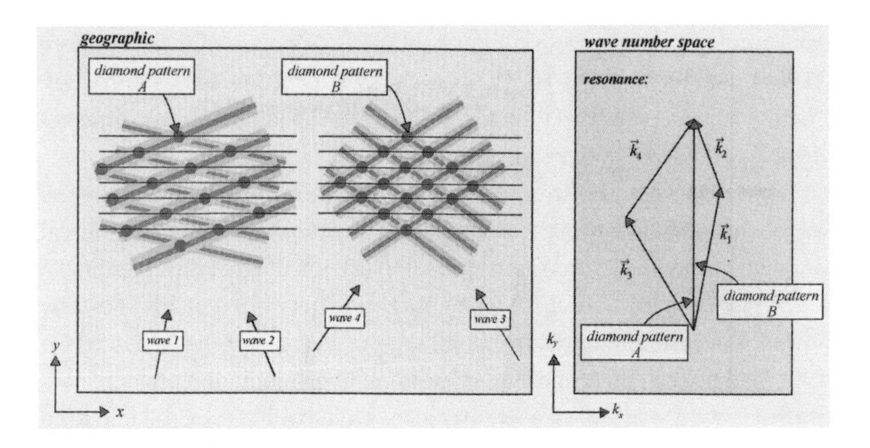

FIGURE 1.8 Diagram of quadruplet wave-wave interactions. From Holthuijsen 2007, 185, fig. 6.20.

Writing about diagrams in the natural sciences, the anthropologist Christos Lynteris (2017, 473) argues: "What is established by the blank backdrop is nothing less than the de-located, de-temporalized universality of the scientific principle graphed on it."

The universalist aspirations of such models issue not only from their abstract, mathematical form, but also from the assumption that time is a durable, homogeneous measure apart from waves themselves. For some Dutch engineers, comfortable calculating more than fifty- to one hundred-year spans, such an assumption may resonate with, or be anchored in, the taken-for-granted longevity of the Netherlands' coastal infrastructure (the name of the newest part of the port of Rotterdam, *FutureLand*, emblematizes the faith)—a stability they narrate, with a mixture of modesty and pride, as continuous with longer Dutch tradition. To be sure, the work of the men with whom I spoke was also informed by international conversations. And the role of computers cannot be underestimated; these researchers had been the generation to move from physical models to computational ones, edging waves from wild and watery to coded and calculated. The next generation younger—people just finishing their doctorates in cohorts that include more women and international scholars—were building on these models, taking them in new directions.

Still, physical models had by no means vanished. My next stop was at an institution called Deltares, a descendant of Delft Hydraulics, where I visited the very newest physical model, the giant Delta Flume.

Station 5: Delft and the Delta Flume

Deltares is a hybrid institution. A descendant of Delft Hydraulics, it is a private entity that works in the Dutch national interest and functions as an academic research facility closely linked with the Delft University of Technology, to which it is adjacent. It sits in the southern part of the city of Delft—whose name means *delve*—at Boussinesqweg 1, a street named after a mathematician responsible for key equations in fluid dynamics. When I visited the Deltares campus, a sleek collection of office buildings and airplane-hangar-like structures hosting large basins of water, I was met by the hydraulic engineer Marcel van Gent, in his late forties, Deltares's manager of coastal structures and waves. He gave me a tour of what I could only think of as a cutting-edge Waterloopkundig Laboratorium, guiding me into a great hall that held wave pools labeled "Atlantic Basin" (650 square meters, hosting a model of a lock to be constructed in Amsterdam harbor), "Pacific Basin" (good for irregular long-crested waves, for modeling "the influence of three-dimensional wave attacks on structures"), and "Delta Basin" (full of a bunch of rocks in breakwater formation, modeling a Moroccan port at one-fortieth scale). Van Gent told me that the previous model in this basin was for Dubai. I saw an engineer walking around in rubber waders in the Pacific.

Van Gent knew what I was most keen to see was the new, as of 2015, Delta Flume (van Gent 2015). The flume is a three hundred-meter-long, nine-and-a-half-meter-deep, and five-meter-wide channel. A ten-meter-high "wave board" can be moved back and forth by hydraulic cylinders at one end of the channel to generate waves of up to four-and-a-half meters in height. This wave flume generates the largest human-made waves in the world (plate 3).

Van Gent had me outfitted with a yellow safety vest in case I fell into the flume. When we arrived at the channel, it was empty of water. Three workmen were building a wall inside, made of a material a Deltares client wanted to test against an "actual size" wave for its durability as a part of a breakwater system. (Waves may be well characterized numerically, formally, but the materiality of structures they *hit*—rocks, trees—can be difficult to abstract.) Disappointed I could not see a wave travel down the flume, I asked when the test would happen and whether I could see that. In two months' time, he told me, and no, no outsiders, no people outside the client's company, were permitted; it was a proprietary wave—or, better, a wave that would generate proprietary knowledge. There would be other waves made in the service of public, Dutch, projects to test the durability of dikes, but those would not happen for a while. Today, the future flume wave was being written as a numerical model, perhaps by postdocs from the Delft University of Technology, who might find slices of it useful for their academic research.

The water that would eventually be pumped into the flume was held underground. In fact, all nine million liters were just beneath us. It was kept out of the sun to protect it from filming over with algae, going green—as I had noticed the canals of Delft seemed to have done in the previous few days, reminding me that Antonie van Leuwenhoek, a founder of microbiology, was from Delft (and that Vermeer, another Delft celebrity, always painted mirror-clean water). Water—and waves—are not just physical forms; they can be biological habitats. I imagined the water below sleeping, waiting to be formed into a wave two months from now—a waterwolf, reanimated just before its crashing death, its height and orbitals calculated and measured, to tame its real-world analogs. The wave flume was a stage for pre-enactment.

Pre-enacted waves help to design wave traps. As the anthropologist Alfred Gell (1996) once argued, animal traps often contain a model of the thing to be trapped, a negative image, even a parody, of the prey's phenomenological world: think of how rat traps lure rats into a darkness humans imagine they find inviting (Lee 2021). Artificial waves made in the Delta Flume are images of the thing to be countered, but they are also negative images, positives to be negated. These artificial waves, reanimated and resurrected again and again so they can be broken, are undead, zombies. They are surrogates, doubles, models for waves domesticated.

Station 6: Henri's Wave

I left Deltares with a question. How did one *program* a wave—either as a template for a physical wave in a wave tank or as a computer model to be used in puzzling through an engineering project? I connected with a young wave engineer named Henri van der Heiden, a mathematician completing his doctorate at the University of Groningen, who offered to show me how he built a wave in simulation (see van der Heiden 2019). I bicycled to his apartment in the north of Utrecht, in a neighborhood split between young Dutch professionals and recent immigrants, including refugees from Somalia.

Van der Heiden told me there were many ways to represent an ocean wave—in language, in drawings, in mathematics, and in computer code—and each representation had its own incompleteness. When he created virtual waves, he was not interested in poetic evocations of waves, or in wave colors, wave sounds, or, really, waves' palpable wateriness. He was interested only in some aspects of a wave's form—those that could tell him something about the shape and force of a particular kind of wave hitting an obstruction (a ship, an oil platform). In the computer program he used, van der Heiden mapped wave shape and force within a 3D Cartesian grid, with x, y, and z coordinates. It was not "waves" in general that he modeled, just some aspects of one kind of a wave's form.

He told me, to begin, that he was interested in modeling a wave with sharp peaks and low and flat troughs. He doodled some pointy waves on a piece of lined paper. This was a "Stokes" wave, named after George Stokes (see the introduction), who made major contributions to fluid dynamics in the nineteenth century. Van der Heiden clarified that we would be modeling not the abstraction of his drawing but, rather, an actual wave—one that once existed in the real world. He had to hand a database that held records of measurements from waves generated in a laboratory wave flume in the city of Wageningen, at a place called the Marin Institute. He called one up but then thought better of the example, since it came from a project for a corporate client. "Someone owns this wave," he said. He found another. The numbers that described this once real Stokes wave outlined a wave in which the peaks were more pronounced than troughs, sharper than they would be in a sinusoidal wave.

The wave van der Heiden pulled from his files and wished digitally to resurrect was one with a period of 9.001 seconds and a height of 7.49 meters. He moved from hand-drawn and computer-sketched visualizations to another species of representation. What he was after in his wave model, he specified, was a wave that could be described by the Navier-Stokes equations, which describe conservation of momentum and mass in fluids in motion, as in this rendition:

$$\rho\left(\frac{\partial \mathrm{v}}{\partial t}+\mathrm{v}\cdot\nabla\mathrm{v}\right)=-\nabla p+\nabla\cdot\mathrm{T}+\mathrm{f},\ \nabla\cdot\mathrm{v}=0.$$

Navier-Stokes equations, named after the engineer Claude-Louis Navier and George Stokes in the 1820s, describe the motion of incompressible fluids, spelling out how the acceleration of fluid particles (on the left side, which includes ρ as density, v as velocity) is shaped by changes in pressure, viscous stresses, and gravitation (on the right side). Van der Heiden's Navier-Stokes was a mathematical formalism creating a set of lines that traced out the peaky wave in which he was interested.

Things got more exact. Van der Heiden wanted a *fifth-order* Stokes wave. What did that mean? Well, if a zero-th order wave describes a flat surface and a first-order wave limns a sinusoidal wave and a second-order wave starts to introduce some pointiness, a fifth order wave cranks up the peakiness by ratcheting up the math to a higher than fifth-order partial differential equation (Fenton 1985). Henri found the math compelling in no small part because he trained originally as a physicist.

Where, I asked him, was our wave here—our notional wave with period 9.001 seconds and height 7.49 meters? Suspended between multiple realities and representations, it floated at the interface of the ghosted outline of the wet Wageningen wave of a few months earlier; the drawings that van der Heiden made to illustrate the undulation; the Navier-Stokes equations that were harnessed to get these forms mathematically described; and—what was next—the FORTRAN computer code that would translate these equations, using values from the Wageningen wave, into actionable computer processes. My eyes drifted to the on-screen segment of the Extensible Markup Language (XML) pointer that van der Heiden said he would use to set up the variables for the wave:

```
<model model="Stokes5" period="9.001" custom_wave_number="0."
height="7.49" angle="0." location_of_crest="-122.63" order="10"
standing="false"/>
```

The computer code was annotated and pointed to publications in which canonical versions of Stokes equations appeared. Doing an archaeology of the code reveals papers and the names of scientists; this wave was full of people and their ideas:

```
!-----------------------------------------
!  Compute parameters (shape, velocities, pressures) for the 5-th
order Stokes wave.
!
!  The nonlinear finite amplitude 5-th order Stokes wave is used.
Given:
!     H -wave height
!     T -wave period
!     d -water depth
!  compute remaining parameters, wave geometry and hydrodynamic
field (velocities and pressure).
!
```

```
!  Publications used:
!  Stokes finite amplitude nonlinear wave of the 5-th order, as
presented in:
!      Skjelbreia, L., and Hendrickson, J.A.,
!      "Fifth Order Gravity Wave Theory,"
!      Proceedings, 7-th Coastal Engineering Conference., The
Hague, 1960, pp. 184-196.
!
!  The so-called sign correction to this theory has been
incorporated, see:
!      Fenton, J.D.,
!      "A Fifth-Order Stokes Theory for Steady Waves,"
!      Journal of Waterway, Port, Coastal and Ocean Engineering,
!      vol. 111, No. 2, 1985, pp. 216-234.
```

I had recently seen wave paintings by the nineteenth-century Russian artist Ivan Aivazovsky (plate 4), who thought of his paintings as composites: "A sea storm seen by me near the coast of Italy might be transferred in my painting to some place in the Crimea or the Caucuses, a beam of moonlight that was reflected in the Bay of Bosporus might light up the citadels of Sevastopol. Such is the nature of my brush and characteristic feature of my artistic mind."[20] While this might be, in addition to an artistic reflection, a Russian imperial fantasy about gathering new territories into the nation, it has something in common with computer models of waves, which are *also* composites of measures and models from more than one place (or, in the case of JONSWAP spectra, just one place over time: the North Sea).

Next came programming. Van der Heiden used a 3D grid whose cubic slices—"cells"—represented both the "container" of the wave and the virtual "water" of the "wave" itself. He told me to think of something like a 3D piece of graph paper—this was the "flow domain"—in which each cube would change state depending on input from its "neighbors." The flow domain permitted an inventory of movements: pressures, speeds, and, when wave height became large compared with wavelength, deformations and breaking. Iteratively computed, this would be a "wave," a 3D digital animation—a diagram, a mash-up of reality and theory—that could be time stepped from one moment to another.

It would also be necessary to assume a constant wave elevation—actually to assume that there was *no wave* at first. In other words, it would be necessary to install a *level*—kind of like all the sea level markers I had seen around the Netherlands, I remarked. Van der Heiden said, "Yes, the level appears as an abstraction of the observation of the wave." Just as sea level is a generalization, one created by factoring out watery oscillations—and

FIGURE 1.9 Computer visualization of Henri van der Heiden's wave.

in its motivating articulation, tailored to specific Atlantic ports—this virtual level was a benchmark required to keep nonwatery infrastructure in stable focus.

Van der Heiden compiled the program. The computer heated up. A visualization appeared (figure 1.9). He started a "moviemaker" program to generate a real-time visualization. By tweaking its elements, one could pick out features of the wave's internal structure—so if one visualized pressure changes, one could pick out the "orbitals" in the wave.

Like the older wave scientists I had met, van der Heiden had an impressive bookshelf, and we fell into talking about Italo Calvino's "Reading a Wave," that tale in which a Mr. Palomar seeks to follow a single wave from an on-the-beach vantage (see the introduction). Palomar eventually becomes so absorbed that he gets lost in time. As he watches the waves, they seem suddenly to be going backward: "Now, in the overlapping of crests moving in various directions, the general pattern seems broken," Calvino writes. "If you concentrate your attention on these backward thrusts, it seems that the true movement is one that begins from the shore and goes out to sea. Is this perhaps the real result that Mr. Palomar is about to achieve? To make the waves run in the opposite direction, to overturn time. . . . ?" (Calvino

[1983] 1985, 7). Here, in van der Heiden's Utrecht apartment, we were also playing a game with time, since the programmed waves we were looking at were resurrections of waves already gone by. This was not just reading waves, but also writing waves—not in books but in programs, abstracted into animated formalisms. Since one aim of this modeling was to help structural engineers design structures to withstand waves in the *future*, there was anticipation here, too; these waves, made of information, were cybernetic oracles.

Station 7: The Datawell Waverider Factory

I have noted that, if "Reading a Wave" is an inquiry into how to "read something that is not written," then the Palomar character already comes to the sea with mathematics and measurement in mind, a grid through which to apprehend the world. One might say the same of the *wave buoy*, an instrument deployed in ocean waters to monitor and report wave heights, periods, and directions to regional and national weather services, private companies that operate oil-drilling platforms, networks of container ships, fishing boats, and surfers. The world's most widely used wave buoy is the Waverider, manufactured since 1961 in the Netherlands by the company Datawell.[21] Like Mr. Palomar, such buoys are tuned to pick out "the pattern of the waves," tracking "forms and sequences that are repeated" (Calvino [1983] 1985, 5, 4). Unlike Mr. Palomar, though, such devices are imagined not as seats of subjective experience but, rather, as mere material recorders and relays of measure, as instruments that *take readings* rather than *do readings* in the sense of offering informed exegeses.

I had been keen, since learning about the Datawell Waverider at the Banff conference in 2013, to get a better sense of what it did, how it *read*. I took a train to the hamlet of Heerhugowaard and visited the site of Datawell Service, Production, and Sales.[22] Dressed in jeans and a plaid shirt, the Datawell sales representative Harry Pannekeet took me to a modest clutch of modern office buildings. Pannekeet had worked for Datawell for thirty-five years, starting out in the company's warehouse. He was a store of immense knowledge, though modest in telling Datawell tales, referring me for elaboration to a book Datawell published on the company's fifty-year anniversary (Joosten 2013). Books of waves seemed to be everywhere.

Pannekeet walked me through a facility that had the look not of a massive production line, but of a small craft factory. Three Dutch women, around middle age, sat at desks next to large windows, chatting while they assembled circuit boards and set them on metal chassis that would be inserted into buoys. Those circuit boards, with parts manufactured somewhere else in the global assembly line, attach to various global positioning system (GPS) and transmission devices.

Pannekeet ushered me into a laboratory to show me the interior plastic orb of a Waverider, a fishbowl-like globe that eventually would be cradled within the metal sphere of a buoy casing (figure 1.10). The globe materializes a design that has been in place (and patented) since the 1960s, when Datawell began making these devices—a task for which the company was founded after the Dutch floods of 1953: "The need for a wave height measuring buoy originated from the tragic floods in the province of Zeeland, in 1953. In order to prevent such catastrophes in the future, the government decided to close the sea-arms and raise the dikes. The required dike height is not only determined by the maximal water level, but also by the height of the waves threatening to wash over the dikes" (Datawell BV 2001, 1). The waves for which the buoy was originally designed were those storm surge waves to which the Dutch coast is vulnerable. The buoy is a sentinel, inheriting a tradition sometimes celebrated in maritime prose and poetry, as we hear in Rudyard Kipling's "The Bell Buoy" (1896, 1), in which a buoy speaks as a loyal colonial servant: "They made me guard of the bay, / And moored me over the shoal. / I rock, I reel, and I roll," or in *Dracula*, in which one character tells of "a buoy with a bell, which swings in bad weather, and sends in a mournful sound on the wind" (Stoker 1897, 65). Unlike these nineteenth-century buoys, which required an auditory reading, the Waverider is, for people who query it, silent. And it does not speak in the first person; insofar as buoys themselves "read" waves, they do so through a mathematical idiom, their measures filtered into a sign system that has no place for an experiencing, speaking subject. (Unlike natural languages, the "language" of mathematics has no "I," "me," or "us" [Rotman 2000].)

The center of the buoy apparatus hosts an accelerometer, a sensor that tracks the changing speed of a thin sideways-mounted platinum-iridium rod waving up and down inside a plastic box filled with an electrically conductive liquid. The rod's waviness, generated by and acting as an interior scale model of the waviness of an animating ocean wave in the buoy's immediate surroundings, is tracked by two electrodes that measure oscillating electric potential difference (voltage). That oscillation (relayed via two electric wires up to a buoy's transmitter) serves as a proxy for vertical acceleration—and therefore for wave height.[23] But measures of wave height can be extracted

FIGURE 1.10 The interior globe of a Datawell Directional Waverider buoy in the factory/lab at Heerhugowaard, Netherlands. Photograph by the author.

only if a buoy can separate vertical acceleration from the pitch, yaw, and roll of "a randomly swaying buoy." The solution to isolating up-and-down movement, the company's literature reports, "was found in introducing a so-called stabilized platform, serving as an *artificial horizon*, on which the accelerometer is mounted" (Datawell BV 2001, 2, emphasis added). The buoy needs to bring the level of the world into its innards. If, as Gilles Deleuze ([1985] 1995, 121) once wrote of surfing, "the form of entering into an existing wave . . . no longer [demands] an origin as starting point, but a sort of putting-into-orbit [of the surfer]," the buoy, borne up by orbiting water molecules, nonetheless retroactively simulates an originary stillness.[24] The buoy materializes a diagram of wave action.

The buoy requires a design tuned to the physics of the sea that is dependent on precisely chosen fluids, plastics, and metals. A key puzzle was to get interior liquid and plastic to have matching thermal expansion coefficients so the "submerged weight of the platform and the accelerometer [would be] virtually nil" (Joosten 2013, 50. The buoy, in other words, would not be *measuring itself*, which would deliver a *misreading*, an undesirable mechanical reflexivity.[25] A combination of polystyrene plastic matched with a mix of distilled water and glycerin offered a path toward stillness. This permits the buoy to be both *in the sea but not of it* and *in the sea* and *of it*. At the heart of the Waverider is an almost alchemical balance. For the Waverider to read waves, its apparatus has to be just so.[26] Pannekeet told me it would take two months of on-the-shelf aging for each globe to be ready to be locked into the metallic, yellow-colored container that houses the device. A set of globes on a shelf sat nearby, aging like fine cheese.[27]

I had read in the company's literature about the ocean conditions to which the device materially had to adapt. Seawater. There was also *pollution*, which would eat "into the stainless steel hull within half a year," a problem addressed by using copper nickel. There was the ice cold: "The raw north proved that rubber cords can freeze, and that ice on the antenna can make the buoy tumble upside down: plunged into the salty water, the ice melts off and the buoy comes up again, resulting in a transmission at intervals" (Datawell BV 2001, 3). Sometimes fish ate into the mooring. The buoy could become an ecosystem for marine creatures—in which case it would be subject to "biofouling," which could influence the "accuracy of the transducers which are integrated into the hull of the buoy." The solution was an anti-fouling paint, which, as a biocide, demanded compliance with environmental regulation.

Waves exist in a vital ocean. That vitality can jam the measurement of processes understood as analytically separable as "physics." Wave buoys read waves as formal structures and less as chemical heaves of swirling hydrogen oxide dosed with sodium, chloride, sulfate, magnesium, calcium, and potassium ions (a.k.a. seawater). Waves are not here apprehended, either, as moving ecologies of microbes, plankton, or fish.

The physics-centered, formalist commitments of buoy measurements have been leveraged by the American artist David Bowen (n.d.), who in a 2011 installation, "Telepresent Water" (figure 1.11), created a piece that

> draws information from the intensity and movement of the water in a remote location. Wave data [are] . . . collected in real-time from National Oceanic and Atmospheric Administration data buoy station 46246, 49.985 N 145.089 W (49°59′7″ N 145°5′20″ W) on the Pacific Ocean. The wave intensity and frequency is scaled and transferred [via marionette-like strings] to [a] mechanical grid structure resulting in a simulation

FIGURE 1.11 David Bowen, "Tele-present Water," 2011, National Museum, Wrocław, Poland.

of the physical effects caused by the movement of water from halfway around the world.

The transmission of data from buoy 46246 (a Waverider sited around Hawai'i) to an art gallery computer underscores the assumption that waves can be known through form alone. That, of course, is a technical and aesthetic simplification. Just as optical character recognition can be confused by blurry documents, handwriting, and more (Mills 2016), automatic wave recognition can be stymied by the information the buoy's designers fashion as separate from its watery substrate.

Some buoys have sunk below the waterline due to water ingress, which has then permitted animals—barnacles, mussels—to grow on them. Sometimes they sink far enough below their pressure rating that they implode. Pannekeet took me through Datawell's repair shop, a hospital of damaged buoys. Some had been run over by ships. Some had been used as moorings. Buoys from US waters sometimes come back with bullet holes—a sign that people are either using them for target practice or seeking to disable them. Some had been taken apart for scrap metal. If buoys are other-than-human sensory organs for apprehending the ocean, these organs point to processes other than the oceanographic; one might read buoys, against the grain, as signs of politics of resistance, appropriation, or resource inequality (as with people who steal buoys to secure ransom from their owners, whose organizational names are often painted on their exteriors). The materials in buoys also hail from widely distributed and changing political economies of mining (platinum, iridium, copper, zinc, nickel), the manufacture of synthetic materials (polystyrene plastic), and plantation agriculture (rubber).[28] Spooling an inventory of elements out from the Waverider provides a snapshot of relations of resource access, regulation, and modification over the past fifty or so years of European industrial manufacture.

When they work as designed, Waveriders provide information about significant wave height, period, and direction sampled from patches of water over thirty-minute spans of time (long enough to discern regularities; short enough to be a "snapshot"). Such accounting does not provide anything like a biography of a single wave, in the way that Mr. Palomar desired—or that the wave scientist Hisashi Mitsuyasu seems to deliver in *Looking Closely at Ocean Waves: From Their Birth to Their Death* (2009), which describes waves as having "lives" that begin and end. Rather, it delivers data points that can be assembled into a population profile, a spectrum of frequencies. Information traveling outward from buoys has, too, to fit into legal frameworks. In early days, the Dutch authority that hosted the Rijkswaterstaat (the Directorate-General for Public Works and Water Management) was the same as the one issuing transmitter licenses, a convenient convergence that made Dutch waters "the great outdoor laboratory of Datawell" (Joosten 2013, 185).[29]

The Waverider thus began its life as a technoscientifically materialized container of Dutch images of waves. It made me think of the work of the Netherlands-trained Arab Javanese painter Raden Saleh (1807–80), whose 1863 "Flood in Java" (plate 5) lithograph shows people crowding onto a raft to escape a flood in the Jawa Tengah Province. Saleh here uses Dutch seascape conventions to array people and horizon. The lithograph, like a buoy, transfers wave images across transnational space—though Saleh, who created it in Java after his twenty-year sojourn in Europe, may have meant it as an indictment of Dutch colonial negligence (Krauss 2005). The Datawell buoy, like artistic images of waves, carries a seascape history and, over the years, has been modified to read across ever more seas, even as it has also retained traces of its Dutch domestic origins.

Station 8: Wave Sensors at Schiermonnikoog

The waves had vanished. "There were waves here," Anita Engelstad affirmed, directing my eye to rippled traces at our feet in the finely sculpted sand of Schiermonnikoog, a low-lying barrier island near the Netherlands' farthest eastern reaches. Engelstad, a doctoral student in geosciences at Utrecht University, had brought me along on a field trip to this island, judging me a competent-enough assistant to help check devices positioned to record wave heights and periods on a stretch of sand on this sixteen-kilometer-long, four-kilometer-wide isle. Schiermonnikoog, at the southern edge of the North Sea and just north of the muddy intertidal zone known as the Wadden Sea, is a Dutch national park noted for wide beaches and for its history as land that has drifted among Frisian, German, and Dutch territorial claims. (It was the last piece of European land to be liberated from the Axis powers during World War II.) That drift has sometimes been literal, as tidal currents have nudged the isle southeastward over the centuries.

Engelstad and I were walking along a deserted, drizzly beach, checking in on pressure sensors her research group had positioned a few months previous. I had been learning from Engelstad's colleagues how the travel and breaking of waves could contour beaches. Marion Tissier, assistant professor of environmental fluid mechanics at the Delft University of Technology, who holds a degree from Bordeaux, told me about a genre of wave called an *infragravity wave*. Wind-generated waves (of sufficient strength to overcome the restoring force of surface tension) are known as *gravity waves* because gravity is the force that pulls them back to equilibrium (i.e., back to level with the undisturbed sea surface). Recall that waves often travel in groups. A group has its own wavelength; when that wavelength resonates with the wind-driven gravity waves the group carries, an additional wave, "bound" to these others can emerge: an infragravity wave (figure 1.12). When these infragravity waves arrive at shore, they may carry quite a bit of energy, even amplifying while the wind-generated waves that helped create them break and dissipate. The "biography" of a wave after its generation, Tissier instructed me, was shaped by its interaction with its

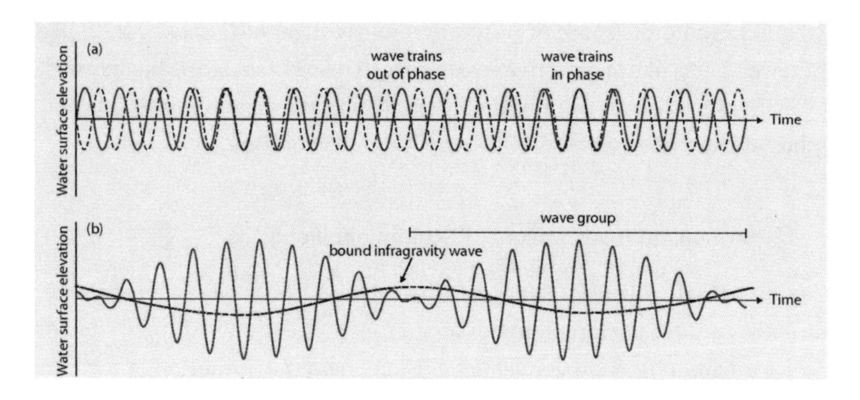

FIGURE 1.12 Diagram of the emergence of a bound infragravity wave. In (a), two wave trains of slightly different wave lengths, but same amplitude, merge; in (b), the point where their amplitudes reinforce one another (constructive interference) becomes the center of a wave group; two groups then induce a long, bound wave. Paraphrased from *Infragravity Waves*, http://www.coastalwiki.org/wiki/Infragravity_waves#cite_ref-OU_5-0, adapted from Open University Course Team 1994.

group and the emergent long infragravity waves that interaction creates in a recursive nonlinear process—one that she thought of visually rather than statistically: "Spectral analysis is a powerful tool, but you lose contact with reality." I learned from one of Tissier's doctoral students, Anouk de Bakker, that infragravity waves are crucial for understanding how sand and sediment are transported on beaches. Why is that important? "It's not only about sea level rise," she told me, "but also about changes in wave characteristics—storm recurrence, different directions they come from" (de Bakker et al. 2016).[30]

At Schiermonnikoog, Engelstad wanted to know whether her sensors had recorded evidence of some parts of the island being *overwashed* by waves, which would mean that waves were arriving at the highest point on the island to then slosh over that boundary rather than recede backward to the sea. If such overwashing waves were transporting sand up the beach as they moved along, they might be *building* rather than eroding the island, at least in the short term (Engelstad et al. 2017). This could mean that, contrary to conventional wisdom, protecting barrier islands such as Schiermonnikoog might best be served by opening up, rather than reinforcing, dikes and artificial dunes. Grappling with sea level rise means thinking about seawater not as a solo act but as a force acting in concert with the worlds of sand and land. And air: Tissier and de Bakker were mentored by the Utrecht physical geographer Gerben Ruessink (also Engelstad's adviser), who is interested in "what happens when Aeolus meets Poseidon"—that is, when wind-sculpted beaches meet wave-driven erosion.

Engelstad and I visited Schiermonnikoog at a time of day when we could be mostly certain waves would not be lapping at our feet. Engelstad needed access to her sensors. These sensors, logging changes in water pressure, recorded the changing heights of waves that traveled over them. Together with measurements from other devices—such as an Acoustic Doppler Velocimeter (for tracking current) that we spent an hour digging out of recent sand cover—these could help Engelstad later figure out things such as whether waves were becoming asymmetrically skewed as they crawled up the beach, a sign that they could be transporting sediment. (That later figuring out was glossed for me by de Bakker this way: "You can just lift out a part, and then study it later behind your desk.") Thinking and acting around a rising and falling sea at Schiermonnikoog was a mix of frustration, brain strain, and dedicated instrument maintenance. (De Bakker told me stories about students' laptops being swept out to sea as they downloaded data from sensors.) Along the way, there was a delegation to sensors, machines, and, later, computer programs, of human questions about what waves might say about erosion, and, perhaps, sea level rise.

Engelstad's project grappled with how to sense the scale of such shifts. Climate transformation will be a story of swirling sand and water both, not only a tale of watery inundation—as, too, will sea level rise be a durational unfolding with sea and scrambled land. It will be local and uneven. Sea level is not a given in nature; it is a heterogeneous measure (and average) that has been conceptualized and created many times over, in many places, with idiosyncratic local definitions (e.g., high-tide marks in Venice, Amsterdam, and Liverpool) keyed to the bounding of national territories, sometimes extended into sea-spanning standards, its origins therefore keyed to what Walter Mignolo (2011, 255) might call an "epistemically imperial" operation. As I learned from Engelstad during fieldwork at Schiermonnikoog, today's notion of the "level" remains a social-technical accomplishment; keeping instruments of measure "still," ready to iterate measures, takes maintenance.[31] It is a question of infrastructure. Whether ocean waves *themselves* might be imagined as infrastructure—a durable pattern of nature, domesticated—leads me to my final station, a place/object called the Sand Motor.

Station 9: The Sand Motor

Around the turn of the millennium, the Dutch state announced a new motto for water planning, "The Netherlands lives with water." This, write Anna Wesselink and her colleagues (2007, 198), "reinforce[es] the idea that the population might enjoy to live near or in water instead of foremost considering it an enemy." Under the optic of what has come to be called Dutch "new nature," rivers would no longer only be channeled into canals; they would be permitted to meander through the grooves and furrows of the landscape (often, it must be said, displacing residents who had long living in the "old," modified nature [see Bulkens et al. 2016; Drenthen 2009]). Such a project is imaginable at the national scale owing to long-established state structures for hydrological planning; the "Netherlands" is the frame within which futures are imagined. Projects of placing dikes along the Dutch coast so they might be modified to "work with" the landscape has demanded a unified state sovereignty. Waves move from foes to friends, becoming part of a controllable, sculptable infrastructure.

Nowhere is this notion more visible than in a large coastal engineering project known as the Sand Motor, located just west of The Hague (figure 1.13). In 2011, the Rijkswaterstaat, the water management arm of the Dutch state, authorized a massive sand "enrichment" along the west coast of Holland. Twenty-one million cubic meters of sand were deposited along a beach, creating a one-kilometer-square, sandbar-shaped peninsula: the Sand Motor or, in Dutch, Zandmotor.[32] As its primary architect, the hydraulic engineer Marcel Stive, told me, the idea would be to "let nature deform it continuously. It can introduce habitats that we don't have yet. It can be interesting to the province and offer more recreational space. It can become internationally attractive." The Sand Motor was imagined as a successor to the Delta Works (see figure 1.5), more attuned to "nature" than those high-modernist dams of Zeeland, a gentler infrastructure made of sand in motion, not concrete and metal against the ocean—a project in what the philosopher of water Irene Klaver and her colleagues (2002, 14) have named "the controlled decontrolling of ecological controls" (see also Keulartz 1999). If the Sand Motor is Delta Works 2.0, it is also an example of reflexive modernization, modern planning recursively taking into account its own effects (Beck et al. 1994).[33] Stive told me that the Sand Motor was meant to have a life span of twenty years, and he spelled out the particulars of what this might mean. The coast sees about 600,000 cubic meters of sand going north every year and 400,000 going south, so a net of 200,000

FIGURE 1.13 Aerial view, Sand Motor, June 13, 2020. http://www.flickr.com/photos/zandmotor/51276240732.

cubic meters is going north, which could be slowed by a properly shaped peninsula. Building the Sand Motor, he said, would "mean that the ecosystem and the people grow with the solution, so that they don't experience it as something forced, initially."

The motto of this enterprise was "Building with Nature" (see de Vriend and van Konigsveld 2012; see also Disco 2002; Onneweer 2006; Temmerman et al. 2013). With waves, I asked? Yes, Stive said, telling me:

> There are two parameters dominating the process of changing beach morphology: wave height and direction. *Transport* is proportional to square of wave height, so a two-meter wave height is four times as powerful. For direction: the sine of two times the incident wave angle is proportional to the longshore transport, with most happening at forty-five degrees. So the whole idea is that waves in the subaqueous realm and wind in the air domain were thought of as being the two engines for making the Sand Motor.[34]

This was a large-scale operationalization of Ruessink's call to fuse the powers of Aeolus and Poseidon. Here, waves move from being figured as *animals* to being imagined as *engines*. I was reminded of Theo Jansen's famous Strandebeests, kinetic beach sculptures made of wood, PVC pipe, airfoils, and zip ties that make use of wind to "walk" on the Dutch shore, perhaps stealing wind from waves to make new kinds of material culture and life (Herzog 2014). But maybe the more fitting machines to recall are those enlisted in massive importations of sand, in what Maarten Hajer has described as "a slow motion ballet méchanique of draglines and bulldozers, excavators and trucks" (quoted in Bulkens et al. 2016, 811). These are the tools—and the politics, which fuse the interests of the sand excavation industry, water engineers, and ecologists—that permit making waves into domesticated machines.

I was eager to see the Sand Motor. I joined a tour of the entity/place led by a group of Hague-based eco-artists who call themselves the Satellietgroep. The artist Jacqueline Heerema, along with a few experts in hydrology and water and dune ecology, would be leading thirty people on a three-hour traipse across the sand. The trip was advertised as a chance to "reflect on the influence of climate change and sea level rise and [on] how we can . . . position . . . ourselves . . . [for] these transitions for the coming future." The members of the Satellietgroep were also hoping to gather data for a projected "Anthropogenic Coastal Atlas," a contribution to what they posed as a project of a Dutch *innovatory heritage*—heritage for the future.

I arrived at the seaside city of Monster and joined an already-in-progress introduction to the Satellietgroep event led by two chefs presenting food they had created from Sand Motor plant life—a mild guacamole-like puree of grasses and a *duneliquor* that they thought exhibited the *terroir* (local flavor) of the place. One scientist told us about the "Building with Nature" ethos. As we set out on the walk, a guide told me she believed one version of the Anthropocene had started in the Netherlands, when someone dug the first ditch for a dike a thousand years ago. The Dutch, she affirmed, had been changing their landscapes—with enduring generation-to-generation effects—for centuries. We walked through a sandscape filled with craggy trees and rewilded heather—with "rewilding" a companion controlled decontrolling of ecological control, working with natural processes to deliver a desired function.[35] I spied kite surfers taking advantage of the waves on this engineered shorescape (waves that might be relatives of those sculpted waves that appear in artificial surf pools, as at Kelly Slater's California-sited facility, which reliably generates long, open-barrel waves

[Goode 2018]). Taken together, the food, the walk, the rewilding, and the tour were all avowedly aimed at making the Sand Motor *Dutch*.[36] The mix of modesty, pride, and measured sensualism that suffused our introduction to the Sand Motor rhymes with Klaver and her colleagues' "controlled decontrolling of ecological controls," a phrasing inspired by the sociologist Norbert Elias's analysis of how tightly reined-in manners have given way, in recent middle-class European culture, to a "controlled decontrolling of emotional controls" (Klaver et al. 2002, 14). Waves of emotion and waves at the shore may be permitted some limited room.

I could not help but hear a Dutch exceptionalism and aspirational universalism. And, indeed, the "Building with Nature" approach is already being launched as a transposable model, particularly as Dutch hydrologists advise engineers in Southeast Asia (Bakker et al. 2017). One graduate student told me that people were so taken by the idea of the Sand Motor that they sometimes imagined transposing it to other ecologies; he showed me an image of the coast of Lima, Peru, with a Sand Motor digitally plopped in using Photoshop. So we are, perhaps, back to the Waterloopbos and its ambition of scaling Dutch things down to scale them back up, both in the Netherlands and in such elsewheres as might still be accessed by this once colonial power, this avatar of wave theory from the North.

The Sand Motor story is one, too, about hope and futurity—something of which I was reminded by another visitor on the tour, a Japanese artist named Nishiko55, who told me about her recent work gathering small household items broken by the Tōhoku tsunami of 2011—items she sought to repair as a work of mourning and remembrance. The Sand Motor does not tarry with past fracture or with the possibility that sand may not roll neatly through time. As Jerry Zee (2017, 216) observes in his anthropology of sand enrichment in China, "Sand renders time into recursions . . . as a material that moves, accretes, holds momentarily steady, or tends to dissolution, it draws together past and future burials." Sand enrichment does not guarantee outcomes but is, rather, he writes, a "chronopolitical experiment," one that may not forestall the calamitous futures it is meant to buffer. Engineered sand, like engineered waves, may be monstrous—an unsettling, boundary-crossing hybrid of the natural and artificial, a mix of forces perilous and promising (Haraway 1992; Lorimer and Driessen 2013).

Waves are here figured as environmental infrastructure, ready to leverage into narratives of national futurity.[37] Take as one Dutch symbolic affirmation of the possibility of controlling waves—and of treating waves as infrastructure—the "Wave" apartment building in the city of Almere

FIGURE 1.14 The Wave apartment building, Almere, Netherlands. Photograph by the author.

(figure 1.14). Waves as animals, waves as engines, waves as infrastructure. Maybe this piece of architecture also suggests that waves could become organized rationally as kinds of nonhuman agents that can be made to hew to Dutch values of reason and measure. Self-consciously expressed Dutch virtues of tolerance for difference are here extended to a world beyond the human, to waves, even as the figure of the "wolf" migrates to far-right Dutch Islamophobic discourse to associate Turkish and Moroccan migrants with "lone wolf" terrorists, in turn to be confronted by populist ideologues such as Geert Wilders, who calls himself a "liberty wolf" (see also Weiner and Carmona Báez 2018).

Beginning in the late nineteenth century (and continuing into the late twentieth), Dutch society was organized into "pillars," or parallel social identity groups—Roman Catholics, Orthodox Reformed Protestants, social democrats, liberals—who had their own schools, unions, social workers, and more, even as their membership nonetheless "crosscut categories of class, hierarchy, region, [and] ethnicity [up to a point; Frisians and Hollanders, for example, remain considered ethnoracially white]" (Taussig

2009, 25). Members of different pillars, "living apart together," were meant to share common national cause; pillarization was a strategy to allow for "the general tolerance of social heterogeneity by bounding difference and minimizing its social threat while containing it in the larger commonality of Dutch society" (Taussig 2009, 28). (In practice, of course, difference was still material for inequality, a fact that has become stark with recent debates about immigration.) Think fancifully of this, then: waves as members of populations—wave trains, groups, sets—that conform to reasoned measurement and pragmatic apprehension; that can be treated as partners, not enemies; and that are most valued when they are not wolfish but ordinary. Think of waves brought into Dutch hydrological discourse as Dutch nonhumans, maybe even as agents domesticating Dutch persons and polities. Think of waves, in this ideology, as occupying their own pillar, playing their own role in structuring Dutch society.

That's probably too much. The wave science in which Dutch practitioners have been active—and to which they have contributed so vitally—has also been an international venture, with American, British, German, French, Japanese, and Russian (at least) traditions crosscutting the field's store of research and theory. The future of Dutch wave science and hydrology may even exemplify such internationalism. Think of Rotterdam, the largest port in Europe, whose Mayor Ahmed Aboutaleb, the first immigrant (and Muslim) mayor of a Dutch city, has been central to organizing international conversations about how low-lying cities might adapt to sea level rise. In 2017, a delegation from Dubai was invited to the city to talk to Dutch hydraulic engineers, as well as to contribute their own expertise, their own modes of controlling water and waves.

Waves will still have uncontrollable animacies of their own. In his short story "The Netherlands Lives with Water" (2009), set in Rotterdam in the year 2024, Jim Shepard delivers an apocalyptic tale in which all the planning in the world cannot stop the inundation of the country. The narrator of Shepard's story summarizes the uncanniness of Dutch visions of waves as frenemies:

> We're raised with the double message that we have to address our worst fears, but that they'll also somehow domesticate themselves nonetheless. Fifteen years ago Rotterdam Climate Proof revived "The Netherlands lives with water" as a slogan, the poster accompanying it featuring a two-panel cartoon in which a towering wave, in the first panel, is breaking over a terrified little boy before its crest, [and] in the

> second, separates into immense foamy fingers so that he can relievedly shake its hand. (208)

I have not been able to discover whether Shepard's tale points to an existing poster or whether he has invented this image—the wave on the edge of breaking, making peace with people—to capture how anthropo-zoo-etc.-morphisms track cultural ambivalences about watery, wavy nature in the Netherlands. But the point is clear: if domestication, which draws on the word *domus* (household) "obtains its dramatic force from the exclusion, control and domination of the wild, the outside" (Hodder 1990, 45), waves are dramatic forces of their own, raising questions, from inside and out, about the conditions under which human houses can hold.

Set One

First Wave

The Genders of Waves

Seeing the sea as a feminine force and flux has a storied history in the cross-currents of Judeo-Christian thought, Enlightenment philosophy, and natural-science epistemology. The ocean has been motherly amnion, fluid matrix, seductive siren, and unruly tide; these castings have opposed principles of masculine monogenetic procreative power, ordering rationality and self-securing independence and dominion over the biophysical world (see Bachelard [1942] 1983; Grosz 1994; Irigaray [1977] 1985; Theweleit 1987). At other moments, the ocean has been masculine, the embodiment of Poseidon or Yahweh or the virile power of storms and vigorous hydrotherapy. Ocean waves—icons of rhythmic and predictable motion, as well as of chaos and destruction—have been similarly gendered, though most often as kinds of women.

The gendering of waves is intriguing to think through at a moment when the humanities have turned toward the material world to make sense of domains beyond the human—the "posthuman," "multispecies," and "nonhuman" (see Barad 2008; Braidotti 2013; Cohen 2015; Grusin 2015; Chao, Bolender, and Kirksey 2022; Wolfe 2009). With the arrival of conversations about "vibrant matter," scholars have written of the ocean as a site of "wet ontologies," a zone of "three-dimensional and turbulent materiality" (Steinberg and Peters 2015, 247).[1] Traces of the human still haunt such a-humanisms, particularly in the time of the Anthropocene. Timothy Morton (2013) calls Anthropocenic objects such as the *globe* and *climate* "hyperobjects"—phenomena that exist on massive scales. Such objects are *viscous, nonlocal, temporally undulating*, and *phasing*.

The scales at which hyperobjects operate would seem far from arenas in which gender is relevant (or even individual waves, though at scale these are certainly viscous, nonlocal, temporally undulating, and phasing). But, as Judith Butler observes in *Bodies That Matter* (1993), the notion of "matter"—at least in languages tangled up with Latin—has been densely figured as female, as *matrix, mater*, an inheritance difficult fully to escape. More, the ocean has often itself been framed as a kind of Mother.

That said, gendered castings of the sea have always been unstable. Gender is a concept created "to contest the naturalization of sexual difference" (Haraway 1991a, 131) and employed, in concert with race, class, and other analytics, to decode how inequalities manifest across human selves. But it may also be adapted to upturn the assignation of sex, sexuality, and other genres of difference to nonhuman entities.[2] More: today's gender-bent waves might be signs of how the analytic of gender may be *viscous, undulating*, and *phasing*; at sea.

The essayist Jonathan Raban (2010, 159) writes, "Of all natural symbols, the breaking wave is the most laden with suggestive meanings. For several thousand years, the waves have been talking power and sex and death to us." It may therefore not be a surprise that waves—at least in the West, the parochial focus here—have spoken in symbolisms of sex and gender. In *Male Fantasies*, Klaus Theweleit (1987, 283) offers examples of how invocations of watery flow have been filtered through masculinist anxieties, forms that have made waves into a curl of subaltern feminine imagery:

> A river without end, enormous and wide, flows through the world's literatures. Over and over again: the women-in-the-water; woman as water, as a stormy, cavorting, cooling ocean, a raging stream, a waterfall; as a limitless body of water that ships pass through . . . woman as the enticing (or perilous) deep, as a cup of bubbling body fluids; the vagina as wave, as foam. . . . [L]ove as the foam from the collision of two waves, as a sea voyage.[3]

Tara Rodgers (2016) argues that waveform inscriptions, which represent waves as curvilinear forms, as "pure," tame what Elizabeth Grosz (1994,

203) names "a formlessness that engulfs all form, a disorder that threatens all order." Drawing on Luce Irigaray (1985, 116), who observes that "historically the properties of fluids have been abandoned to the feminine," Rodgers (2016, 202–3) suggests that natural philosophy ideologies in the West have had "waves [as] both form-giving . . . and perpetually in excess of formal representation."

Where to begin a historical investigation of the gendering of waves? Perhaps, following the feminist theologian Catherine Keller, with the Bible. In *The Face of the Deep: A Theology of Becoming* (2003), Keller argues that patristic Christian readings of Genesis have interpreted the void introduced in verse 1:2, "And the earth was without form and void; and darkness was upon the face of the deep," as either a nothingness from which God created the cosmos *ex nihilo* or as a chaotic, fluid, and maternal principle tamed by a masculine God. The "deep" (*tehom* in Hebrew), Keller observes, derives from Tiamat, the Babylonian goddess of the ocean, a motherly principle subsumed into the syncretic biblical creation tale. "What happened," asks Keller, "to the chaos of Genesis 1:2? . . . Was it murdered? Was it a 'she'?" She writes that the "void evinces fullness, its waters, viscosity . . . the second verse sends a mysterious tremor through the whole narrative of creation." For early Christian writers, the "*tehomic alterity* which has been relegated to the outer darkness threatens to flow back monstrously: the *flux*, repressed, returns as *the flood*" (Keller 2003, xix, 9–10).

Noah's flood survives as the excess of Creation, with waves the echoing substance of that inundation, a flood God keeps at bay. In the book of Job, Yahweh declares His power over the seas: "Here you may come, but no farther, here your proud waves break" (quoted in Keller 2003: 130). The fourth-century bishop Basil of Caesarea wrote that, at the moment when the sea meets the land, it "withdraws out of respect, *bowing its waves*, as if to worship the Lord who has appointed its limits" (quoted in Corbin [1988] 1995, 27). Waves are at one moment defiant agents of disorder and at another pious believers.

Waves gather a gendered agency in another tradition of thinking the supernatural. In Old Norse mythology, the nine daughters of Ægir, god of the sea, and Rán, goddess of the drowned, are ocean waves, and each manifests a different wavy form. In the *Prose Edda*, by the Icelandic poet Snorri Sturluson, compiled circa 1200 CE, these daughters are named: Himinglæva, "that through which one can see the heaven"; Dúfa, "the pitching one"; Blódughadda, "bloody hair"; Hefring, "riser"; Udr, "frothing wave"; Hrönn, "welling wave"; Bylgja, "billow"; Dröfn, "foam fleck"; and Kolga, "the cool one" (Sturluson [ca. 1200] 1916, 219).

FIGURE 1A.1 W. G. Collingwood, *Heimdal and His Nine Mothers*, woodcut, 1908.

In Norse stories, these "wave maidens" afflict seafarers by embodying their father's force and their mother's mercilessness. Judy Quinn (2014, 92–96) writes that in *Edda* stanzas in which "names are deployed, the personified wave is ascribed agency, and often rather willful agency . . . female personifications of the sea [are] imagined as nubile, alluring, self-willed, determined, and destructive." Quinn (2014, 96) gives a sample in which wave maidens vex male sailors: "Who are those brides who go along in the surf-skerries and have their journey along the fjord? They have a hard bed, the white-hooded ones, and they play little in the calm. . . . Who are those women who go around together? . . . Seldom are they gentle with the band of men, and they are awakened in the wind. . . . They have pale hair, the white-hooded ones, and those women do not have husbands."

These are scary women. To Barbara Ehrenreich (1987, xv), such manifestations give body to a masculinist dread of "a nameless [feminine] force that seeks to engulf—described over and over as a 'flood,' a 'tide,' a threat that comes in 'waves.'" In a later tale, the wave women are mothers of Heimdallr, a Norse god who keeps watch for Ragnarök, the world-ending battle among the gods in Eddic eschatology (see also Lindow 2001). W. G. Collingwood's woodcut *Heimdal and His Nine Mothers* (1908) envisions the maidens as embodied water (figure 1a.1). The wave maidens move from stern temptresses to maternal, protective figures.

"The Tide Rises, the Tide Falls," a poem by Henry Wadsworth Longfellow (1885, 289), poses waves as delicate carriers of mournful news about the evanescence of life:

The little waves, with their soft, white hands
Efface the footprints in the sands.
And the tide rises, the tide falls.

Waves offer a durable deliquescence—one supported, here, by a whiteness that may also animate a racialized femininity. Compare this with Hortense Spillers (1987, 72, 68), on one oceanic trajectory for the making of Blackness, in which she writes, "African persons in the 'Middle Passage' were literally suspended in the 'oceanic,'" captured in an "undifferentiated identity" that made of them "ungendered" flesh." For audiences at the start of the twentieth century, the *whiteness* of wave women went hand in glove with their anthropomorphization as individuals.

White wave women steer me toward a perhaps unexpected connection: the permanent wave hairstyle. After World War I, many white American middle-class women found their lives transformed as new social possibilities emerged, a result of the shortage of male labor after the war. On the path to a new independence supported by well-paid office jobs and a sense of the possibility of suffrage, many women signaled modern status with shorter hairstyles, such as the bob, at once practical and radical. But "simple haircuts for women never gained acceptance in the 1920s, at least among the men and women who publicly expressed their opinions" (Søland 2000, 38). Commentators worried that short hair risked blurring the differences between the sexes. But in the late 1920s, a new technology emerged to offer "feminine and graceful styles with curls and waves" to short hair (quoted in Søland 2000, 38; see also Willet 2000). The permanent wave, delivering "new curlier versions of bobbed hair, . . . marked the reestablishment of gender distinctions in fashionable self-presentation" (Søland 2000, 40). Wavy short hair could "refeminize" white women. A joke in a 1926 issue of the farmers' magazine *Northwest Poultry Journal* quipped, "The ocean must be feminine. Anyway, it early adopted the permanent wave."

World War II saw a state-led rescripting of women's possible identities in line with the figure of the soldier-citizen. With many young men away at the front, the military turned to women to fill jobs as engineers, doctors, attorneys, cryptographers, clerical workers, and officers. In 1942, the US Navy founded the Women Accepted for Volunteer Emergency Service (WAVES), placing women in temporary shoreside naval jobs so that men in these positions might be sent to sea.

The association of waves with evanescence and femininity would not have been lost on many publics. But the WAVES also summoned the image of a wave of attack, a more battle-ready meaning; they called up symbolism both feminine and masculine. Indeed, women in this organization were meant, in the most heteronormative way, to be both feminine and masculine, idealized in 1943 as "trim girls in the smart navy blue uniform, with its mixture of the romantic and functional" (Ross 1943, 1). The WAVES, numbering eighty-six thousand at war's end, were overwhelmingly white, meant to stand as a vanguard for traditional American femininity—though also, at moments, for sea changes in gender relations.

Consider the figure of "waves of feminism," meant to point to fronts of social change. In the US context, the first wave came to refer to suffrage movements; the second, to 1960s women's liberation; and the third, to women of color and queer feminism of the 1980s and 1990s. Many critics, however, have pointed to limitations of the wave metaphor—not least, its original anchoring in the experience of mostly middle-class white women (see Ahmad 2015; Bailey 1997; Delap 2011; Graff 2003; Hewitt 2010; Siegel 1997; Slovic 2016; Spencer 2007; Spigel 2004; Taylor 1998). Ednie Kaeh Garrison (2005) argues that the *feminist oceanography* (a term coined in 1997 by Deborah Siegel) of the wave narrative homogenizes women, linearizes movement, and posits times of lulls, mismeasuring histories of activism. Alison Wylie (2006, 173) is more sanguine about the possibilities of the metaphor, suggesting that "waves do not so much overtake and succeed . . . one another as rise and fall again and again in the same place, transmitting and diffusing energy in complicated ways."

Waves, like genders, are malleable things. The defiance of waves gathers a secular cast in Friedrich Nietzsche's *The Gay Science* ([1882] 1974, 247), in which they become emissaries of unrepentant modernity: "How greedily this wave approaches, as if it were after something! How it crawls with terrifying haste into the inmost nooks of this labyrinthine cliff! . . . But already another wave is approaching, still more greedily and savagely than the first, and its soul, too, seems to be full of secrets and the lust to dig up treasures. Thus live waves—thus live we who will." Eugene Victor Wolfenstein (2000) argues that these waves embody a Nietzschean valorization of a masculine will to power. Nietzsche's waves have an affinity, too, with Greek mythological renderings of waves as horses. As Tamra Andrews (2000, 223) writes, "White crests resembled the horses' billowing manes. Breakers, or particularly strong waves, were sometimes referred to as wild bulls."

Wave horses are usually masculine and white, but not always. Here are becoming-animal waves in Virginia Woolf's *The Waves* (1931, 75–108): "The waves drummed on the shore, like turbaned warriors, like turbaned men with poisoned assegais who, whirling their arms on high, advance upon the feeding flocks, the white sheep. . . . They fell with a regular thud. They fell with the concussion of horses' hooves on the turf" (see also Beer 2014). Robin Hackett (2004, 69) argues that this moment in Woolf is an anti-imperialist parody of colonial fantasies about precious white womanhood under threat from dark alterity: "The words 'assegais' and 'turbaned' mark these figures as dark-skinned others; the use of warriors as opposed to soldiers primitivizes them." Waves become dark, animal, masculine—other to a domestic, imperiled, sheepish white femininity.

Flip the script, though, and look to Nnedi Okorafor's Africanfuturist novel *Lagoon* (2014), set on the oil-poisoned shores of Nigeria, where a shapeshifting crew of formidable aliens arrive to use their powers of material transformation and reorganization to cleanse the sea of petro-pollution. The extraterrestrials' first envoy, Ayodele, comes to a beach in Lagos in a form both science-fictiony and folklorically familiar to the people she meets—as a molecular machine and, phasing in and out of being a "dark-skinned African woman with long black braids" diving "like Mami Wata" (13), a West African water deity. Ayodele arrives as a *wave*, barreling toward a trio who will be our main characters: "Something was happening to the ocean. The waves were roiling irregularly. Each time the waves broke on the beach, they reached farther and farther up the sand. Then a four-foot wave rose up. . . . The wave was heading right for them. . . . *PLASH*!" It was a "wave that looked like the hand of a powerful water spirit" (11, 13). Melody Jue (2017a, 177) argues that this wave condenses Afro-diasporic figurations of the sea as both a harbinger of kidnapping terror and part of Africanfuturist "speculations that figure the ocean as both a means of cultural survival and a catalyst for future evolution, a familiar alterity."

Where might this itinerary from Genesis to these many wave anthropomorphisms, zoomorphisms, and theomorphisms arrive? Let me return to Keller's theological project, tuned to recapturing, remaking, the "face of the deep"—not as essentialist feminine energy or as the reanimation of Tiamat, but as what Keller (2003, 193) calls, with a nod to Donna Haraway, a "trickster matrix." Committed to a "queer, postcolonial, polymorphous and possibly perverse feminism" (34), Keller writes about a reimagined *tehom*: "As the wave rolls into realization, it may with an uncomfortable passion fold its relations into its future. . . . The relations, the waves of our possibility, comprise the real potentiality from which we emerge" (227). Keller moves waves into an animal idiom, now calling not on the equine but the cetacean: "We are drops of an oceanic impersonality. We arch like waves, like porpoises" (218). This swerve to porpoises may spiral back to old-school gender symbolism, since these creatures are sometimes described as distillations of hydrodynamic femininity (see Bryld 1996).[4] But it may open up a different becoming-animate materiality. Keller writes, too, that "waves [are] membranes of energy from which matter forms and stabilizes" (Keller 2003, 232). Waves transition between states of energy, operating across matter, gender, and species.

Mel Chen's (2012, 136–37) writings on transgender theory suggest a reading of the prefix *trans-*: "*Trans-* is not a linear space of mediation between two monolithic, autonomous poles. . . . I wish to highlight a *prefixal trans-* not primarily limited to gender." Thinking of waves' entanglement with gender and animality, we can call waves "tranimals," adapting a term defined by Lindsay Kelley (2014) to describe species-, sexuality-, and semiosis-crossing creatures (see also Hayward 2014). If attention to the materiality of the nonhuman world—including minerals, toxins, and water—may be key to accounts of the animacy of the world, then waves exhibit a *tranimacy,* becoming *tranimate objects.* Astrida Neimanis (2012, 94) suggests in "Hydrofeminism: Or, On Becoming a Body of Water" that feminist analyses of water politics should step away from essentialist identifications of women with water but not ignore such sex/gender/water eddies as the travel of chemical estrogens through oceans and bodies or, to take an allied example, that "North American breast milk . . . likely harbors DDT, PCBs, dioxin, trichloroethylene" (see also Ah-King and Hayward 2014). The waves of *tehom* may be good for thinking *with* and *against* gender. In "Has the Queer Ever Been Human?" Dana Luciano and Chen (2015, 189) offer that "the encounter with the inhuman expands the term *queer* past its conventional resonance as a container for human sexual nonnormativities, forcing us to ask, once again, what 'sex' and 'gender' might look like apart from the anthropocentric forms with which we have become perhaps too familiar." Thinking about "sex" and "gender" as they are assigned and undone with respect to nonhumans—as chaotic nature, trickster God, waves, flows of marine chemistries—might render the analytic of gender freshly "at sea," as Mother Ocean becomes not Other Ocean, but Ocean Otherwise (cf. McTighe and Raschig 2019), a genre-bending tranimate transoceanic.

Set One

Second Wave

Venice Hologram

In 1972, the artist and designer Judith Munk, with her oceanographer husband, Walter, wrote an essay entitled, "Venice Hologram." Guiding readers of the *Proceedings of the American Philosophical Society* through the centuries-long relation between Venice and its sheltering, surrounding lagoon, the Munks narrated a *longue durée* story of seawater sustaining but also, increasingly, soaking the city. They reviewed the chronicles of those city planners who had struggled with one another to stay or amplify the embayment's tidal flows and revisited accounts of civic-minded antiquarians seeking to conserve the city's architecture against the wear of waves. In an unconventional conclusion, Munk and Munk called for efforts to capture the city's inventory of statuary using *holography*, a photographic technique aimed at recording three-dimensional images of objects. They offered, as an animating warrant, a quotation from the nineteenth-century English art critic John Ruskin, who, calling Venice "a ghost upon the sands of the sea," worried about the city's future: "I would endeavor to trace the lines of this image [of the city] before it be for ever [*sic*] lost, and record, as far as I may, the warning which seems to be uttered by every one of the fast-gaining waves, that beat, like passing bells, against the Stones of Venice" (1851, 2). For the Munks, drawing on the etymology of *hologram* as *holo* (complete) and *gram* (message), Venice was itself a hologram—a complete message. And holography, a technology for making a three-dimensional image by exposing an object to a field of laser light to record diffraction patterns that might be conjured into an image under proper later illumination, was a technique that might preserve the auras of Venetian art without the presence of the objects themselves—a form of immaterial materiality that answers to Ruskin's message to fight against the time of gaining, beating waves.

Venice has long been visited by waves from the Adriatic Sea. In its earliest days, waves were attenuated by wetlands that dotted the lagoon, wetlands created by dynamics of silting from rivers that flow into the bay and by the tidal flushing of the waters of the wider harbor. Munk and Munk reported on contests in the mid-sixteenth century between the hydrographer Cristoforo Sabbadino and Alvise Cornaro, a believer in land reclamation. Sabbadino argued: "Let us divert the rivers and at the same time allow the tides the widest possible expansion in the area of the lagoon and the deepest possible penetration into the mainland. We will obtain the result of a more powerful flow in the inward phase and a greater pressure in that of the ebb" (quoted in Munk and Munk 1972, 428). Munk and Munk observed that Sabbadino "likened the Lagoon to a lung that, when filled, needed the marshland in which to expand" (428). Cornaro, countering Sabbadino, viewed swamps as undesirable homes to disease. "Reclaim the land and plant corn," he declared (428). Both of these figures offered, in different voices, something that the anthropologist Caterina Scaramelli (2023) calls a *swamp theory of history*, with swamps the past but persistent presence organizing and disorganizing human enterprise. As sea level rises and Venice continues to subside (owing, earlier in the twentieth century, to groundwater extraction but now also to continued compaction from construction and the movement of the Adriatic tectonic plate [Bock et al. 2012]), the figure of the swamp—as inheritance (Venice was founded on marshland), as enemy, and as savior (with wetlands as reengineerable "soft infrastructure" [Russi et al. 2013])—looms importantly. Think of swamps as not so much infrastructure as *inframuddle*, or that which messes up infrastructure, renders it choppy.

Waves, like swamps, are watery, flowing others to Venice's marbled walkways, piazzas, and canal boundaries. The storm waves of the Adriatic have long been perceived as enemies to be protected against—a vision behind the construction of the sea walls (Murazzi) of Pellestrina, started in 1751. There are also the flood waves consequent on storms and tide, which increasingly swamp the city itself. *Acqua alta*—high water at high tide in the Adriatic—rises between knee- and waist-high in locales such as Piazza San Marco, covering open areas of the city with small lapping waves.

Walter Munk was likely the author of those scattered passages in the "Venice Hologram" essay that called attention to the power of waves (while Judith Munk, with her training in art history, was likely behind the art and architecture angle).[1] Walter had had a big picture intervention in mind. In fact, he spent his sabbatical in 1971, taking time away from his post at the Scripps Institution of Oceanography, explicitly to consult about what might be done about flooding in Venice. He advocated for a system of gates that might control the flow of water. Such a construction, he urged, would be more flexible and pliable to incoming Adriatic waves than the rigid breakwaters that had previously been installed. Such a system, while not fully in the style of "soft infrastructure," would permit more give and take—and might also permit the nearby marshes to endure. What eventually became the Modulo Sperimentale Elettromeccanico (Experimental Electromechanical Module [MOSE]) project—an envisioned system of movable gates at the Lido, Malamocc, and Chioggia inlets (under construction since 2003 and first used in 2020) aimed at protecting the city from a tidal rise of three meters—is a descendant of Munk's idea (see Aguilera 2007; Fletcher and Spencer 2005). It is also, of course, dependent on wave-prediction work done by Italian hydrographers, civil engineers, and wave scientists such as Luigi Cavaleri, who was instrumental in the research behind the *Acqua Alta* oceanographic tower, an instrument and observation station posted in the Adriatic Sea twenty kilometers from Venice and first built in 1970. Cavaleri (2000, 58, 60) reports on

> a strong focusing of wave energy to the west of the Lido entrance, connected to the complex bathymetry in front of the inlet due to the interaction of the tidal flows and the littoral sediment transport. The focusing area shifts to west while the incoming direction moves more to east. . . . Apart from refraction and shoaling, the evolution of waves while moving towards the coast is dominated by generation by wind, dissipation by bottom friction and breaking, and nonlinear wave–wave interactions.

These are the wave conditions with which MOSE reckons and that it is built to confront and reshape.

Return to Judith and Walter Munk: they were interested not only in the waves coming into the city, but also in those in the city's canals—and particularly those created by canal-going vessels. They wrote, "Of the many remaining 'water problems' evident even to the casual visitor is the wave roughness in the Grand Canal, a result of the motor-boat traffic. This constant wave action withdraws the mortar and thus destroys the walls of the palazzos lining the Grand Canal (and irritates the dwindling gondolaphiles)" (Munk and Munk 1972, 437). The Munks had a solution, one that sought to beat the waves at their own game: "Perhaps one can reduce the wave activity by lining the banks with absorptive material, somewhat analogous to the use of acoustic tile. F. Rizzoli and W. Munk have tried a design that takes advantage of the fact that the wave spectrum is highly peaked, at a frequency corresponding to a phase velocity equal to the legal speed limit plus twenty per cent!" (437). Here waves become a kind of infrastructure that can be modified to operate against themselves. Waves as inframuddle, oscillations that disturb and confuse, become infrastructure—and in this instance, infrastructure that is finely tuned to the aesthetic desires of tourists with Romantic visions of Venice as "The Floating City."

Is Venice a hologram? If a hologram is a record of a past event, a pattern of memory, then yes: insofar as Venice can be experienced as a representation of itself (perhaps one of its primary manifestations, ever since the days of the Grand Tour [see also Davis and Marvin 2004]), it *is* a hologram, what the anthropologist Roy Wagner (2001, xx), in his writings on the "holographic world view," would call a "near-life experience." But think, perhaps, rather, of Venice *not* as a replicable aura that can be detached and then reattached to itself—a diagram like those of waves in chapter 1—but as a *muddle*; a mixed-up liquid turbidity; an oscillation of turbulent, swampy, silty, and wavy forms that sometimes undoes the city's neat infrastructure of canals and sometimes is the very condition of the city's existence at all. "Muddling through," write the science studies scholars Mike Fortun and Herbert Bernstein (1998, xv–xvi), is "marked by perseverance: a dogged pursuit and relentless enactment of both pointed inquiry ('How do you know that?') and thoughtful social and political work ('Given *that*, let's try this') . . . adequate to the monstrous, rapidly changing worlds of nature and society." With swamps and waves both the past and the future of the city, Venice is the message in the muddle.

Set One

Third Wave

Wave Navigation, Sea of Islands

When I first met my key guide into Dutch wave science, Gerbrant van Vledder, at a canal-side restaurant in Utrecht, he made a point of showing me his copy of David Lewis's *We, the Navigators: The Ancient Art of Landfinding in the Pacific*, from 1972. Lewis, a British doctor and sailor, reported on sailing from Tahiti to New Zealand under the guidance of two men from the Santa Cruz and Caroline Islands in Oceania, navigators versed in traditional wayfinding techniques, orienting themselves using a combination of star sighting, watching for seabird paths, and tracking, using eye and balance, patterns of waves.

In 2015, van Vledder had accompanied a wayfinding voyage in the Pacific, around the Marshall Islands, a journey that had been led by the Marshallese navigator in training Alson Kelen, who a few years earlier had become the director of Waan Aelõn in Majel (Canoes of the These Islands [see Tingley 2016]). Kelen had been engaged in multidisciplinary projects to revive traditional wave piloting, the practice of using visual, haptic, and proprioceptive sensing to discern, from sailing canoes, the remote presence of islands by attuning to patterns of swells and currents. That tradition had been ruptured by Protestant missionizing that started in the

FIGURE 1C.1 Marshallese stick chart made of pandanus roots called a *wapepe*, often read to represent wave patterns around a central island, 2005. Created by the Rongelapese navigation expert Isao Eknilang. Photograph by Joseph Genz.

1850s; German colonial administration of market networks for coconut oil from 1885 to 1919; and Japanese occupation between the wars, which had prohibited voyaging and had seen Indigenous craft replaced by schooners. The US administration of the Marshall Islands starting after World War II led to further, catastrophic disruption, as the detonation of sixty-seven nuclear bombs from 1946 to 1958 on Bikini and Enewetak sent radiation through the archipelago. Many residents sickened; some were exploited by the United States as unconsented biomedical subjects. Many left their islands, some becoming refugees (Barker 2004).

Kelen had, since the early 2000s, been in apprenticeship with Korent Joel, from the island of Rongelap, who was born in 1948 and had left the islands after his grandfather died of radiation sickness. Joel then lived in Hawai‘i from 1968 to the early 2000s, working as a captain of government cargo ships. Joel had been inspired, during his time in Hawai‘i, by the efforts of the Polynesian Voyaging Society, which had built *Hōkūle‘a*, a reconstruction of a traditional double-hulled canoe that, while sailing from Hawai‘i to Tahiti, generated new and persuasive evidence that the Pacific Islands had been settled through purposeful transoceanic voyaging (Finney 2003). In 2005, he inaugurated a collaboration with Ben Finney and Joseph Genz, anthropologists at the University of Hawai‘i, as well as the oceanographer Mark Merrifield, as part of a project to revive Marshallese navigation. Among the tools ready to Joel's hands were customary diagrams made of coconut fronds and shells that represented knowledge of swell patterns. One such diagram, a *wapepe* that represented a top view of wave patterns around a central island, had been created for him by a Rongelapese elder, Isao Eknilang, one of a few people willing to share such artifacts beyond close confidants (figure 1c.1) (see Genz 2016, 2017).

To students in physical oceanography, what may leap out are how the four curves suggest how wave fronts bend—refract—around an island. At the same time, tools such as the *wapepe* are not so much scaled cartographic models as they are illustrations of analytic principles to be used in preparation for voyages. They also materialize initiate, often secret, lineage-specific navigation knowledge; part of reviving the wayfinding tradition would mean figuring out how to make the work of these charts more widely available to Marshallese youth.

I learned more when Kelen visited Cambridge, Massachusetts, in 2017 to join with van Vledder, along with Genz and the Harvard University particle physicist John Huth, to give a lecture at Harvard on the voyage they had taken, a one hundred-kilometer journey from the capital of the Marshall Islands, Majuro, north to the island of Aur. Kelen told the audience he had wanted a team to help him study a sort of wave pattern on which wave pilots depend called the *dilep* ("backbone" in Marshallese and marked on the *wapepe* in figure 1c.1 by vertical and horizontal lines). The *dilep* is a kind of *wave path* between atolls created either, according to one school of thought, as "an extension of reflected swells emanating from the destination island" or, according to another, as "the crossing of opposing or nearly opposing swells within a bimodal seastate that forms nodes of intersection . . . between islands" (Genz 2016, 20–21). It is represented on tools such as Eknilang's *wapepe* as a reed pointing from one node to another. The *dilep* is apprehended not only visually but also proprioceptively, in one's stomach. Kelen, following Joel's lead, had become interested in whether scientists could use computer models to capture something of what *dilep* might be.

Wayfinding techniques in Oceania have long been of interest to European mariners. German ethnographers of the nineteenth century called objects of material culture such as the *wapepe* "stick charts" (*Stabkarten*), comparing them with, and in some sense assimilating them to, the graphic maritime mapping techniques with which they were familiar (Genz et al. 2009). In 1901, a German naval officer, Captain Raimund Winkler, published the article "On Sea Charts Formerly Used in the Marshall Islands, with Notices on the Navigation of these Islanders in General," reporting that they were not charts but, rather, heuristic devices for teaching about the direction of swells. His placement of the devices into a receding past ("formerly used") became a self-fulfilling prophecy for much of the twentieth century, as occupying and radioactive colonialism continued to undo Marshallese life. Later ethnographers, in efforts to acknowledge wave piloting as a long tradition, often ended up suggesting that it belonged to a deep, general human history, effectively spiriting its practitioners into an ancient, timeless, and ahistorical past. That analytic operation then staged the tradition as simply parallel to, for example, European nautical systems, readying it for comparison, often by translating Marshallese approaches into putatively more universal Western formalisms (e.g., to do with diffraction) while simultaneously positioning it as entirely other to dominant Western epistemology.

Meanwhile, streams of scholarship from cognitive anthropology—dedicated to understanding category systems that structure human thought—often envisioned wave piloting as a window onto universal human orientational operations, a move that had the unhappy effect of rendering Marshallese pilots as exhibiting some kind of bare human nature (see Frake 1995; Hutchins 1995). The anthropologist John Mack (2011, 118) offered that Marshallese charts are "less representations of space" than "representations of *experience* of space"—experience emanating from and embodied in the habits of practiced wayfinders who are trained to feel, in their bodies and boats, the analog materiality of wavy motion. Such an opposition—between representations of space and experience of space—is too stark. Latitude-longitude charts, sextant-enabled navigation, and radar readings of seascapes *also* enable and assume specific sorts of mariner experience—conjunctures of perception and cognition. More, European-oriented interpretations overlook the possibility that these traditions may have met before. Captain Cook, for example, was in dialogue with navigators from other traditions; the Raiatean navigator Tupaia, who traveled with Cook, was instrumental in making charts with wind directions, charts that were based on extensive wayfinding experience in Oceania, from Sāmoa to Tonga and Rapa Nui (Eckstein and Schwarz 2019). Modern Western seafaring is inextricably entangled with and dependent on other traditions, much as canonical European histories might seek to purify their technical lineages of Indigenous cultural tributaries. That purification has often animated a "salvage" approach to wave piloting, which, as Elizabeth DeLoughrey (2019, 168–69) writes, risks "recuperat[ing] the pernicious colonial fantasy of the 'vanishing native' in which the white westerner tries to salvage—and mourn—the loss of what the global north has effectively destroyed."[1]

Hands-on projects of cultural revival, especially if shaped by Euro-diasporic participation, may sound, at first hearing, more like projects in retrieval than futurity, but they are also more multivalent. A strategic syncretism seems the order of the day.[2] Genz (2016, 15), for example, writing about the "vestibular knowledge" of wave pilots, suggests that mixing *not quite* translations from traditional knowledge systems with those from Western-ish seafaring is necessarily how any revival of wave piloting might function.

Any such revitalization must also be understood in the context of late twentieth-century projects of anticolonial resistance in Oceania, projects that have included everything from antinuclear activism in sites of French colonialism and American imperialism to forms such as the Hawaiian Renaissance and post- and decolonial literary movements. In 1993, the Tongan anthropologist Epeli Hauʻofa famously called for what were once commonly known as "Pacific Islands" to be reimagined, to be reconstrued, as a "Sea of Islands"—a vision of Oceania unified, not divided, by water. Wayfinding is one element in this reframing (see Hauʻofa 1994).

And such a reframing need not be seen as simply activating nostalgia for ways gone by. The waves of the Sea of Islands exist in a geopolitical present. Writing some twenty years before Hauʻofa, the Kanaky (Indigenous Melanesian, New Caledonia) writer Déwé Gorodé narrated waves as connecting not just Indigenous Oceania but also a wider Global South world. The year 1974 found Gorodé incarcerated for activism with the Kanaky self-determination movement, and from prison in Nouméa, New Caledonia, she wrote the poem "Wave-Song" (2004, 42), which narrates her hearing waves over the compound walls that surrounded her:

Only the endless wave-song
    beyond the barb-wire
is a lullaby that rocks our enclosed and watchful sleep

The waves deliver news of "the huge white mushroom cloud infecting the sky over Mururoa" (43). And they carry, from the Eastern Pacific, the cries of people tortured in Chile after the fall of Allende in 1973. Gorodé entreats the waves to carry peoples of Indigenous Oceania into a future

in dignity
    stronger and more serene
    more timeless than that of the ageless majestic
    stone guardians of Rapanui
(2004, 43)

These transoceanic connections are more than revitalization. These are waves as messengers of resistance (Keown 2018, 590; see also Scanlan and Wilson 2018; Walker 2011).

Such resistance continues to call for reparations for the damage of radioactive colonialism and now enfolds the matter of climate change. I learned, when I met Kelen, that he works not only in navigation but also in activism that calls attention to the Runit dome, a concrete hemisphere 377 feet in diameter that was built to contain radioactive waste left by the United States in the wake of bomb detonations on Enewetak. With sea level rise, the dome may be at risk of breach. Kelen observes that his activism bridges "the nuclear age and the climate change age" (quoted in Barad 2019, 526).

Climate change sees the Marshallese concerned with waves in a new way—with waves that follow from sea level rise, a consequence of ocean warming. If the middle of the twentieth century visited nuclear catastrophe on the islands, the twenty-first century has already seen encroaching new disaster, as island settlements and dwellings are inundated by waves. In a poem entitled "Two Degrees"—a reference to an end-of-century planetary temperature rise that would cause seas to cover low lying atolls (Keown 2018, 586)—the Marshallese activist Kathy Jetñil-Kijiner (2017, 78) links Oceania's history of atomic injustice to the present, with a vignette of patients at an island hospital inundated by an underestimated tide:

a nuclear history threaded
    into their bloodlines woke
    to a wild water world
a rushing rapid of salt.

Jetñil-Kijiner and others have described the inundation of their islands as a pointer to what is to come for wider parts of the ocean world, as rearranged sea level submerges land and amplifies climate migration. As DeLoughrey (2019, 169) notes, "Fittingly, the figure of the tropical island, particularly the low-laying atolls and islands of Tokelau, Tuvalu, Kiribati, and the Marshall Islands, are now gaining attention for the ways in which the threat of sea-level rise, which disproportionately affects the tropics, anticipates a planetary future." Decolonial struggle in the Pacific is struggle against climate change (see also Anderson et al. 2018; Bahng 2019; Kirsch 2020).[3] And imperial and nuclear history needs constantly to be kept in view. The waves that lap at the islands are not timeless waves but waves in history—as well as waves that carry the future, in motion.

Chapter Two

Flipping the Ship

Oriented Knowledge, Media, and Waves in the Field, Scripps Institution of Oceanography

COMING INTO SIGHT over the horizon, visible from the launch I have hired from Marina Del Rey, California on this October 2017 day, is the FLoating Instrument Platform (FLIP), a research vessel operated by the Scripps Institution of Oceanography (SIO) at the University of California, San Diego (UCSD). After two hours thwacking over choppy seas, TowBoatUS Pilot Dan and I find FLIP moored thirty-five miles off the coast of Malibu, positioned in six hundred-meter-deep water, floating at coordinates I was emailed a week previous by the vessel's chief scientist. A singular vessel, FLIP offers an off-kilter vantage on the sea (plate 6). In its horizontal orientation, FLIP can be towed like an ordinary oceangoing craft. But by "flipping" 90 degrees into a vertical position once it reaches its at-sea destination, it can become an enormous spar buoy, more-or-less stationary in the wave field, looking like nothing so much as a maritime metal treehouse. With seven-eighths of the platform's 108 meter length below the water's surface—and all of the furniture and instrumentation inside its top one-eighth having swiveled on gimbals by 90 degrees—scientists can study ocean dynamics from a field station with only a slow bob (figure 2.1).

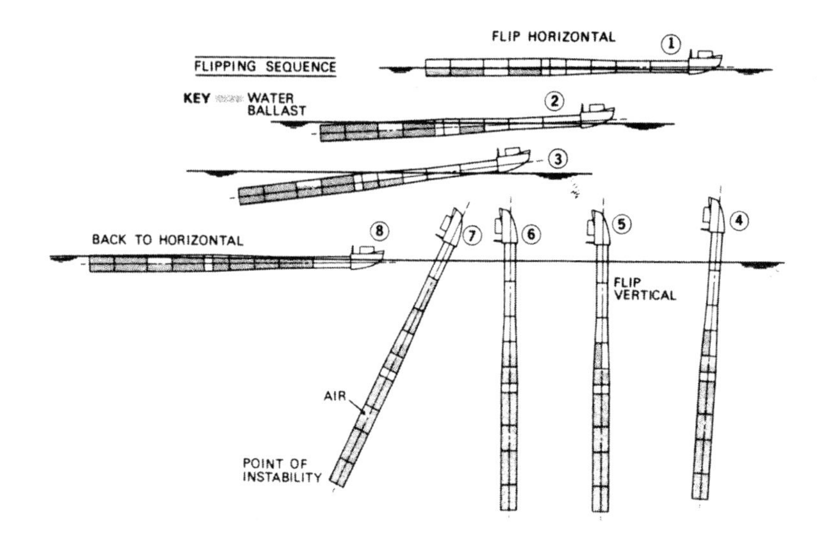

FIGURE 2.1 FLIP's flipping sequence. From Bronson and Glosten 1985, 17, fig. 2, adapted from Fisher and Spiess 1963. Courtesy of Scripps Institution of Oceanography, University of California, San Diego.

After a few years interviewing researchers at Scripps, I have secured a berth on FLIP to do anthropological fieldwork on oceanographic fieldwork, joining scientists examining the interface where water meets air; where waves broker the turbulent exchange of heat, gas, and momentum between sea and sky, modulating weather and climate (see, e.g., Sutherland and Melville 2013). After I leap from my boat onto the neck of FLIP, clambering up its slippery ladder, I meet the nine researchers and five crew with whom I will spend the next three days. They have already settled into the platform's verticality, having been on board a few weeks.

The "FLIP ship" is a legend in oceanography, a "one-of-a-kind wonder," says the Smithsonian Institution (Waters 2012). Its capacity to flip by 90 degrees has made it an object of astonishment. The platform—not technically a "ship," since it holds no propulsion power of its own—puts me in mind of the gravity-bending lithographs of the Dutch artist M. C. Escher. It is full of sideways doors, swiveling sinks, and tables in dual orientations. Having toured FLIP back in port, on its side, in San Diego, I compare my photos of what rooms look like *this way* rather than *that* (figures 2.2a–b). It is impossible not to recall Ludwig Wittgenstein's storied cartoon duck,

FIGURE 2.2A FLIP computer lab, in dock. Photograph by the author.

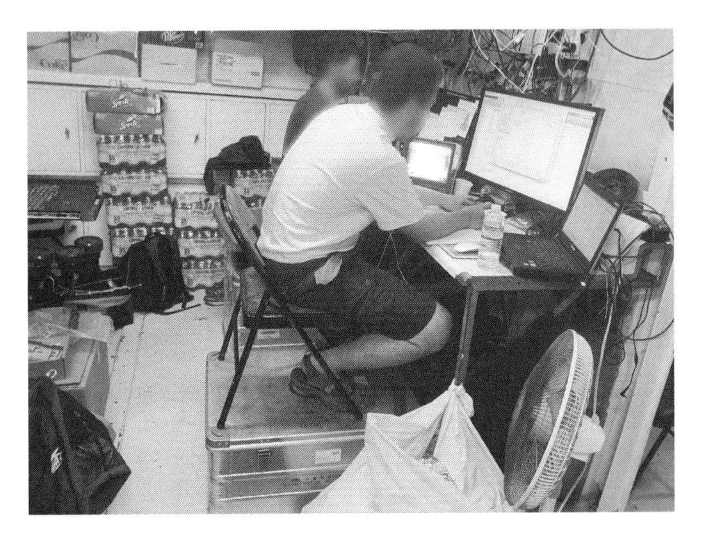

FIGURE 2.2B FLIP computer lab, at sea. Photograph by the author.

which becomes a rabbit when one flips one's perception of it, seeing what was, in one frame of reference a duckbill becoming rabbit ears in another (figures 2.3a–b).[1] During my first night on FLIP, I realize that my top bunk faces a ceiling that is, on sideways days, a door. The chief scientist remarks that after FLIP flipped, she was "surprised by how much *bigger* the science space got. I had tried, before we flipped, to look at everything sideways to get ready, but that didn't work." One scientist tells me it was a relief to "kiss the waves goodbye," to leave seasickness behind as the platform angled upward.

Built in 1962 by the Marine Physical Laboratory at Scripps, with funds from the US Office of Naval Research, and through its life largely its original Tri-Ten steel self, FLIP can be read as an artifact marked by the history of Scripps wave science. It came into service a couple of decades after research on wave prediction for World War II, which mostly unfolded at or near beaches and aimed at forecasting wave arrival for Allied amphibious landings at Nazi-held shores. During the Cold War, Scripps oceanographers turned their attention to the open sea, employing FLIP in studies of how storms transmit waves and in research tuned to submarine acoustics, of interest in scenario planning for possible submarine war with the Soviet Union. Oceanographers listened to underwater sound through the undulations of massive subsurface waves—*internal waves*—that emerged at the interface between deep water layers of different densities and temperatures.

The Scripps oceanographers Fred Fisher and Fred Spiess commissioned FLIP around 1960, after a colleague at the Woods Hole Oceanographic Institution, Allyn Vine (for whom the famed *Alvin* submersible is named), "suggested upending a submarine to make a stable platform" after "observing how stable a navy mop floated in choppy water."[2] Unconventional vessels both, FLIP and *Alvin* were built just after the Soviet Union launched the Sputnik satellite, which motivated US technoscience to undertake particularly spectacular, often "futuristic" constructions. The historian of oceanography Helen Rozwadowski (2004) suggests that such "strange and imaginative platforms and vehicles" were of a piece with an American Cold War moment that envisaged the sea, like outer space, as a frontier inviting bold technological innovation. Fisher and Spiess worked with naval architects on a vessel that could turn upright by leveraging the controlled flooding of eight ballast compartments arranged down a thin stem (6.5 meters in diameter), a shaft Fisher and Spiess wanted lengthy enough to attach hydrophones to down to 90 meters (see Bronson and Glosten 1985; Fisher and Spiess 1963). "A stable platform from which to perform experiments at sea" (Fisher and Spiess 1963, 1633), FLIP became

FIGURE 2.3A "Kaninchen und Ente" (Rabbit and Duck), *Fliegende Blätter*, October 23, 1892.

FIGURE 2.3B FLIP in mid-flip. Photograph by John F. Williams, US Navy/Creative Commons.

central to making visible to oceanographers, but also to the military (though operated by Scripps, the vessel is owned by the US Navy) unseen sea territories, surface and submarine. In the early 2000s, FLIP became a site for studying why and how waves break in the open ocean, an unanswered question (see Babanin 2011).

In this chapter, I chronicle my stay on FLIP—a platform that, as of 2020, was retired from service (about which more later)—interleaving this narrative with a reading of the history of wave science at Scripps from World War II through the Cold War to today. I detail how contexts of discovery, motivation, and justification in wave research have shifted from military to ecological concerns, with one effect being a rapprochement between physical and biological oceanography, even as ecological questions are still contoured by the wake of military framings.[3] I argue that ocean waves have become readable to scientists through conceptual schemes made available by twentieth-century technologies and media, including time-stepped aerial photography, visualizations of wavelengths in sound and electronic communications (e.g., optical sound in film), models of the flow of air over aircraft wings, infrared readings of turbulence, and more. Scientists apprehend waves through *mediations*, viewing waves—to adapt terms from media theory—as forces of *transmission*, as patterns that materialize at *interfaces* (e.g., of air and water, as well as of buoys and computer screens), and as mechanisms of *inscription* (e.g., producing traces that can be read for evidence of distant winds and storms). Media forms have imprinted on how ocean waves are apprehended as objects of knowledge and on how they are represented as processes in space whose futures in time might be foretold. That claim is not to gainsay the use value of these formulations but to place them in cultural history—or, better, to *reorient*, or *flip*, a dominant angle of vision on the science, which would discern a bright line between what is *discovered* and what is *formalized* about waves. What scientists take waves *to be* is impressed by the media through which waves become legible. Wave science, to draw on the literary theorist Roland Barthes ([1970] 1974, 4), is a practice of interpretation in which "the reader [is] no longer a consumer, but a producer of text"—that is, someone who makes a text what it is through the act of reading it.

I attend not only to the *mediations* that have made Scripps wave science possible, but also to the *rotations*, *reorientations*, and *flips* in scientists' frames of reference that have formatted their views of waves and wave science. Inspired by the clever backronym that gives FLIP its name, I draw on a range of meanings offered by the verb *to flip*: to cause to turn; to turn over; to turn on or off with a switch; to change from one state to another; to become enthusiastic. The title of this chapter, "Flipping the Ship," aligns not only with the sideways turn of thinking that led to the making of the platform—and with the sometimes perception of FLIP as a flipped ship (passing ships have occasionally radioed to make sure the vessel is not sinking)—but also with the notion of "flipping the script" or "making an unexpected or dramatic change, reversing the usual or preexisting positions in a situation."[4]

In torquing received accounts of wave science as simply the cumulative layering of discoveries one upon the other, I ask after scientists' attitudes toward their work and, in particular, how wave science, so invested in prediction, offers to its practitioners *orientations* toward possible futures—for their science, for their careers, for the ocean world they study. "Orientations," writes Sara Ahmed (2006, 3, 8), "shape not only how we inhabit space, but how we apprehend this world, . . . how we proceed from 'here,' which affects how what is 'there' appears." At stake in researchers' orientations are how they think about the relation between knowledge acquisition and application—between, otherwise put, matters of fact and matters of concern, the pursuit of scientific objectivity and of sociopolitical objectives.[5] Such *oriented knowledge*, embedded in institutions and history, shapes how researchers judge what counts as good science, a good scientific life, and desirable uses of scientific knowledge.

Aboard FLIP

When I arrived at FLIP, climbing onto its metal-meshed deck, I was given a hand through a trapdoor by Tom Golfinos, FLIP's captain of thirty years, having come to the job after a maritime career in his native Greece. Chief Scientist Qing Wang, a professor of meteorology at the Naval Postgraduate School in Monterey, California, directed me for an orientation to her assistant, Ryan Yamaguchi, an engineer with a bachelor's degree from the University of California, Irvine. He drew me a diagram of instruments arrayed on FLIP, keyed to their scientist operators (figure 2.4). This stationing of FLIP was part of a sprawling project that involved a nearby array of buoys; autonomous underwater vehicles; a twin turboprop plane; and another Scripps ship, the research vessel *Sally Ride* (named after the late UCSD physics professor who was the first female US astronaut in space). The assemblage aimed to understand how ocean swells and whitecaps modulate temperature and humidity across the sea surface, modulations

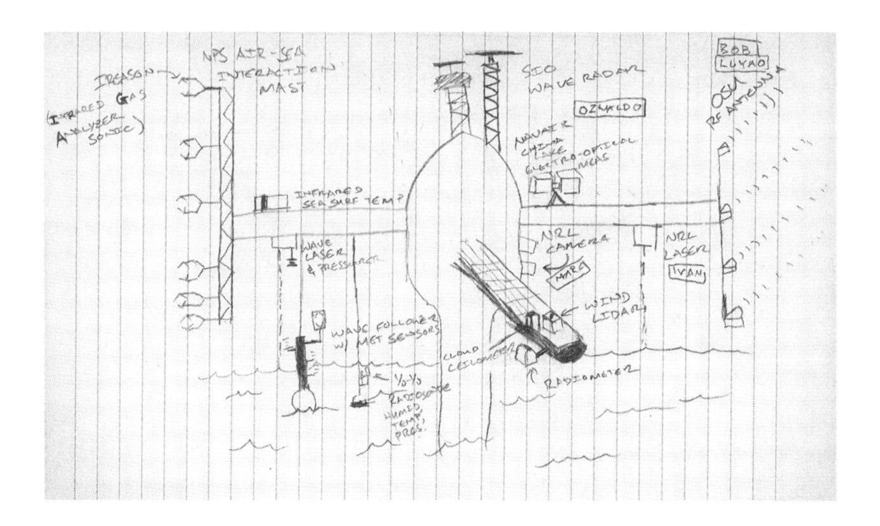

FIGURE 2.4 Yamaguchi's diagram of FLIP research projects, fall 2017.

that might both facilitate and interfere with radio waves. The science crew's research would inform the quality of near-sea surface electromagnetic communications—including radar, one reason some Navy monies were behind this research—as well as, for some of these (and later) scientists, climate models.

Wang checked Yamaguchi's diagram, ticking off the names of the team she was leading: one other professor, three oceanography postdocs, two graduate students, and two technicians. Of these nine, four, including Wang, born in China, grew up outside the United States. With scientists hailing from Russia, France, and (a second from) China (and the Americans Latino, Euro-American, and Japanese American), the demographic cast of the science party reflected trends in international scientific training, collaboration, and migration (and contrasted with the mostly white working-class crew), even as the science plan was tethered to ongoing US state and military projects. Given the perpetual and ambient war footing of the United States, this ongoingness could make the work seem suspended in historical time (aided, here, by the antique character of the vessel), with implications that, since they were not fully formed, kept the future in partial abeyance, partially deferred, for many researchers, making their work basic science, with later uses framed as applications.

Wang was leading the project because of her expertise in tracking how changes in temperature and humidity over the sea are shaped by waves and swell. She had completed her undergraduate degree in atmospheric physics in Beijing, examining air quality over soybean fields, a study that employed a model derived from US surveys over farms in Kansas conducted in the 1960s. She came to the United States in the 1980s, her path paved by her father, a professor of mining in Beijing who was a keen student of English; when China opened up, he spent time in Colorado, where he met a Taiwanese friend whose wife worked as a programmer at the National Center for Atmospheric Research. That person connected Wang to a scientist in her field who pointed her to Pennsylvania State University as a site for doctoral study.[6] When Wang received her degree, the job market was tight. She was already married with a child when she saw an ad for the Naval Postgraduate School. "I didn't know too much about the Navy," she told me, "but this was a tenure-track job."

The Navy connection is no surprise. Much Scripps oceanography since World War II has been funded by a mix of monies from the National Science Foundation (NSF) and the US Navy (e.g., the Office of Naval Research, the Bureau of Ships), even as scientists themselves are not in uniform or, usually, much interested in military work.[7] While generations of Scripps oceanographers after World War II arrived motivated by the adjacency of naval priorities, that is not now necessarily the first thing on their minds. Many of the people I came to know at the institution were attracted to the field because of a love of the ocean (surfing, swimming, boating) and a concern about the ecological health of the sea.[8]

I could relate, having grown up in the surf of southern California as a white middle-class East Coast émigré, more tuned to beach life than to naval contexts. Waves, for me, materialized as shimmering promises, hyperreal translations of the virtual (the perfect wave) into the actual; they stood as invitations to self-actualization across an open future. I encountered the Pacific, I now realize, through my German and New Englander parents' optimistic eyes as a future-facing horizon of possibility rather than as a domain shaped by the settler colonial dispossession of Indigenous territory; the ocean's history as a theater of US-Mexico conflict; and its use as stage for hydrocolonial expansion—US, Japanese, and more.[9]

When I first received instructions about how to get to FLIP, I was provided with a map revealing a seascape of which I had been ignorant: the Point Mugu US Navy Sea Range, off the coast of Los Angeles, dedicated to testing naval maneuvers and technologies. Just outside this space was the instrumented area within which FLIP, along with all those buoys, the *Sally Ride*, and the plane would map wavespace. The geographical sidling up of

FLIP to the Sea Range was a spatial correlate of the contiguity of American oceanography with naval worlds. It also fixed this swath of ocean "field" as a kind of experimental space, a "lab" (distinct from the "sanctuary" of the Channel Islands, or the Heritage National Marine Sanctuary, proposed in 2015 by the Northern Chumash Tribal Council just to the north). The military map prompted for me what the anthropologist Michael Scott (2017, 217) might catalog as "awe, marvel, astonishment, shock, dread, amazement, and horror," an uneasy genre of wonder in the face of this long-in-the-making mathematical, computational, military-industrial, and oceanic sublime. I knew I was headed into, and writing from, a privileged position in "the belly of the monster," that nexus of US scientific and military enterprise that, Donna Haraway (1991b, 188) has named as the matrix for so much contemporary technology—though working at the Massachusetts Institute of Technology, an institution with a strong dose of defense funding, I knew I was also already partway digested by this leviathan, grappling, like the scientists with whom I worked, with how to navigate my career and politics within its contexts.

This FLIP trip, I learned, meant to examine *atmospheric ducting*, a process through which electromagnetic signals travel through a kind of invisible channel in the lower atmosphere, a horizontal segment of air just above the ground or sea through which radio signals can be transmitted long distances, following the curvature of the Earth. Such ducts are well characterized over land but less so over waves. There is some folk knowledge about the effects of ducts in the optical domain. Take the Fata Morgana, a floating mirage over a calm sea created by flipped layers of cold and warm air (a thermal inversion), which act as a lens that conjures to seafarers surreal, upside-down, flipped images of faraway ships that are sometimes taken, in folkloric accounts, for magical cities. Or the "green flash," on occasion visible as the sun sets over the ocean, its light sliced into different colors of the spectrum by the lensing effect of water. While thinking of all of these kinds of waves, watery and electromagnetic, rippling alongside, through, and with one another, recall that genre of water wave/radio wave interference called *wave clutter* (see Bell et al. 2011), a fidgeting disorder across lots of wave fields, each with its own spectra.

To understand how oceanography at Scripps became so crosscut with military matters, we need to revisit World War II. We also need to shift our frame of reference back to the southern California shoreline, where questions about how waves operate on the beach, not so much in the open sea, were crucial early on to weaving together Scripps and US Navy imperatives.

Waves of War, Hot and Cold

American wave science was forged in the crucible of World War II as researchers at Scripps—in alliance with other US institutions, as well as with, particularly, British researchers—turned their attention to predicting patterns of waves arriving on shores held by Axis powers.[10] That work turned waves into strategic, if temperamental, environmental infrastructure for the amphibious landings of Allied troops in North Africa, Europe, and Oceania.

Scripps had not always featured physical oceanography so centrally in its mission. The organization was founded in 1903 as a station for the Marine Biological Association of San Diego and became the Scripps Institution for Biological Research in 1912, when it joined the University of California, becoming the SIO only in 1925. In the early days, Scripps hosted a cadre of gentleman scientists doing work on re-outfitted yachts provided by wealthy San Diego benefactors such as the newspaperwoman Ellen B. Scripps, who named the institution after her late younger brother, George. Research was dedicated to topics such as the travel of sardine populations along the California coast. Its founder, William Ritter, of Wisconsin by way of Harvard, deemed La Jolla an ideal site for marine work since it had "proximity to oceanic depths and other truly oceanic conditions; a favorable climate; a large variety of shore line; and accessibility through sea ports and railroads" (quoted in Raitt and Moulton 1967, 31; see also Hlebica and Cook 1993). This narrative of nature and of a nascent networked nation posed the place as a kind of Eden. Connections to historical and contemporary indigenous Kumeyaay communities or to nearby Mexico were not in the spotlight, and researchers' visions of the Pacific came from their biographical points of departure from the East Coast of the United States. Collaborations with Mexican institutions over the years, such as nearby Ensenada's Autonomous University, while occasional (there have been a few joint surveys of the Gulf of California/Sea of Cortez), were few. Scripps's horizons opened up to an imagined frontier Pacific.

The early ethos of the place was shaped by a sense of La Jolla as a fresh site for science. Ellen Scripps's brother E. W. described it in 1915 as "An Odd Place: A New Town Where High Thinking and Modest Living Is to Be the Rule" (Raitt and Moulton 1967, 77). Ritter wrote books in biological philosophy, including *The Probable Infinity of Nature and Life* (1918), and declared toward the end of World War I that "the sea and atmosphere are preeminent among the features of the earth as the common physical heritage and bond among peoples and nations" (quoted in Raitt and Moulton

1967, 93). Bonds among nations would be tested by the wars of the twentieth century, as would hopes for international, cooperative oceanography (Adler 2019; Hamblin 2005).

The Scripps wave story as usually told centers on the oceanographer Walter Munk, born in Vienna in 1917. Munk finished a bachelor's degree in physics at the California Institute of Technology (Caltech) in 1939 and applied for American citizenship that same year, after the Nazis took Austria. He moved to Scripps to start a doctorate under Harald Sverdrup, another expatriate European who had moved from Norway to become the director of Scripps in 1936 after leading the science team of Roald Amundsen's North Polar Expedition in the 1920s. When Germany invaded Norway, Sverdrup extended his stay in the United States. In the early 1940s, at the request of the US Army Air Corps, Munk and Sverdrup began working to determine whether wave weather might be predicted to time Allied amphibious invasions. Munk and Sverdrup's work became decisive for the landings of "duck boats" at Normandy on D-Day (June 6, 1944). To generate their approach to wave prediction—to improve on the in-place Steere Surf Code, which, like the Beaufort scale of wind speeds, was a "rule of thumb" set of guidelines—Munk and Sverdrup first needed to gather data: spatial data.[11]

While dozens of "wave stations" would eventually exist around the world from which observers would report incoming wave heights and periods, much proof-of-concept work was done at Scripps, which hosted a long pier from which observations could be made. In one Scripps document, "Height of Breakers and Depth at Breaking," from March 1944, the authors report on a preliminary project in which "four or five [aerial] photographs were taken of a single well-defined wave as it advanced from the outer end of the pier to the point of breaking."[12] Such photos were synchronized with underwater measurements of pressure, which corresponded to the rise and fall of waves at surface.[13] Munk argued that one could use photographically captured changes in wavelength and speed as waves traveled to shore to infer the changing depth of the water beneath. A once confidential document titled "Effect of Bottom Slope on Breaker Characteristics as Observed along the Scripps Institution Pier" provided an illustration—wavy lines ranged against a grid backdrop (figure 2.5)—underscoring waves' amenability to measurement.[14] To secure confidence in diagrammatic accounts such as this, Scripps oceanographers also recruited lay observers to gather more data.

In January 1944, the US Navy conducted a study of "waves in shallow water."[15] The Navy's Bureau of Ships contracted the Woods Hole Oceanographic Institution and the University of California to make "observations

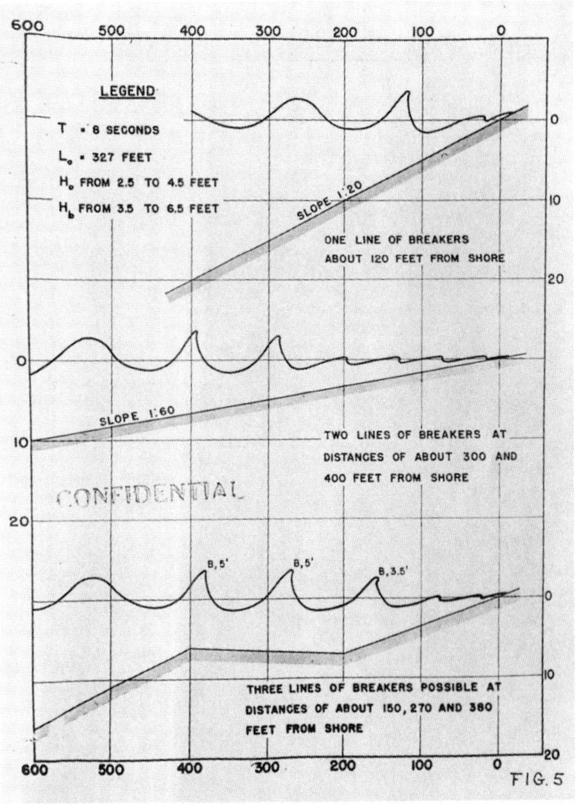

FIGURE 2.5 "Effect of Bottom Slope on Breaker Characteristics as Observed along the Scripps Institution Pier," SIO Wave Project, report 24, October 23, 1944, Walter Heinrich Munk Papers, 1944–2002, accession no. 87-35, Scientific Papers, Manuscripts and Talks, Scripps Institution of Oceanography Archives, box 23.

of waves and breakers along the Scripps Institution pier and other piers in Southern California" for "the purpose of preparing an urgently needed manual for use in amphibious operations."[16] A man named John C. Hammond was deputized to locate people living near piers to conduct observations. Hammond's letters to Sverdrup document him running around providing people—what we might now call "citizen scientists"—with stopwatches. Hammond's report of April 16, 1944 barely contains his exasperation with organizing lay people into a project of scientific observation, getting them to adopt a proper frame of reference.[17] His amateur sociology of observers offers a glimpse of some residents of California's 1940s shores, as well as a Rorschach of what Hammond judged—drawing on his attitudes about gender and class—to be proper and improper scientific orientation:

> The problem consisted of locating observers for seven piers and maintaining their interest 7 days a week over a period of 6 weeks, at compensation

of $3 a day, in wartime. While this was accomplished apparently easily, there were a few times when there was a threatened breakdown. . . .

> 1 The original observer at Oceanside left after five days because of an eye injury, but helped to locate a successor. This lady is the wife of a civil engineer and lives next to the pier. She was an ideal choice as she has little at home to engage her and this work offers something to narrow the gap between her housework and the highly technical work of her husband—a gap keenly felt by the more intelligent housewives in the later years of such marriages. . . .
>
> 4 The life guard at Huntington Beach was located thru the city officials and has done a good job. He had had some experience in studying wave heights for surf-board enthusiasts. He is above average in intelligence. . . .
>
> 5 The Redondo observer was obtained from one of the little fish markets on the pier. His interest has been keen and, outside of the necessity of absorbing much irrelevant and unimpressive amateur meteorological data in order to maintain his sense of importance, no difficulty has been experienced. . . .
>
> 6 At El Segundo the Standard Oil pier is under strict military guard. It was necessary to obtain the pier manager and his assistant, and it was impossible to allow them compensation on account of company rules. The manager is a highly paid man with heavy responsibilities. He found the observations to be an annoyance and desired at times to be relieved but his cooperation was held by stressing the serious nature of the project and its implications. This pier is believed by the writer to be well located and these observers of high caliber. . . .
>
> 7 Mr. Weir at Ocean Park was obtained through his employer, who owns the amusement pier. This arrangement works well and is to be recommended where the employer is a reasonable person.

The difficulty, for Hammond, was getting people—housewives, lifeguards, fish sellers, oil men—to enter the frame of reference, both in terms of disciplining their observation and submitting to what he took as the reasonable demands of science and national security. It may be no wonder that the next steps in wave observation were in trying to automate it, to standardize or turn off the human element.

FIGURE 2.6 *Eugene Cecil LaFond [ca. 1938] Standing by His Wave Recorder Installed near the End of the Scripps Pier*, Eugene LaFond Papers, SMC 102, Special Collections and Archives, University of California, San Diego, Library, box 18, folder 16.

Record keeping—and standards for understanding what waves *were*—became essential to wave science. In "Proposed Uniform Procedure for Observing Waves and Interpreting Instrument Records," from 1944, Munk was already offering a rubric for such features as *wave height*. He wrote, "The wave height shall be taken as the average of the highest one-third of the waves observed during a time interval of at least ten minutes," spelling out the measure he would later call *significant wave height*.[18] Such measures came increasingly to be embedded in wave-recording devices. The unfolding story of wave inscription devices at Scripps can be tracked through a run of papers and diagrams in the archives of the SIO, from early examples such as Eugene LaFond's 1937–38 wave recorder (figure 2.6) through a long, crumply scroll of wriggly wave-height measurements from October 3, 1966, that I found one day rolled up in a cardboard box, a dry archival trace of a once watery wave.

Waves were becoming knowable at this mid-twentieth-century moment not only through time-and-motion studies at piers and beaches, but also through something relatively new to wave science: aerial photography (figure 2.7), a representation of the sea flipped from looking across the horizon—the seascape—to looking down on the sea as a kind of map of

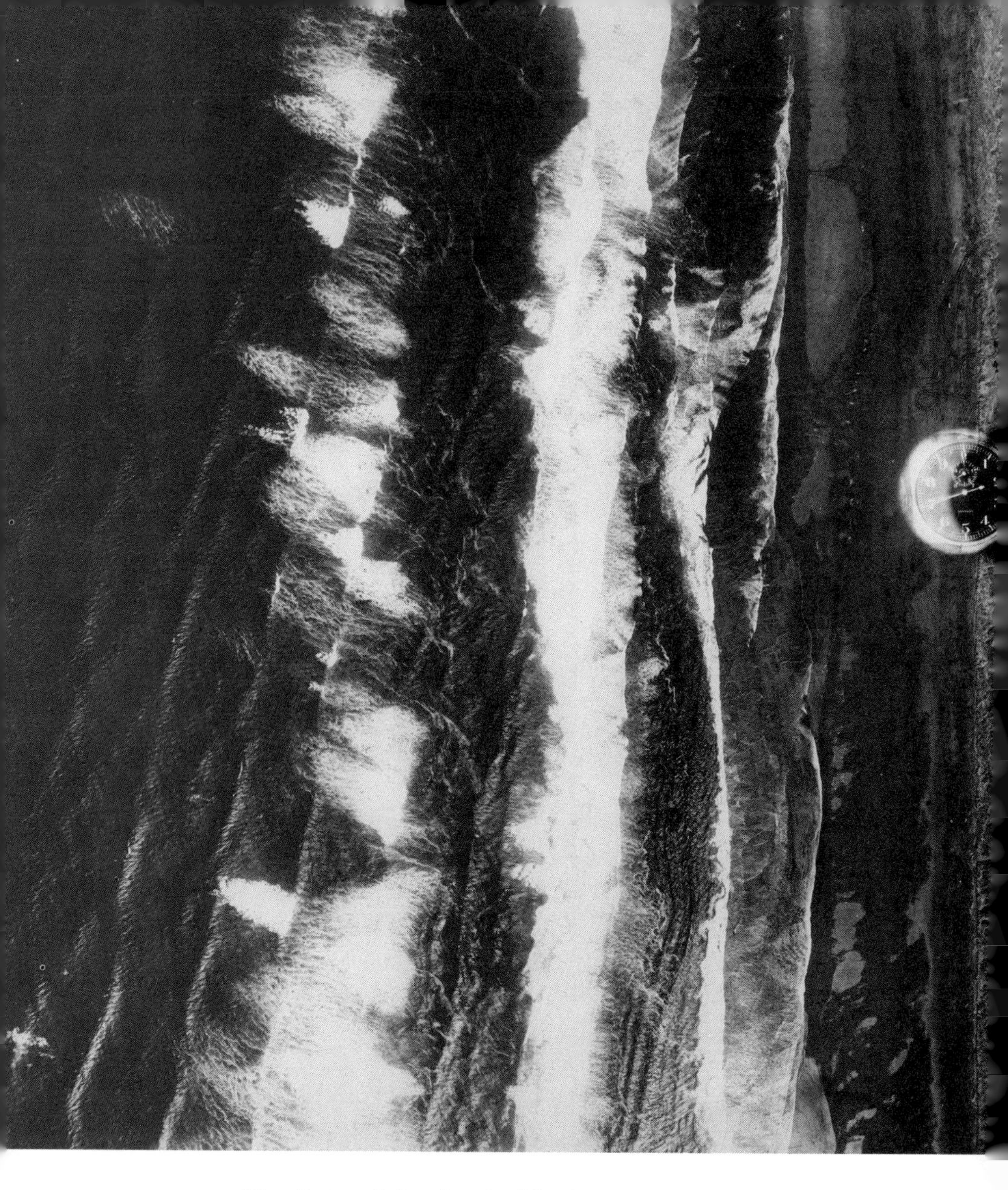

FIGURE 2.7 Wave Project, University of California, Berkeley, *Swell from NW, May 6, 1950*. Willard Bascom Papers, SMC 62, Special Collections and Archives, University of California, San Diego, box 12, folder 2. The waves should be read from left to right.

itself, permitting a view from above (Kaplan 2018). With sequences of shots lined up to create a flip-book cinema, such views could foretell futures by turning wave time into wave space. Earlier, nineteenth-century viewings of waves from above had developed in concert with the surveying of ocean space from cliffs, a vantage from which scientists often described the sea as a flux to be brought into calculative order by an objective observer. Hermann von Helmholtz wrote that a "great multitude of different systems of waves mutually overtopping and crossing each other . . . is best seen on the surface of the sea, viewed from a lofty cliff. . . . I must own that whenever I attentively observe this spectacle it awakens in me a peculiar kind of intellectual pleasure" (quoted in Rodgers 2016, 204). Recall and contrast Willard Bascom's outdoorsy stance (described in the introduction) as he sought to measure waves from a point of view *in* the water, here fusing a dedication to disinterested accounting with an in-the-thick-of-it heroism, one that the historian of science Naomi Oreskes (1996) has argued crafted mid-century oceanographic objectivity as a potent mix of masculine detachment and passion (often erasing the labor of those many women who were tabulating, calculating, and theorizing ocean data from sea and sky). Twentieth-century aerial photography doubled down on detachment, in the name of military aims, seeking to annihilate the horizon itself.

Predicting waves for Normandy and beyond was a project about strategic planning for near-term futures. The next phase of militarily sponsored work on waves at Scripps turned to a different future—a *possible* future, conjured by the specter of nuclear war. In 1946, the United States detonated a twenty-kiloton fission bomb in the Bikini Atoll, in the South Pacific, calling the explosion a "test." "Operation Crossroads" enrolled not only US military officers, but also oceanographers, led by the Scripps scientist Roger Revelle, who wrote that the Navy "wanted to learn about the waves that would be produced by the air and underwater explosions and about the dispersion of radioactive materials in the lagoon and ocean waters" (quoted in Shor 1978, 380) and later declared that "the atomic bomb is a wonderful oceanographic tool" (quoted in Rainger 2004, 93) (figure 2.8). The aim of Operation Crossroads was to investigate the effects atom bombs might have on ships and ocean environments. Munk was signed on for "assistance in mathematical and theoretical problems of ocean diffusion" (Raitt and Moulton 1967, 143). Revelle recounted:

> The waves from the underwater explosion were about what we had predicted from our model studies, producing 12-foot high surf on Bikini

FIGURE 2.8 Mushroom cloud consequent on underwater explosion of nuclear device by United States during Operation Crossroads, Bikini Atoll, July 25, 1946.

> Island. But completely unexpected was the behavior of the giant jet of foaming seawater that shot a mile into the air, and then fell back in a huge doughnut-shaped mass of spray nearly 1,000 feet high which rapidly spread out and enveloped the entire array of target ships. The initial outward velocity of this "base surge" was so great that for a few moments we thought it was the water wave produced by the explosion, and that it would break right over the rim of the atoll. (Quoted in Shor 1978, 381–82)

The year 1946 was the same one the Office of Naval Research (ONR) was founded, and Operation Crossroads was the beginning of the ONR's relationship with Scripps. When Revelle became the director of Scripps in 1950, the connection strengthened. With Bikini, the immediate emergencies of a hot war were replaced with speculative terrors—speculative for the military and scientists, that is, and not at all for people living in Oceania (see Masco 2014).[19] The script of wave science had been flipped, from emergency forecasting to preparedness simulation, even as real bombs inflicted enduring injury on peoples of the Pacific.

Nuclear detonations in the Pacific saw oceanographers from Scripps participating again in 1952, when the United States dropped a hydrogen bomb on Enewetak Atoll, in the Marshall Islands (see Rainger 2004, 98). Wearing only bathing suits, Munk and Bascom were positioned on three-by-three-foot rafts floating within sight of the impending explosion. Munk had "become concerned that the H-bomb would trigger a tsunami with distant outreach. . . . [W]e were concerned that the shock of a magnitude 7 earthquake (the thermo-nuclear explosion) would trigger an underwater landslide. Such landslides are good generators of tsunamis" (quoted in von Storch and Hasselmann 2010, 26). Munk and Bascom scanned readouts of bottom pressure to see whether there were any heralds of tsunami activity, which there were not. If wave predictions for Normandy sought to know a near-future wavescape, Cold War events in the western Pacific figured nuclear-weapon shock waves as virtual proxies for not-yet-arrived waves of war, even though the test did produce radioactive fallout, as the Japanese oceanographer Yasuo Miyake argued in 1954, in the process doing fast and slow violence (Nixon 2013) to people and environment. French detonations in the Pacific were part of the story, too; they began in 1966 and continued into the 1990s (see Rainger 2004, 105–6).[20] US-based scientists then came to live in what anthropologist Rebecca Bryant has called an "uncanny present," acquiring a "sense that what [they would do in the] . . . present [would] be decisive for both the past and the future, giving to the present the status of a threshold" (2016, 20). That turned out to be correct, though not so much for preventing the irradiation of Earth's oceans as for enacting it—dosing the ocean and "the bones and teeth of all subsequent generations" (DeLoughrey 2019, 71–72) with radioactive isotopes of carbon, cesium, strontium, and plutonium so that now, at millennium's opening, some geologists point to nuclear tests as key stratigraphic markers of the Anthropocene.[21]

It is not clear how many scientists or naval participants later developed cancer from this event. (Munk wondered whether a thyroid removal could be traced to this moment; Bascom, by contrast, credited radiation with *curing* a putatively terminal cancer that he had used as a rationale for exposing himself to radiation.[22]) What is certain, however, is that this detonation, with many others, was devastating to Marshall Islanders, who developed cancers and passed on a legacy of birth defects to their descendants (Parsons and Zaballa 2017).[23] Their islands and the waters around, which the United States captured from Japan in 1944, had become available for military experiment when Micronesia became a US Trust Territory in 1947. Bikini and Enewetak were oceanic territories captured by what Winona LaDuke (1983) would call "radioactive colonialism." What Kevin Hamilton and Ned O'Gorman (2021) name "experimental imperialism in the nuclear Pacific" was subtended by "laboratory" analogies, shored up with photos of nuclear detonations as simple (and aesthetically arresting) records of data rather than as killing forces unleashed in specific places, on people's homes. American visions of the South Pacific as a tropical paradise of isolated islands populated by racially subordinate natives, coupled with the extension of American hegemony over the Pacific Ocean (rendering the expanse into a theater for *war*, not at all as a "pacific" space), are the flip (but enabling) side of a vision of the sea as a site for military wave science. Against persistent stagings of the Pacific Ocean as a frontier or space of optimistic futurity, Marshallese activists such as Kathy Jetñil-Kijiner have worked to keep visible the Pacific as a space of loss and decolonial struggle (see Anderson et al. 2018; Bahng 2019; see also "Set One, Third Wave: Wave Navigation, Sea of Islands").

The Cold War ocean realm was not only a military world; at sea, it was also a male world, reinforced by the gender segregation of the US Navy. Scripps oceanographers were overwhelmingly white men. The writer Helen Raitt, who was married to the geophysicist Russell Raitt, was a rare exception to the all-male cast of scientists on Scripps trips. She joined the return voyage from Enewetak, writing about it in *Exploring the Deep Pacific* ("My whole sex would be on trial" [Raitt 1956, 34]). Raitt was also the founder of the Scripps wives service organization, the Oceanids, which offered practical support to Scripps families. Raitt was the eventual coauthor of a major Scripps history and, earlier, coauthor of a play staged to entertain the Scripps community. "Endless Holiday," with lyrics by Raitt with Ellen Revelle (wife of Roger Revelle), offers a scene in which male scientists on a cruise meet mermaids who complain about the aggressiveness of scientific investigations:

> SECOND MERMAID And do you have any idea how many times you've lowered a bottle, or a bucket?
> THIRD MERMAID Or that silly long tube of yours to the bottom?
> ROGER About four hundred times maybe?
> ALL MERMAIDS Four hundred and thirty-three times.
> FIRST MERMAID We want you to quit.
> SECOND MERMAID It's not safe down there anymore!
> THIRD MERMAID You are making us all nervous wrecks.
> SECOND MERMAID You've giving us ulcers.
> FIRST MERMAID You don't seem to realize what you're doing to our community life. This continual bombing day and night!
> ROGER We're sorry, but we're oceanographers.
> FIRST MERMAID We know you're oceanographers[!] Oh dear God how we know about oceanographers. Why, when the sardines de-

> cided to move out in '46 and you started that abominable survey program . . . , we all had to move west. Life just wasn't worth living in close to shore any more.[24]

Though the "bombing" of which the mermaids speak refers to oceanographers lowering equipment into the sea and is meant as a flippant exaggeration, poking fun at boys and their toys, the text can also be read as a critique of the violence done during this period of oceanography to living community and natural resources—violence that was possible in part because of how Cold War masculinity operated by compartmentalizing work and home, science and ethics. Pitching "mermaids" as the flipped image of "oceanographers"—figures of feminine concern versus masculine reason—reveals how masculine and feminine gender formation and heteronormative orientation worked as resources for shoring up projects of ocean objectivity during the early Cold War period.[25]

Aboard FLIP: Fixing Horizon and Level

We live now in a different period, even as the wake of earlier social trends still tells on the present. Fast forward to the 2010s and one of the first things one might notice on oceanographic vessels is sociological: the scientists are no longer all men. The change has been gradual, though beginning in 2005 the founding of organizations such as Mentoring Physical Oceanography Women to Increase Retention (MPOWIR) has generated significant transformation. (Some 80 percent of participants in the program with doctorates prior to 2012 are now in faculty, government, or nonprofit research positions [Mouw et al. 2018, 171]). The MPOWIR, sponsored by NSF and the ONR, has acted against sexual harassment at sea and put mentoring in place to advocate for women scientists navigating family-career tensions, which can make it difficult to schedule the trips at sea that are essential for building a research profile. (The MPOWIR has recently reframed its charge to speak more directly to LGBTQ+ oceanographers and to histories of structural racism in the field [see Keisling et al. 2021]). Another interest group, #BlackInMarineScience, inaugurated in August 2020, has gathered mostly marine biologists, though some physical oceanographers have lately signed on.[26] The horizon of possibilities for people in oceanography—and on Scripps ships—has been expanding.

This is so even as the military bounding of such horizons retains a continuing force. The larger project in which my FLIP science party's oceanographic work was embedded was the Navy-funded Coupled Air Sea Processes and Electromagnetic Ducting Research (CASPER), aimed at investigating how atmospheric conditions just above the sea surface might affect the propagation of radio waves and microwaves from one site to another (Wang et al. 2018; Wang et al. 2019). Understanding how the spatial distribution of water vapor (as moisture, humidity) hovering just above the waves might channel telecommunication—might, that is, enable "electromagnetic ducting"—could improve the transmission of signals across the sea surface, minimizing distortion and clutter.

So channeling at-sea communication for a mix of civil, defense, and recreational purposes demands a kind of holding steady and taming of the horizon, even as it might also permit the possibility of "over-the-horizon radar," a prospect that seeks to overcome the very notion of the "horizon" as a limit to knowledge (see Maleuvre 2011). If the horizon is a line beyond which lie spaces unknown at a time-space of observation, over-the-horizon radar suggests an overcoming of this liminality for those who hold its power, reeling the future into the present. What was sought after by scientists on this FLIP trip, then, was not so much what Bryant (2016) has dubbed the "uncanny present"—a sense of the present as a turning point—but a present smoothly coextensive with a foretold, and flexibly controllable, future. An onboard notebook with information about CASPER featured an image of the cartoon character Casper the Friendly Ghost, fitting for a project designed to make sensing invisible messages efficient but a bit disconcerting, even unsettlingly flip, given its military applications. I knew that there were nonmilitary projects nestled within this FLIP trip, too—research programs that had to do with the kind of weather prediction and coastal safety work I had seen in the Netherlands. Much of the knowledge many of us get—on TV, online—about weather comes through channels wound through with military infrastructures.

I used the cartoon Yamaguchi drew for me to locate scientists around FLIP and immediately recognized that people were executing tasks in time-bound loops, repeating observations at regular intervals, the time of science keyed into a cyclical time discipline meant to track the nonhuman time-space of the sea. I also got on the ship's Google hangouts; if, once upon a time, oceanographers and crew at sea communicated with one another through pulling on ropes or winches or, later, coordinated their activity on walkie-talkies, now more is done on screens (see Goodwin 1995; Hutchins 1995). I saw a rolling script of scientists worrying about wave heights, complications with instruments, and more.

A master's student at the Moss Landing Marine Lab studying ocean chemistry asked if I might help secure devices on one of the booms. Having watched videos of people walking the two-foot-wide catwalks that hang from FLIP some thirty-five feet above the water, with rope handholds on only one side, I had thought: yikes! I'm going to flip out if I have to walk out there! To interview scientists, though, I have to follow their tightrope acts. I joined the student in positioning humidity, temperature, and pressure sensors on the boom one night. Hovering above a black sea we could not see, it was like we were on the scaffold of a space station.

The next day, I came across Dave Ortiz-Suslow, fresh from his doctorate at the University of Miami's Rosenstiel School of Marine and Atmospheric Science (Ortiz-Suslow et al. 2016).[27] He stood on a grated metal walkway studying the waves. We fell into talking about his dissertation, and when he learned that I had done research in the Netherlands, he told me that a wave scientist from Delft (someone I had interviewed) served on his dissertation committee. He was fascinated, during his time in Holland, by how *engineered* the landscape was and how different this was from the United States, where, he said, greater value seemed placed on the "wild"—like the open ocean around us. When I asked him what he saw when he looked at this sea, he looked to the horizon and said:

> Young waves developing. Westerlies turning on—it's the afternoon sea breeze. It seems like the swell has died a little bit—but you can still see a longer oscillation coming in from the Pacific Ocean. And the short waves we see here are going to come and mess things up for everyone trying to surf. You're seeing white-capping a bit—10 knots is about the cutoff. Nice classic southern California wave field; this is what we've had every single day. Except during the Santa Ana winds.

Ortiz-Suslow was in charge of a "surface-following meteorological mast" that he called "FLOP," for "FLIP's lowest observation point" (the implied call out to flip-flops, California beach footwear descended from the Japanese *zōri* that American servicemen brought back after World War II, placed us firmly in a Southern California surf milieu). Tethered to one of the booms, this was a pole that had mounted onto it not only temperature, humidity, and pressure sensors but also a "motion pack," which measures "linear acceleration and rotation, from which we can extract wave information. We can directly relate that to wave height, and hopefully, with some tinkering, wave direction." Back inside FLIP, Ortiz-Suslow showed me a computer screen translating that data into a computer visualization that displayed, in on-screen squiggles, the spectrum of wave energy over the course of the day—the quantitative correlate of his qualitative remarks.

Some squiggles on the rolling graph represented the unhurried yawing of FLIP—a side-to-side wobble with a minute-long period. I had registered it subconsciously. This motion would need to be digitally removed when Ortiz-Suslow cleaned up the data, he explained, saying it might take "some time back on shore, in front of the computer, because right now we're *also* very slightly rotating—back and forth by about 20 degrees every thirty seconds—and that's because the mooring is slightly off. And the keel is pitching by 5 degrees because of currents. The trick will be canceling all this out. You go out for five weeks, then spend twenty-four months fixing the data!" FLIP was mostly stationary, but that needed to be further curated in the record-keeping process. As with buoys, stillness—and sea level—was necessary to install in this infrastructure of observation.

Watching the spectrum of wave frequencies scroll, I asked Ortiz-Suslow what led him to waves. He said:

> I started by working on a Gulf of Mexico Research Initiative–funded project that came about as a result of the 2010 BP oil spill. This project combined many different disciplines, and you could really see the impact this work had on the public because our goal was to better understand and predict surface oil plume transport after the Deepwater Horizon disaster. Waves were a key component of this work, critical to modeling currents. My dissertation was on the mechanism by which the atmosphere and ocean are coupled and on how sea spray affects the airflow over waves. Waves are the intermediary between air and sea, and they affect the lower atmosphere.

He got into surfing when he arrived at UCSD: "Surfing fed into the interest. There are pockets of wave scientists whom you know are really into surfing. You can pick them out at conferences. More casually dressed. People my age—twenty-eight—tend to be surfers. Vans shoes and jeans." Ortiz-Suslow pointed to the T-shirt I was wearing, celebrating a beach in my hometown in North County San Diego, and remarked on how waves were rendered on the shirt, reminded that back in high school, he used to doodle waves on his notebooks, trying—like many surfer types—to represent the ideal wave ("mind surfing"). Waves are entities that people understand in idealized ways—through equations, images, T-shirt cartoons—as well as through their demographic location; though Ortiz-Suslow is some twenty years younger than I, we had overlapping attitudes to the sea. Its military history was far from our attention, in part because the military entanglements of ocean science are so distant from the experiences of many middle-class, often white surfers in the days of a volunteer military marked by class and racial inequality.

Meanwhile, the people sending X-band signals from FLIP to *Sally Ride* were testing how far they could get these electromagnetic waves to propagate, using different altitudes of air ("atmospheric ducts") as "wave guides" that could carry radio waves long distances. All sorts of waves out here seemed to be unfolding across a spectrum. Where did the idea of the wave spectrum come from?

Waves across the Pacific, a Media History

Wave observations during World War II and just after were not yet animated by a sense that waves arriving at shore might be sorted by frequency—into that framework that would, after a watershed conference in 1961 (see Irvine 2002), commonly come to be called a *wave spectrum*, an analogy to spectra of light, sound, or radio waves. Wave scientists had long known that waves were generated by ocean storms, which impart energy to water and create swells that move across the sea with definite wavelengths and periods. In *The Sea around Us*, Rachel Carson (1951, 113–14), summarized the dominant scientific wisdom (and observational practice) at the mid-century:

> As the waves roll in toward Lands End on the westernmost tip of England they bring the feel of the distant places of the Atlantic. . . . As they approach the rocky tip of Lands End, they pass over a strange instrument lying on the sea bottom. By the fluctuating pressure of their rise and fall they tell this instrument many things of the distant Atlantic water from which they have come, and their messages are translated by its mechanisms into symbols understandable to the human mind. If you visited this place and talked to the meteorologist in charge, he could tell you the life histories of the waves that are rolling in. . . . He could tell you where the waves were created by the action of wind on water, the strength of the winds that produced them, how fast the storm is moving. . . . Most of the waves . . . , he would tell you, are born in the stormy North Atlantic eastward from Newfoundland and south of Greenland.

Such waves, it began to be surmised in the 1940s, might be sorted by their frequency (the inverse of their periods) into a spectrum. The intuition

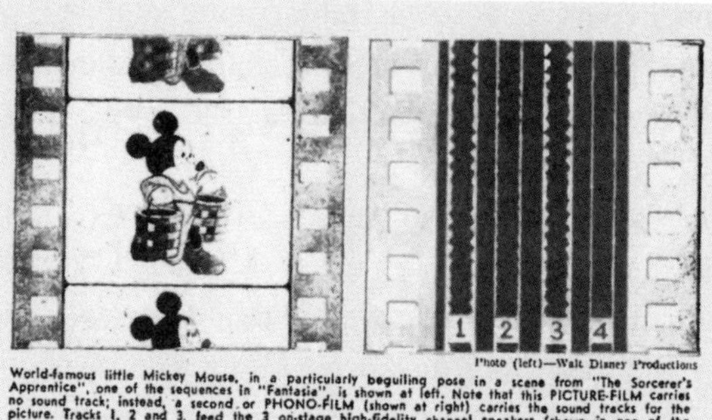

World-famous little Mickey Mouse, in a particularly beguiling pose in a scene from "The Sorcerer's Apprentice", one of the sequences in "Fantasia", is shown at left. Note that this PICTURE-FILM carries no sound track; instead, a second or PHONO-FILM (shown at right) carries the sound tracks for the picture. Tracks 1, 2 and 3, feed the 3 on-stage high-fidelity channel speakers (shown in one of the accompanying photographs) and the series of auxiliary loudspeakers (shown in the block diagram of the theatre sound installation). Channel No. 4, extreme right, shows only slight variations, but these are sufficient to control the dynamic range of the reproduced sound. (The phono-film shown here was clipped for *Radio-Craft* from Beethoven's "Pastoral" [6th] Symphony.)

A new era in talking motion pictures is being introduced by the $2,200,000 production "Fantasia." This fantastic color picturization of musical masterpieces heard in new interpretations of tone and volume includes a $200,000 sound-recording set-up and a $30,000 theatre 3-dimension sound system. Amplifiers, loudspeakers, photocells, an interphone, and a disc recording system, are used as here described.

•

"FANTASIA" INTRODUCES "FANTASOUND"

FIGURE 2.9 Illustration of optical sound. "'Fantasia' introduces 'Fantasound,'" *Radio Craft* 12, no. 8, February 1941, 495. The film image on the left was meant to be projected simultaneously with the playback of wavy black-and-white silhouettes on "phono-film."

came from new communications media that had become part of scientists' everyday life, and from looking at these media from unusual angles.

British wave scientists, gathered by the British Admiralty during World War II into a body called Group W (*W* for waves) lit on the idea that it might be possible to "measure the variation of frequency with time" (mathematician Fritz Ursell, quoted in Longuet-Higgins 2010, 43) of wave records from a Lands End observing station and infer something about the many geographical origins of waves with varied frequencies. Twenty-minute records of wave pressure, taken by undersea gauges (the "strange instrument[s]" of Carson), would be relayed to a rotating drum. But where a more usual wave record might have been rendered by a pen zigzagging lines on a scrolling cylinder, here researchers turned to a method inspired by Hollywood movies. The director of Group W, the oceanographer George Deacon,

> had a friend in the film industry and we learned that the Walt Disney film, "Fantasia," which was then a recent success, had the sound part of the film as black and white wavy silhouettes along the side of the picture frames [see figure 2.9]. It therefore occurred to us that we could do the same thing. . . . We could use [photographic] paper . . . , and instead of printing lines we would have a big block of light which would move about, generating a silhouette of the waves [figure 2.10]. We could then put this on a wheel and, as the wheel spun round, the variations

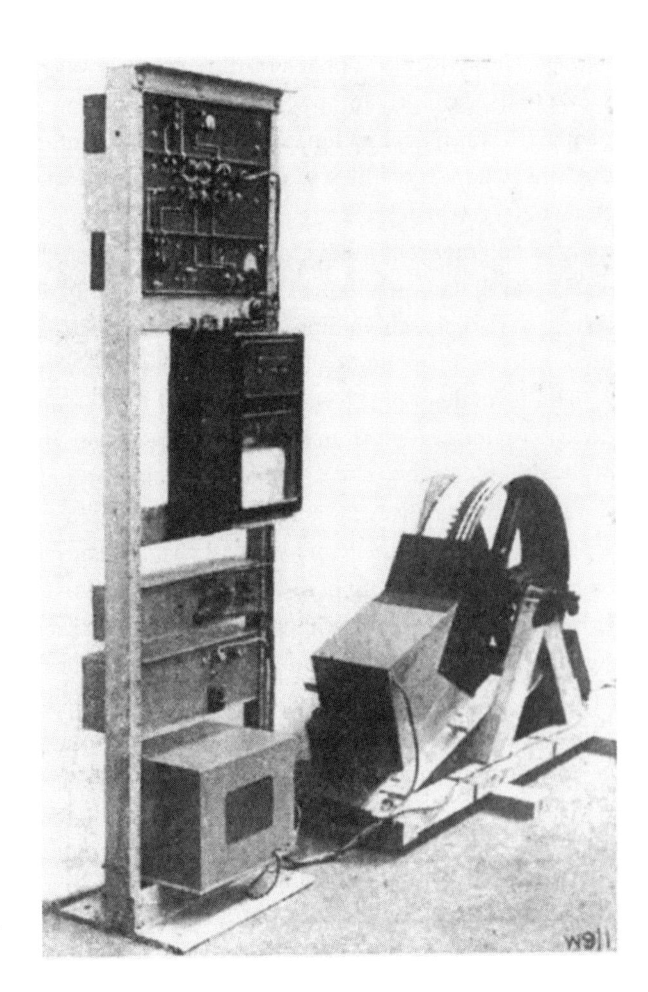

FIGURE 2.10 Frequency analyzer. From Barber et al. 1946, fig. 2.

> in black and white could be detected by a photocell. (Mathematician and oceanographer Jack Darbyshire, quoted in Longuet-Higgins 2010, 50; see also Barber et al. 1946)

Fantasia itself had made much of its optical soundtrack, even featuring a "meet the soundtrack" segment in which animated frequency signatures shimmied onto the screen (though the technique of optical sound had, in fact, already been in use for some years).

Optical film soundtracks were not the only media to inspire the spectral model. The oceanographer Willard Pierson adapted the spectrum model to wave records from the work of the Bell Laboratories statistician John Tukey, who had used it to examine the statistical properties of noise ("noise color") in electronic circuits (see Pierson and Marks 1952; Tukey and Hamming 1949).[28] Tukey's frame rendered waves as populations, not individuals, similarly to how sociology at the time was coming to study people. (It was no coincidence, then, that Tukey later cowrote a review of the Kinsey Report on the statistics of American sexual behavior [Cochran et al. 1953; see also Igo 2007].) In these borrowings, scientists turned to tools from emergent media to understand the medium of the ocean; if water waves had once been the model for sound and electromagnetic waves, here researchers flipped the script, inflecting ocean science with models from the technological apparatuses of electromagnetic measure.

The formalism of the wave spectrum, then—derived from visualizations of sound waves, from electronics—reads ocean waves in the idiom of twentieth-century media; indeed, it makes ocean waves, materialized in the medium of seawater, thinkable *as media* in a more technological sense. As John Durham Peters (2015, 3) writes, "The old idea that media are environments can be flipped: environments are also media." Melody Jue (2020) argues that oceanic phenomena, knowable through media (cameras, sonar, scuba, satellites, equations in hydrodynamics), can themselves be theorized as media, forces that crystallize as well as query conceptions about how information manifests as inscription, across interfaces, and in storage and transmission.

The story of wave spectra came operationally into focus in an ocean-spanning project that unfolded during the summer of 1963 when Walter Munk led Waves across the Pacific, a venture aimed at tracking trains of waves from their origins in Antarctic storms across ten thousand miles of Pacific Ocean toward their eventual arrival on the shores of Alaska. Munk has reported that he was inspired by studies in radio astronomy that used diffraction patterns to locate the source of interstellar radiation (von Storch and Hasselmann 2010, 7). Munk, with eight other oceanographers, including Klaus Hasselmann (see chapter 1), arrayed wave sensors at six sites across the Pacific. A documentary about the project released in 1967 lists the full roster of locations (figure 2.11):[29]

> Cape Palliser Light in New Zealand, a rugged storm-battered point where the arrival of the great waves from an Antarctic storm could be expected
>
> Tutuila, one of the volcanic islands of Samoa, 2,100 miles to the northeast
>
> The uninhabited equatorial atoll of Palmyra, 1,600 miles beyond Samoa, only two miles wide with no point of land more than six feet above sea level

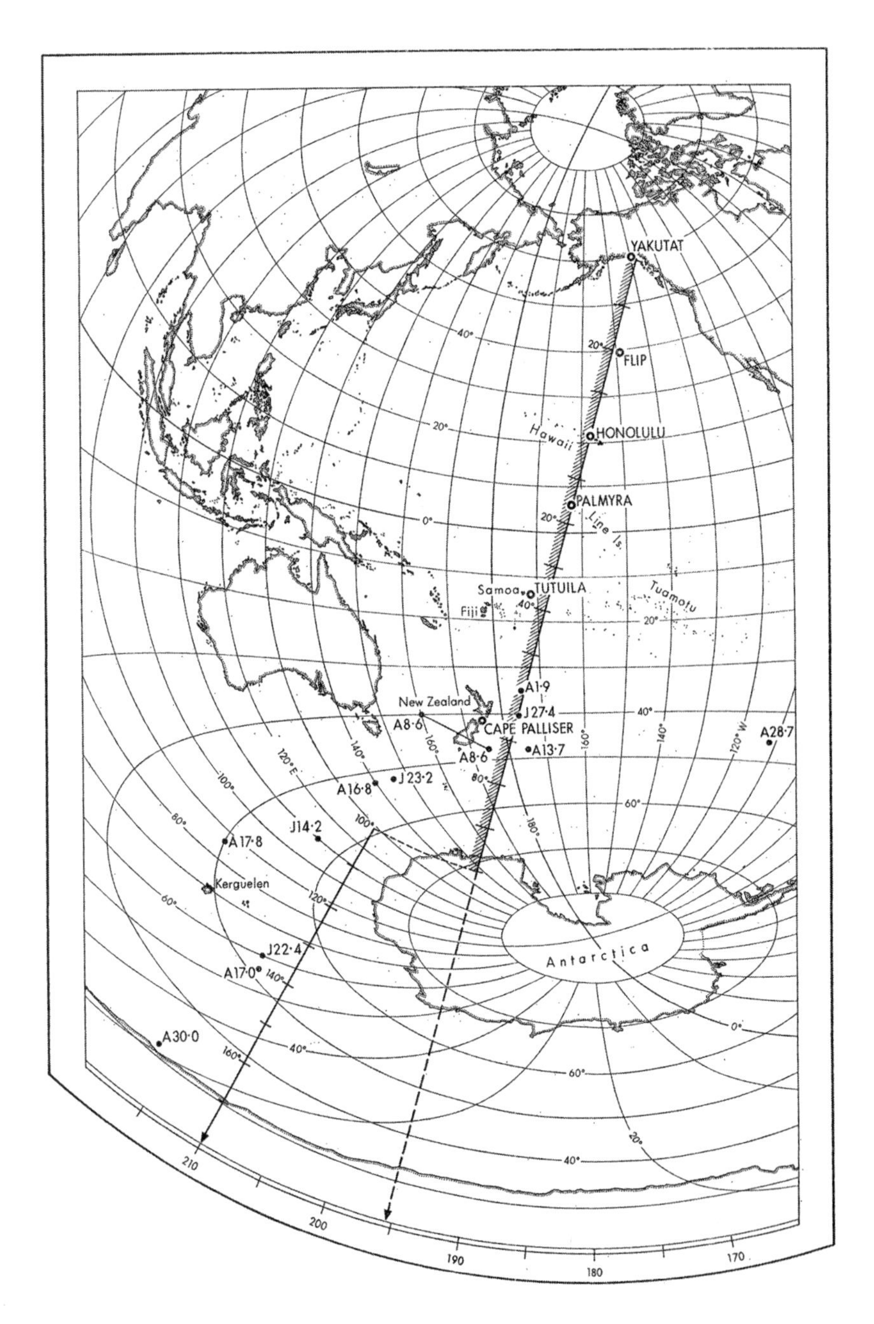

FIGURE 2.11 Field locations for Waves across the Pacific project. From Snodgrass et al. 1966, fig. 1.

The easily accessible Kewalo basin in downtown Honolulu, selected for the central wave station and expedition headquarters

The islandless North Pacific, where the US Navy's mobile island FLIP (FLoating Instrument Platform) was stationed at 45 degrees north and 150 degrees west, and

The final recording site an Alaskan beach, the end of the line for the trains of waves[30]

The task was to follow wave forms and energies as they traveled over a great arc spanning the Southern and Northern Hemispheres, investigating whether swells would be scrambled by crossing the equator. (The answer was no.) The project required observing stations that could permit scientists to position pressure sensors on a not-too-deep seafloor (or, in the case of FLIP, on a portion of its stem just below the surface). Sensors would need to be sensitive enough to measure waves that could be as extensive as a mile long and as squat as a tenth of a millimeter high. Sensors translated changing water pressure into electrical signals, which were relayed to spooling computer punch tape. The film shows scientists unrolling punch-tape records onto graph paper and poking pencils through the holes to make dots representing oscillating wave heights, producing graphs of the travel of waves. Such graphs permitted scientists to map patterns they hypothesized would become readable to stations later in the chain. Munk, at station on American Samoa, is shown in the film radioing ahead to Palmyra.

The tape data themselves were sent back to Scripps to be processed. As Munk put it in the film, spectral analysis could reveal how groups of waves traveled, since waves in the wild "are of all sizes and come from all directions. They are mixed and piled atop one another in lovely confusion." Translating punch tape into graphs, radioing, phoning, shipping, and more could generate spectra to sort it all out. The film shows the deployment of multiple sorts of media—tungsten wire, water, magnets, bronze cylinders, globes, rulers, typewriters, filing cabinets, paper, pencils, punch tape, tape computers, telephones—all of which underscore the mediations through which waves come to be known. (Again, all the oceanographers were men; women—mostly secretaries, record keepers, some scientists—were on land, a division of labor shaped by oceanography's use of sex-segregated Navy infrastructure, dominant US gender roles, and masculinized notions of heroic sea science.)[31]

Taken as a whole, the Waves across the Pacific project *renders the Pacific Ocean into a giant transmitter of wave signals*, the travel of its swells miming the relay of information by scientists from one station to another.

This ocean becomes like a radio that stores and sends information about storms. Technological and media forms imprint on how these ocean waves are brought into readability.

The Pacific, of course, is and was also a geopolitical medium. In the film *Waves across the Pacific*, the script has Munk remark that, for this project, "the sea itself was our laboratory." This laboratory was available to Munk and company because, with the exception of New Zealand, most points along the great circle since World War II had come to be under American rule, either as territories of the United States (Tutuila, on American Samoa, and Palmyra) or, more recently, as states (Hawaiʻi and Alaska). As the historians of oceanography Michael Reidy and Helen Rozwadowski (2014, 351) argue, "Knowledge of the ocean was—and remains—inextricably connected to midcentury geopolitics and the growth of modern science." They note, too, that "the expansion of empire enabled scientists . . . to study topics that required the accumulation and subsequent reduction, tabulation, and graphing of large amounts of observational data from all over the globe" (350). Ruth Oldenziel (2011, 16), writing about US island territories as an array of "naval nodes for the control of ocean space," reads the Pacific Ocean as layering communications infrastructures, military outposts, and colonial enterprise (of Japan, the United States, more). The geopolitical context that made the Waves across the Pacific project feasible is left out of the film. Its narrator paints a picture of the Pacific as a space of nature, with scientists tracking "waves that propagate across the entire Pacific, the largest body of water on our planet, waves that are interrupted only occasionally by the upthrust of a lonely coral atoll, waves that smash endlessly against the rock of high volcanic islands like Samoa, waves which ten days later die quietly on the beaches of distant coasts, the end of a journey half around the world." These waves move through geopolitical space. The "lonely coral atoll" of Palmyra was occupied by the US Navy from 1939 to 1959, during which time the Navy built a runway and dredged a ship channel. After that, in 1962, the island was used as a site from which the Department of Defense observed nuclear tests above nearby Johnston Atoll. The waves that "smash endlessly against the rock of high volcanic islands like Samoa" meet not just a shore but a Marine Corps outpost. And the "distant coast" of the experiment was in Alaska, which in 1959 had become a state of the United States. When the script (by the freelance writer Harry Miles) has Munk casually say about the experiment, "We preferred islands we could get to," it leaves to one side the historical contexts that made these places available.

FIGURE 2.12 "Measuring ocean swell from a Fale in Tutuila, American Samoa (1963). Walter had persuaded Judith to take the 0400 daily watch of swell recording. (Edie, Kendall, Judith, Walter, and Silau playing the guitar.)" From von Storch and Hasselmann 2010, 8, fig. 1.2.

Let me flip the script, then, and make explicit how the film represents island environments as zones of wild nature, emphasized through images of fieldworking oceanographers—white men—often in shorts, wearing no shirts, sometimes sporting tropical bead necklaces. A scene of Munk in American Samoa—"along the exotic southwestern shore of Tutuila"—shows him aided by shirtless local men placing a sensor underwater (figure 2.12). He works with tape records in a *fale*, a "house . . . built entirely of coconut palm, an ideal place to work with waves," and he tells us, later in the film as we see him reading a book on a verandah: "Data taking became quite routine. My wife and some of the Samoans learned to operate the recording equipment." The film's positioning of local people (and some American women) as ready help for male scientists appears elsewhere. In Alaska, Gaylord Miller, working at a Coast Guard station and "the only one of us who sometimes met a bear on his way to work," employs "a fishing boat operated by an Eskimo crew" to aid in positioning a sensor. As if to underscore the tropical vacation that the South Pacific sites represented for white scientists, we see many with newly grown beards.[32] Munk and other participants in the project had more complex relations in person and on the ground than were captured on camera.[33] However, the filmmakers drew on tropes representing Pacific Islanders in Cold War Hollywood films as docile "natives" (Hereniko 1999). The film gives viewers scientists and waves with lives (recall Carson's "life histories of the waves"), but not local or Indigenous people. The rendering both lively and technical of waves was made possible by an American orientation to the Pacific enabled by what Kerry Bystrom and Isabel Hofmeyr (2017) have named *hydrocolonialism*, the scripting of the sea into projects of extending colonial or imperial range.

Meanwhile, FLIP, "at her lonely north Pacific station," is, in the film, a space of pure technology: "Wave measurements aboard FLIP were comparable and in some ways better than those on land. FLIP floated with negligible drifting taking wave measurements for three hours twice a day." FLIP becomes what Rozwadowski and David van Keuren (2004) might call "a machine in Neptune's garden," drawing here on Leo Marx's analysis in *The Machine in the Garden* (1964) of how American pastoral landscapes have come to enfold and include modern technology (think, for example, of bucolic scenes of railroad trains in the countryside). As an avatar of science, FLIP comes to fit right in to wild ocean space, both within and apart from it, bobbing at the elemental interface of air and water.

Aboard FLIP: Turning to Wonder

But FLIP is a weird sort of machine in the garden, its serious side always doubled by the sideways flippiness of the thing. It awakens in scientists a sense of wonder.[34] That sense is in part amazement about what engineers can build—and, perhaps, too, astonishment that funding could have been gathered to build such a bizarre device. But it also derives from a disorienting but welcome affirmation, for those invested in an ethos of objectivity, that the embodied perspective of humans is conventional—that answering questions always requires an explicit frame of reference and that that frame can, under the right circumstances, be shifted to reveal unexpected facts.

Otherwise put, FLIP offered for scientists an instantiation of a meta-objective standpoint in which the relative coordinates of human experience nest within a grid of absolute values in which "up" and "down" are orientations, not ontologies. The anthropologist David Valentine, in his ethnographic work on space-travel enthusiasts, observes that the aspirational space travelers he knows think relativistically about gravity: in zero gravity, on other planets, on hypothetical circular space stations that might produce artificial gravity through the centrifugal force of rotation. For them, he says, gravity, suspended and reorganized, puts its status as a "*universal* universal"—as a ground—into question (Valentine 2017, 191). On FLIP, of course, where gravity never changes magnitude but instead shifts experiential direction, disorientation highlights the limited perceptual tuning of humans—a tuning that can be outsourced to instruments able to flip x-, y-, and z- axes without getting confused.

Still, the moment of the "flip" has been, for scientists, a time-space of suspended expectation, an interstitial moment of wonder. Some of the science party were sorry for me having missed the flip. "You missed the fun part," one said. "It was like an amusement park ride." Another declared, "It was really awesome. It starts off slow; they fill in lower tanks, and then, in the last fifteen minutes, it starts going faster, and then the last two to three minutes there is *roar* of air coming out of the vents, and then it's silence." Still another commented that it was "pretty cool! The coolest thing was realizing how much more *space* there is out here. Everything just transforms. The actual flipping was holding on to the rail—a little bit like a rollercoaster, bobbing up and down." One of the postdocs pronounced, "It was just amazing. You put your feet on the wall and your back on the floor and the wall becomes the floor and the floor becomes the wall." A student worker remarked, "We were giggling like kids. The momentum when it righted itself—it's like a mixture of excitement and 'Oh, shit! This could be the time that FLIP has a catastrophic accident.' It became a big metal tree house, like the Lost Boys had in Peter Pan."

Indeterminate, shifting gravity, observes the anthropologist Debbora Battaglia (2012) in her reading of astronauts' diaries, is often experienced as a zone of levity, clowning. Examining the diary of a Soviet cosmonaut perplexed when he "came in at an unusual angle" upon arriving (through an airlock) at a space station, Battaglia notes that perspectival shifts are often managed through joking, suspending seriousness. FLIP operated just so among scientists. (Crew members, particularly the cook, were less enchanted. The able-bodied seamen crew, who had to make sure everything went properly during the flip, had less license to give themselves to giddiness.) As if to confirm that analysis, scientists on FLIP received a "certificate of flipping," complete with campy calligraphy suggesting an antique age of sail—something like those tongue-in-cheek documents that sailors get when they cross the equator.

But wonder does not occupy some space outside objectivity; it may be *part* of it, a resource and oscillating opposite to it—perhaps in the way the emotional mermaids in the play excerpted earlier were the flip sides of rational oceanographers, or in the way that an "expeditionary heroism" (Oreskes 1996) has been part of the narrative of making objective knowledge at sea. This is not the full upside-down turn that would have wonder as the flip side of a deracinated objectivity. Rather, it is a turn that poses wonder as ever coming into and out of alignment with empirical accounting, as that which makes measure meaningful. Whether FLIP is rabbit or duck depends on what scientists ask to see. Fixing what counts as "objective" often entails *having an objective*, a goal, an orientation. Think of the flip, then, as an *epistemological turn*, with the imaginative and the empirical conjoined, though also swiveling in and out of alignment.[35]

For oceanographers on FLIP, the axis that stays constant is the idea of a frame of reference—and a frame of reference, ultimately, for humans, since the flip is about keeping observing humans breathing and working in the air just above sea surface, even if their instruments duck back and forth between water and air. The platform is not, initially, a vantage for what the historian Lorraine Daston (1992) has called *aperspectival objectivity*. Rather, it is an aerie offering a provisional frame of reference, a view not from nowhere, but from a close-by and controlled somewhere (though this doesn't prohibit scientists from later trying to factor that somewhere *out*, to leave only waves against a universalized level). Keeping the specifics of its location (oceanographic, political, technical) in view, think of FLIP as a platform that generates what Haraway (1991b) has called "situated knowledges"—claims made from embodied *somewheres*—though think of it, too, as a setting that makes *oriented knowledges*, knowledges aimed at particular futures, at specific objects and objectives.

In a place where human senses are pressed to orientational limits, much sensing is delegated to machinic devices, offering readouts in numbers, false colors, and sounds. One of the older technologies on the platform is radar, although its protocols are now often translated into up-to-the-minute computer visualizations. A screen displaying an azure circle representing the output of the ship's radar tower during my trip pointed down at a circle of sea around FLIP. Across this circle moved wormy wriggles of green—a digital highlighting of the poking-up crests of *internal waves*, huge waves with half-kilometer wavelengths that were traveling underneath us with a period of about fifteen minutes. A scientist took me outside to see real-world evidence: more-or-less regularly spaced smoothed bands of water. Internal waves are a reminder of an earlier set of questions people asked from FLIP—and a pointer to an area of wave science that has become of new interest not only to physical oceanographers, but also to biologists, who now consider waves not only as formal processes of ever recurring cyclicity, but also as carriers of ecological process. And history.

Waves beneath the Pacific, Internal Accounts

Rob Pinkel is a Scripps physical oceanographer who began doing science on FLIP in 1969, when he was directed to work on the platform by his adviser, Carl Eckart, a physicist who trained with the quantum theorists Erwin Schrödinger and Werner Karl Heisenberg and who in 1946 became the first director of the Marine Physical Laboratory at Scripps, following a wartime post as leader of the University of California's Division of War Research Laboratory. Pinkel was tasked with studying internal waves, enormous waves that travel underwater, with upper boundaries rising and falling just beneath a stratum of surface water that betrays only a hint of the massive movement beneath (see Thorpe 2010). Caused by both surface winds and submarine tidal flows over large lumps of seabed, such interior ocean waves can be as much as 150 meters tall. They can distort sonar communications. During the Cold War, that made them of sharp interest to the Navy.[36]

When I met Pinkel in 2017, he gave me a crash course on internal waves (see Pinkel 1975, 2016; Pinkel et al. 2015; Pinkel et al. 1987). He pointed to a desktop wave machine—one of those seesawing lava-lamp-like devices made of clear plastic through which one can see a layer of gluey blue fluid rolling beneath a clear liquid, like oil under vinegar. This, he said, was a side view of an internal wave. "The difference between surface and internal waves," he continued, "is that you can't *see* internal waves, so there's no intuition to go on." And from a platform like FLIP, it is also impossible to *feel* them; they are kinds of hyperobjects (Morton 2013) that operate at a scale beyond embodied human perception. Knowledge about internal waves has been largely won through underwater sonar data. Pinkel said that collections of internal waves, rolling beneath the surface, could be envisioned as akin to "the rolling hills of Kentucky, except that all of the hills are kind of moving in different directions at walking speed or a little slower—and, if you go beneath their surfaces, it's like there would be different layers of soil with different topographies on them." Getting an image of this wavescape from FLIP in early days was a matter of some suspense, since in the field "early sensing systems could not convey a good sense of the beautiful wave propagation patterns that were unfolding deep in the sea. We were challenged to create sensing systems that collected data across a large range of scales, such that we could watch these patterns evolve." Scientists had to depend on improvised, often kludged together, instruments. Now, with improved technology, internal waves are easily detectable from above, including through synthetic aperture radar images from satellites. When Pinkel started, such apprehensions were not available.[37] Patches of "dead water"—known for ages by seafarers—became legible only later as the crests of internal waves peeking up from below, slowing the flow of surface water (see Baehr 2014).

What was research like in early days? Pinkel zoomed out:

> In 1968 to 1975, the Vietnam War was full-on. And there were differences among scientists about whether to do certain kinds of research—differences at every level, and it was the right sort of discussion and dialogue to be having. Obviously, the students, myself included, were not fans of going to participate. We had different political opinions about it, and about the rectitude of working on Navy-related items in general. I was a student, and I had long shaggy hair—my mother said I looked like a young version of Mark Twain—and the people in the outside world just assumed I was a liberal hippie. Because I was doing this work, some of my student colleagues assumed I was a warlike hawk. Image-wise, I didn't fit in at all.

One of Pinkel's contemporaries, Douglas Inman, reported in an interview with the historian Ronald Rainger:

> My own personal approach had been the fact that I certainly was against what we were doing in Vietnam, but on the other hand not sufficiently so to deny what expertise I had to the country to facilitate what our guys were doing over there. So I sighted the harbors that were used in the Vietnam War, did it from small craft. . . . We had generated wave data for the entire coast. So to an extent, Vietnam became a testing point for some of our concepts and theories, and this, in retrospect, was a very useful and well-funded approach to science.[38]

Inman became so invested in keeping a boundary between oceanographic research and its politics—which politics he saw as irrelevant and not, as others did, enabling—that he sought to diffuse protests at Scripps, including by the new leftist social theorist and UCSD professor Herbert Marcuse, whom he said was

> egging on the student revolts. . . . [T]here were several attempts to have marches, student marches, down to Scripps, and our geographic

separation prevented this. . . . But we played this in a rather smart way, I think. We finally invited Marcuse down to give a lecture in Sverdrup Hall, and we made sure that the graduate student population at Scripps filled up at least half of the audience, and then when Marcuse gave his talk and his followers would start to yell and holler and so forth, we had sufficient people in the audience that it was controllable.

Flipping the script on Marcuse was a tactic for orienting the Scripps student body toward a frame of reference in which science and politics were seen as separate. Science was meant to be understood through its internal logic, not with respect to putatively "external" forces. Not everyone was convinced, as Munk reported in 2000 in an interview, naming two pacifist Quaker colleagues at Scripps who avoided ONR-funded projects.[39]

Pinkel viewed relations of oceanography and the military from a later vantage, saying: "I think right now the people actually doing military research are extremely non-hawkish, because if any of this technology started to be used, it would not look like a John Wayne movie, it would be just devastating to civilization. And so the technology game gets played in the hope that the war game never actualizes." This is an orientation to a future that is suspended, held in abeyance, a horizon that may never come (or, in the case of Inman, one that is justified "in retrospect").

Internal waves continue to be of interest in oceanography. I met up with Matthew Alford, once a student of Pinkel's (see Alford et al. 2015). When I arrived at his office, Alford was getting back from surfing, board under his arm. Next to his standing desk an arresting image floated on his flat-screen TV: a false-color digital visualization of internal waves propagating away from Hawai'i across the Pacific, headlined by the phrase "Wave Chasers," the name of a research group Alford once led at the University of Washington (plate 7). Internal waves can radiate far across the ocean, and in 2007 Alford and his colleagues published a paper titled, "Internal Waves across the Pacific" in homage to the 1960s project. Understanding the global paths of internal waves is a work in progress, owing to the effects of sea floor ridges, the reflection of waves off continents, and the generation by storms of wind-animated internal waves (created in part from the Earth's rotation).[40] Making these waves legible requires (to begin) ships, sonar, and conductivity-temperature-depth measuring instruments, and representing them visually draws on conventions that fuse the realist with the diagrammatic, often through computer graphics that invite envisioning the actual through the aesthetics of the high-tech virtual model.

Internal waves may also host unique ecologies, brokering new relations between physical and biological oceanography.[41] The Scripps biological oceanographer Peter Franks explained to me that internal waves can gradually transport animal larvae, plankton, and chemical mixtures, as waves compress and rarefy seawater and nudge substances along (see Lennert-Cody and Franks 2002). Advising me that "physics is simple to model; biology is messy and complex," he showed me an aerial photo of a dinoflagellate bloom arriving off the shore of Scripps. Phytoplankton were caught up in a nonlinear wave traveling toward the shore, moving in "patches," the internal wave forming an aggregation that extended and compressed like an accordion.[42]

Indeed, the story about internal waves today is less about submarine contests for underwater territory and, because internal waves are bound up with processes of ocean mixing (exchanges of heat, salt, nutrients, carbon dioxide, and more between layers of the sea), much more about ecology and climate (see MacKinnon et al. 2017). The physical oceanographer Amy Waterhouse, who earned her doctorate in 2000 from the University of Florida, told me more about internal wave-marine biology interactions, informing me that the La Jolla submarine canyon hosts an in-water observatory, a three hundred-meter-long mooring that hangs down into the water, dotted with instruments along much of its length. It measures physical and biological oceanographic process that range from temperature to conductivity and chlorophyll concentration. Waterhouse and her colleagues use this infrastructure to study how energy from internal waves breaking in the canyon—breaking that jumbles up temperature-salinity-density stratifications—shapes the flow of nutrients, influencing which creatures live where and shaping local and distributed climate (Waterhouse et al. 2017). Waterhouse did her postdoctoral work at Scripps with Jennifer MacKinnon (2014, 164), whose research program looks at turbulent mixing, "crucially important for everything from regional pollutant dispersal and nutrient budgets to global patterns of heat distribution, which is especially important in a changing climate" (see also McPhee-Shaw 2006).[43] Undersea breaks, if not signs of a broken ocean—an ocean wrecked by sociogenic process—become signs of compromised pasts and uncertain futures. Waterhouse came to her concern about climate "after working as an environmental consultant for the mining industry, who were disposing of mine tailings in the oceans," she said. "Our job was determining the best place for those. At some point, I asked myself 'What am I doing?' I decided that one way to make a difference was to ask questions about climate and climate change. I applied to oceanography graduate school to study coastal oceanography." Narratives of scientists struggling with the application of their work were offered to me time and again, with many researchers ambivalent, and sometimes quite upset, about the deleterious wakes—of radiation, pollution—left by some of the research histories that had put in place the questions they inherited.

I sometimes thought of Gregory Benford's science-fiction novel *Timescape* (1980), in which scientists at Scripps, working in the early 1960s, receive a message from 1998 warning them about a future in which oceans have been smothered by algal blooms whose origins lie in 1960s failures to think about the long-term ecological effects of research on the ocean. In the novel, the messages, sent in Morse code using hypothetical faster-than-light particles called tachyons, forestall an apocalyptic future as scientists are able to warn their predecessors, effectively erasing the wake—and its biological and ecological effects—that has been left for them. The orientation to time here is one in which generations of scientists come to know the effects of their science only in retrospect, positioning themselves as always already incompletely aware of the meaning of their science in the present. Today, however, many scientists are less likely to accept such claims of "unintended consequences" (see Parvin and Pollock 2020), seeking instead to conduct their science in the service of ecological and social values they support, to apprehend cause, effect, and consequence in an ocean now chronically under threat.

Aboard FLIP: Breaking Waves, the Lab in the Field, Paradigm Scripts

If models of ocean waves often bear the traces of the media with which they have been measured, imagined, and compared—becoming something like doubles or afterimages of, say, radio waves or flip-book animations—it is when waves *break* that analogies to other kinds of waves often break down. Why *do* surface waves break in the open ocean? When they are not in contact with an underwater slope, which makes waves "feel" the bottom and curl up, what initiates breaking?[44]

Understanding breaking waves at sea was a question that kept scientists on FLIP up at night, because it was a question they often asked *at night*, when they did not have to deal with the turbulence caused by sunlight on the water-air interface, where waves break. "The sun for me is noise," Ivan Savelyev, a Moscow-born and Florida-trained oceanographer, told me.

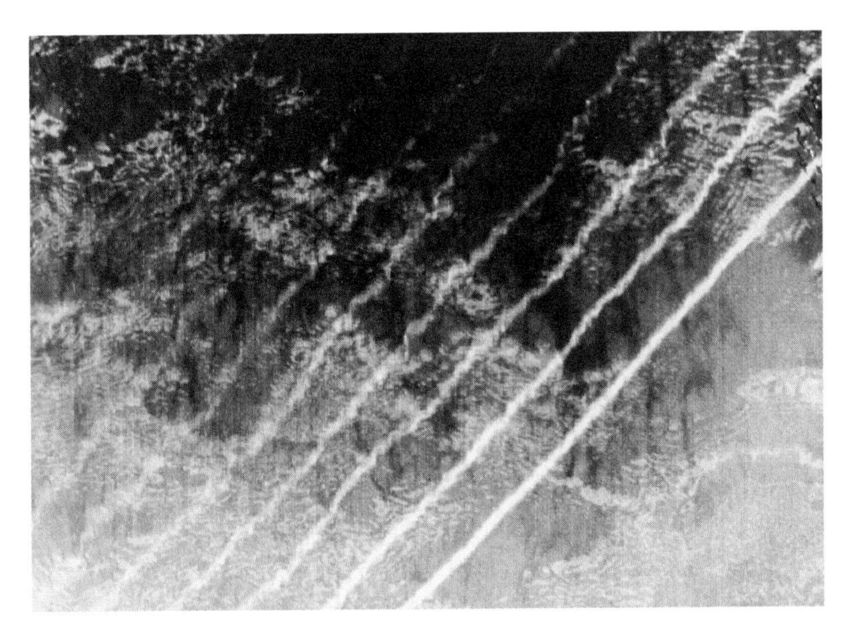

FIGURE 2.13 An infrared image, from above, of waves on 1.5 square meters of sea surface, striped with laser lines projected to track patterns of turbulence. From the research of Ivan Savelyev.

Working at night, he averred, made the open-ocean "field" more like a "lab," where variables could be controlled, kept within parameters—and, indeed, the very flip of the platform from horizontal to vertical made the vessel not so much immersed in a field as floating above it, moving from the tossed time of the sea to a time outside and above, objectifying the ocean. Over its history of use, FLIP often flipped between what Deborah Coen (2018, 93) identifies in her history of climate science as "a 'replicative' approach that treats the local atmosphere [or, for this case, ocean] as a laboratory for the investigation of universal laws and a 'chorological' approach," studying field environments as unique pieces of a global picture. Think of FLIP as a field-lab hologram.

I had run into Savelyev just as he was starting his day at 5 p.m. His instrument, on the starboard boom, featured two midwave infrared cameras pointed down at the ocean surface onto an area 1.5 meters square. Onto this patch of sea, Savelyev projected beams of infrared laser light—light just below the visible spectrum—that he had arranged into grids, points, and lines. He imaged those lines through two cameras (to compose a stereo image) whose feeds he could translate into visibility on his laptop in FLIP's computer lab (figure 2.13; see also Savelyev and Fuchs 2018). The lines squirmed on the sea surface, revealing patterns of turbulence. Infrared light, which discloses heat signatures, lingers for a moment after it hits a surface, leaving lagging traces that track the motion of water. It was almost as though Savelyev's apparatus was enacting an ancient Greek theory of vision that has eyes shooting beams out at the world to know, to *be-hold* it. The anthropologist Veronica Strang suggests, "Wonder comes from all sorts of light. People tend to appreciate unusual light phenomena such as sunrises and sunsets, rays from behind clouds, exciting color contrasts, and rainbows" (quoted in C. Henderson 2014). And infrared. The images, which slurred like molasses, relayed the material otherness of the sea, even as their eeriness emerged from algorithmic data conversions developed for night-vision surveillance (Parks 2014). But while these looked like otherworldly waves, with invisible lives, they were also waves that, known through the thermoception of infrared, were very much of our moment, when tracking the exchange of temperature across water and air may be key to understanding what the media theorist Nicole Starosielski (2019) calls the "intense thermal volatility" of climate change.

What was Savelyev trying to find out? He dictated: "to resolve the fluctuations of wave and turbulence fields on the smallest spatial scales, and to see how they respond to atmospheric and wave forcing as well as larger scale oceanographic conditions, such as currents and internal waves. Ultimately the results of this research should improve our understanding of mixing processes across the air sea interface, with implications for weather and climate forecasts."

One of Savelyev's colleagues on board, Marc Buckley, a French American postdoc, was also interested in wind-wave interactions. He had stationed cameras pointing obliquely at the sea surface below the boom. Nearby, a fog generation system sent water particles dancing in the air above the waves. To take freeze-frame photos of this dance, he installed a laser-beam generator that, through beam manipulation, created a sheet of green light that sliced down through the air, reflecting off fog particles, permitting him to capture ten-nanosecond cross-section snapshots of waves. Figures 2.14 a–b show two sequential images, 0.5 milliseconds apart.[45] Read as a movie, they record the motion of fog particles. Air-sea interaction at the fluid boundaries of waves is of concern to research communities tracking the circuiting of greenhouse gases. Close-up work such as Savelyev's and Buckley's might later be joined with data from airborne oceanography, the use of Light Detection and Ranging (lidar) from aircraft to measure ocean waves from above, flying at kilometer altitude, and measuring swaths a

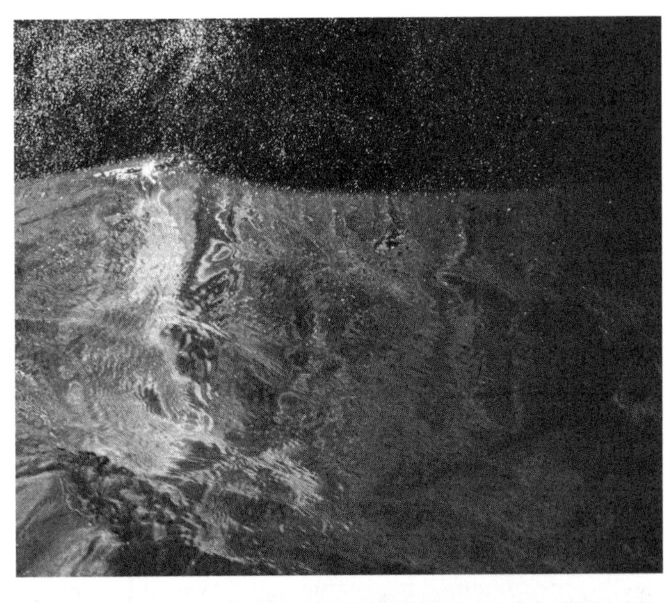

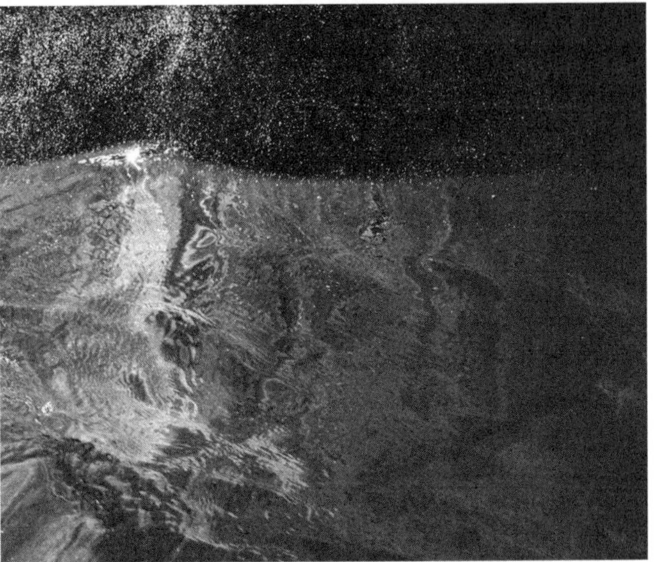

FIGURES 2.14A AND 2.14B Two sequential images of cross-sectioned waves, imaged with laser light by Marc Buckley.

kilometer wide of surface, down to centimeter wave heights and tens of centimeter wavelength. A multimedia sea.

Savelyev told me he was interested, too, in the complicating effects of internal waves, which he could also study from FLIP: "That's the thing about fieldwork, you realize that there are things going on you didn't know about—questions you didn't know you should be asking. You come home with kinds of data you didn't know existed." Just as with anthropological fieldwork, "being there" can reveal the incompleteness of textbook accounts, reorienting one's questions.

The next day, a plume of smoke wafted off the coast of California, a trace of the Orange County Canyon Fire 2 of October 2017, generating smoke that required some of FLIP's research to adjust. *History*, in the form of effects of climate change, was now intervening in the snapshot time of wave science. The adjustment FLIP made, though, was mostly to pause research, not to attune immediately to the changed air, which would require a differently instrumented field campaign.[46] Fire and smoke were not within this trip's "horizon of expectation," that frame of comprehension that the literary theorist Hans Robert Jauss (1970) names as structuring readers' apprehensions of what a narrative—or, I would say here for scientists, a research plan—promises and portends (cf. Petryna 2022).

What was all this FLIP research *for*? To make unseen sea territories readable. Oreskes (2020) argues that military and civil motives in mid-twentieth-century oceanography became symbiotic. Scientists accommodated to the alternately loose and tightly coupled-with-the-Navy funding structure, to the social form of scientific research that may lead to "dual-use technology"—that is, technology with both civilian and military application.[47] I asked people about that relation today. One told me:

> I don't find it relevant whether research is funded by NSF or ONR. What I mean by that is that ONR funds basic research that benefits science. Any basic physics research can have military applications. Any basic research, actually. In a way, I'm quite thankful that NSF and ONR are funding basic research. I'm more of a pacifist. If people are funding basic research in a nice way and giving scientists freedom to do things, it is a good thing. It's public research. It's not classified. Our results are published.[48]

Another said that although their work was funded by the ONR, "What I understand and what I research is pretty far from what can be used offensively. Most of it will be weather, and it will be public." The sociologist Chandra Mukerji argues in *A Fragile Power: Scientists and the State* (1989) that many military-funded civilian scientists starting in the 1960s began to have a sense of their work as too complex and unpredictable for application. The structure of feeling this created, Mukerji contended, was one in which scientists secured a sense of themselves as *objective*, not beholden to the instrumental aims of their funding. On her argument, this subjectification paradoxically generates "objectivity" and "disinterestedness" as resources on which the state can call in warranting science in the name of sovereignty. In a way, it permits scientists to flip frames of reference between pure and applied—think again of Wittgenstein's duck and rabbit, of FLIP's flip from levity to seriousness, from field to lab, of oscillations between mundanity and expeditionary heroism. Think of these as *paradigm scripts*: modes of thinking that take for granted the possibility of seeing things one way and then another. These flipped scripts generate an orientation to the present and future and to scientific time as operating at a different tempo from political time.[49] Scientific time, for wave science, unfolds in at least two registers here: first, in the *longue durée* time of accumulating scientific knowledge production; but second, in the ever repeating, looping, and repeating time of putatively timeless wave time (measured in milliseconds or minutes, in the time it takes for wave cycles to repeat or dynamics to unfold). It is that second kind of time—cyclical time—that can make waves so difficult to bring into a historical view.

Waves of Warning, Chemical and Climatic

Back on shore, Nick Pizzo, one of the oceanographers who had given me a preparatory orientation to work on FLIP, had been thinking about waves closer to the beach. In summer 2017, he published "Surfing Surface Gravity Waves" (Pizzo 2017), a paper in the *Journal of Fluid Mechanics* that got splashy press from Scripps, which glossed it as about the physics of surfing. Pizzo told me he was trying to figure out where particles on the face of a wave might move most quickly:

> The reason I said it was *surfing* is basically drawing an analogy. . . . When you accelerate with the wave, I call that surfing. And it makes sense. If you're in a motor boat and you're trying to catch a wave, you try to go at the same speed as the wave. For whatever reason it's never been mathematically shown that that was the case. Despite the breezy connotations associated with the word *surfing*, my motivations were purely oceanographic.

It is notable that the Scripps office explains wave phenomena to the public through surfing; the institution is in a different moment from that of World War II, at which time amphibious landing was part of what captured a national imagination. At the same time, Navy funding remains a conditioning context. Interestingly, Navy funding may shield scientists from the political winds of the moment, which more directly contour priorities from the NSF (see van Keuren 2000). With the rise of an anti-science Republican administration in the late 2010s, military funding—ironically or not—may have been more effective, scientists told me, in funding basic research. Pizzo told me, "You know, I have no direct interest in science with military applications. But I think pragmatically in the context of modern research funding: if you're also able to get basic science out of funding that was directed toward a project with more practical application, I think it's a good thing." Pizzo, like many others, is concerned about the looming effects of climate change. What, I asked him, is the relevance of wave processes for climate? He answered:

> Well, the atmosphere and the ocean interact. They exchange masses, gases. They exchange momentum; wind blows on the water and creates currents and transfers energy. The winds create swells and drive mixing. And all of those interactions are affected by what's happening at the surface, and *those* dynamics are dictated by the waves. For example, when waves break, they create bubbles and spray and you *greatly* increase the surface area between the air and sea—so, if you want to understand how carbon dioxide is exchanged between the atmosphere and the ocean, you need to understand what the waves are doing.
>
> Waves are a crucial driving factor for gas exchange. But these effects are really, really hard to constrain because the scales you must resolve to fully understand the problem go from millimeter parasitic capillary waves to swells, which are ten kilometers, to large-scale mesoscale processes, which are one hundred kilometers, to internal waves. I mean, how do you *model* something like that? How do you *measure* something like that?

It requires lots of data. Pizzo said, in words that echoed what I had heard from Munk, "You can't do any wave science axiomatically. . . . You *need* the fieldwork." Pizzo, no surprise, is a surfer, and as our conversation closed, he said, "There's a hurricane swell coming in. So I'm going to go surf at some point. There are a couple of sites just south of here, in Coronado, and it's going to be one of those really short-lived things where you drop everything that you're doing and try to get a few waves."

We flip back to thinking about ocean waves breaking on the shore, back where Scripps wave science began. Now, however, questions had to do not with war but with sea level rise, pollution, and climate. From waves of war to waves of warning.

And waves of warming. The Scripps researcher Falk Feddersen studies "how waves come in to the beach, break, and dissipate their energy." In addition to investigating coastal erosion, sea level rise, and the effects of urban infrastructure on the dispersal of effluents, he is interested in how breaking waves carry and generate heat, a process to keep an eye on as climate change increases the strength of storms and wave action. When I met him, I asked why he had gotten into waves. He answered, "Well, because I surf. I'm not gonna lie. But, really, I want to study processes that I can experience myself. I can *experience* a rip current. I can *experience* turbulence under a breaking wave. But standing on the deck of a ship, profiling something one thousand to two thousand meters below the surface—getting a bunch of digital measurements—you can never *experience* that. It's so abstract." Feddersen's experience had presented him with his key research question, about heat generated by breaking waves:

> One thing we always noticed as surfers is that if you just paddle around, you can sit up on your board just outside the surf zone, and you can feel that the water's colder on your toes than at your waist. You can paddle across temperature gradients. What *that* means is that there are all these vortices and eddies swirling up the water around you. What sets the temperatures in these regions? Obviously when the sun shines, it's gonna get warmer, and then at nighttime it cools. But if you just think of energy conservation, and you look at the amount of energy flux, sorta like the power being delivered by waves, it's actually really large. And some of it goes into currents, but when waves crash at the shore, they *heat up*. (See Sinnett and Feddersen 2014)

Knowing those heat profiles could have implications, Feddersen said, for planning coastal infrastructures, for thinking about vegetation as a buffer for the shore, for tracking the health of coral reefs (on reefs and waves, see Becker et al. 2014). Feddersen connected bodily experience to larger scales of social significance, the subjective to the objective. Bodies emplaced at and in sea, not bodies suspended above the sea, as on FLIP, become significant media for making knowledge about waves. And wave breaking becomes a process to be described not by physics alone, but also by chemistry, biology, ecology.

When FLIP was fashioned in the early 1960s, in the post-Sputnik intensity of the Cold War, US oceanographers imagined the seas as a frontier akin to outer space. Oceans were a new battlefield, but also a realm that might offer farmed bounties of fish and plankton, deep sea-sourced mineral wealth, offshore oil, and desalinized drinking water. Just after FLIP was built, Scripps scientists envisioned creating a research island in La Jolla that might contribute to all of these projects, an artificial island to be sited a mile offshore, operating as a utopia for interdisciplinary science, one that would even include habitats for long-term underwater living and working (Rozwadowski 2004).

In the end, Scripps Island was never built. (Sky-high budgets, pushback from the environmental movement, and skepticism about its utility scuttled the plans.) The ocean future that Cold War science had forecast turned out to carry the legacy of nuclear testing, damage from oil drilling, and algal blooms, leaving in its wake a different sea from that imaged by the island's framers. The future now looks like the compromised time about which scientists in Benford's *Timescape* novel warned their predecessors, more like the rising and polluted sea that Feddersen and colleagues now study.

In the days of this flipped future, Scripps scientists now look at waves as avatars of anthropogenic damage. On my last research visit, in 2019, I met up with the atmospheric chemistry professor Kimberly Prather, founder and director of the NSF's Center for Aerosol Impacts on Chemistry of the Environment at UCSD. She told me that, when waves crash, they release telling traces of the bodies of water through which they travel. In areas such as the coast of California, which sees significant exhaust from transportation emissions as well as contaminants from agricultural runoff, sea spray will contain signs of such pollution. Breaking waves therefore contain what Prather calls a "chemical signature" of human impacts. The wake of industrial and agricultural processes is legible in the wave break.

There is more than chemistry in the mix, too. Wave bubbles pop and transfer a range of organic materials—from lipids in small droplets to heterotrophic bacteria in the larger droplets—so that sea spray carries a complement of microbial bits and bodies. Prather offered a comparison: "Microbes are in your body. You're comprised of more microbe DNA than human DNA—and some of what they're trying to do is keep our bodies in a narrow temperature range. Microbes in the environment, to make an analogy, are similarly trying to keep the temperature of the Earth's climate in a healthy range." Exactly how well microbes control the temperature of our planet depends on which ones make their way into the atmosphere, to be cycled by humans, animals, and plants and to be incorporated as seeds for clouds. That process is modulated by transformations in ocean biochemistry, which include changes due to phytoplankton blooms.

If waves and their effects have been apprehended through and taken on qualities of the media used to understand and model them—from animated film to spectrally sorted radio transmission—here, wave knowledge is filtered through a medium that has only lately come into scientific legibility: the microbiome. Breaking waves convey fossil fuel histories and shifting microbial ecologies in their bubbling swirls (see plate 8), and now tell a story of a rapidly warming ocean, with cascade effects on marine food chains, coral and kelp ecologies, and more. This is what I earlier called the *oceanic churn*, the watery convolving together of multiple materials and time streams. Prather calls microbes "embedded sensors" that deliver accounts of the health of the sea, of possible ocean futures, futures to which her work, funded by the NSF, seeks to orient—and, importantly, orient students, preparing them to tackle complex environmental problems.[50]

Prather's research group, which often gathers under the curved, wave-like roof of Scripps's outsize hydraulics lab—a Douglas fir-posted and redwood-sided building that is host to a thirty-three-meter-long wave tank, a wind-wave tunnel, and a dedicated maker space—answers in a new way the interdisciplinary aspirations that once shaped dreams of Scripps Island. Prather's center consists of about one hundred people—undergraduates, graduates, postdocs, professors—funded by a multimillion-dollar NSF grant. Its demographics represent a flip from earlier Scripps days; some nine out of ten of the undergraduate scientists are women, with many from underrepresented groups—a fact vital to Prather, who, breaking a career-long reluctance to speak up about gender, had recently lectured at Stanford University's Women in Science and Engineering speaker series to urge the necessity of building an inclusive, interdisciplinary approach to ocean science.

Just before my meeting with Prather, I presented a version of this chapter at Scripps's physical oceanography colloquium series. I learned that the future of FLIP was unclear, its captain having retired and the ship sent to dry dock; as it turned out, I had joined its final voyage. Around the same time, I came across a Japanese comic book, *7 Seeds*, set in a postapocalyptic future in which a group of disaster survivors discover a derelict ship modeled on FLIP (figure 2.15). When they board the abandoned vessel, the *Fuji*, it flips to vertical and inaugurates a countdown, and the explorers realize it is setting up to launch nuclear missiles. They find the control room corroded by iron-eating bacteria and speculate that an epidemic did

FIGURE 2.15 Control room of *Fuji*, a flipped ship in Yumi Tamura's *7 Seeds* (セブン・シーズ), vol. 18 (Flower Comics, 2010).

in the human occupants of the ship. This spectral comic-book FLIP scripts the craft as a relic of the Cold War—though one that radiates hazard into a future it does not see—and stages the platform as a metaphor for a human world undone by environmental catastrophe.[51]

FLIP was built in 1962, the year Thomas Kuhn's *Structure of Scientific Revolutions* was published. FLIP was meant to be a revolution of sorts—though its 90 degree rotation, preserving *x*, *y*, *z* space, stopped short of a full torquing of received ideas. Still, it offered unexpected angles. In spite of its popular nickname, FLIP was not a ship. It was more like a *-ship*, that conceptual operator (used as a word-forming element in such terms as friendship, kinship) that points to a "relation between." Flipping the *-ship* reveals those relations to be orientations to history and the future, and many scientists now orient to foretelling and, if possible, forestalling coming disaster. They are caught in back-and-forth ocean time, a tidalectic, that has the past surging into the present to upend expectations of steadiness and balance.

Set Two

First Wave

Being the Wave

During the American presidential election of 2008, close watchers of the news learned that Barack Obama was a bodysurfer, riding waves without aid of a board, propelled only by arms and swim-finned feet. The aim of bodysurfing, writes the enthusiast Robert Gardner (1972, 42), "is to cut diagonally across the face of the wave, trying to slide along just under the breaking curl." The sport lacks the glamour of board surfing, or anything like its promotional industry. It is a solitary enterprise, requiring a feel for how to launch one's body across the arc of a breaking wave. Obama arrived at the activity biographically, growing up in Hawai'i. A first-generation African American, Obama was not, like his University of Hawai'i anthropology graduate student mother, Ann Dunham, a *haole*—a Kānaka Maoli (Native Hawaiian) term that refers to white people from the mainland. He stood apart, too, from Chinese, Japanese, and Filipino immigrant communities, as well as from Native Hawaiians (Obama 1995, 25). Obama cut a unique figure in the water, especially since so much mainland American history—think segregated beaches—has worked against African American recreational swimming.

This figure of Obama in motion (plate 9) impels this meditation on bodysurfing, which asks how a politics might be read from the sport. Drawing on an account of a bodysurfing competition in California I entered, I contend that embodied aquatic techniques shape not only bodysurfing bodies and selves, but also what bodysurfers take a wave *to be*.

What does it take to "be the wave"? It takes work, getting a feel for the curve of seawater and navigating social worlds surrounding beach and breakers. Cultural studies of surfing have mapped such navigations, tracking how the aesthetics of the sublime—of individual union with scary phenomena—have contoured risk-taking leisure (see Booth 1999; Farmer 1992; Finney and Houston 1996; Ford and Brown 2006; Gregory 2013; Henderson 2001; Lanagan 2003; Pearson 1979; Stranger 1999; Waitt 2008). Krista Comer, in *Surfer Girls in the New World Order* (2010), reports on how women have struggled against their sidelining in surf culture, as successive genres of surfer masculinity—strapping, slacking—have claimed social space. Racial politics have shaped the sport, too. The legendary Native Hawaiian surfer Eddie Aikau traveled to apartheid-era South Africa with *haoles* (who, like him, had been inspired by the 1966 movie *The Endless Summer*), only to be turned away from a segregated hotel—an affront his white friends did nothing to address, which then newly racialized existing ethnic divides when the group returned to Hawai'i.[1] The geographer Arun Saldanha documents how shoreside tourism in Goa, India, ripples with race. His *Psychedelic White* proposes that racial identities materialize not through the enforcement of rigid boundaries but, rather, through "*viscosity* . . . how an aggregate of bodies holds together, how relatively fast or slow they are, and how they collectively shape the aggregate. . . . Flows of people are at once open-ended and gradually thickened by recurring, allegedly conscious decision making. . . . What bodies do—swim, sunbathe, shout, hold hands, drink . . . —determines who and where they are on the beach" (Saldanha 2007, 50–51, 121, emphasis added).

What kinds of viscosities shape bodysurfing? To seek to "be the wave" is to be in the pull of the transforming affects and demographics of beach recreation, in the drift of shifting shorelines and practiced apprehensions of seawater as force and substance. Such apprehensions are *phenomenotechniques* (Bachelard [1934] 1984), processes that fashion experiences of the world in calibration with techniques of the body. How does a politics—possibilities for movement and stasis, inclusion and exclusion—inform bodysurfers' encounters with waves?

These questions take me to Oceanside, California, in August 2012 for the thirty-sixth Annual World Bodysurfing Championship. I have entered as a competitor, old enough, as of last year, to join the Men's 45–54 group. I am here, as Comer (2010, 128) says in *Surfer Girls*, to "return to the scene of [my] coming-of-age and take stock of all that has transpired since." I went to high school nearby. A friend from those days, Jeff, with whom I spent a sunburned summer in the 1980s, is in the contest—and has been every year since way back. He asks whether I have practiced. Not much, I admit, having just flown in from Massachusetts. He asks whether I like going left or right on waves. I don't remember. Jeff points to the water and says, "If you go right, you'll have to watch out not to run into the pier."

The event, which has drawn three hundred people, features bodysurfers from the mainland United States, Hawai'i, France, Brazil, Australia—though that hardly covers the "world" promised by the competition's name. A survey sees mostly white bodies (like mine) and more men than women. Beyond the boundaries of the contest are mostly Latino families from Oceanside. The proximity of Camp Pendleton, a US Marine Corps base, means many beachgoers are working-class military. The racial and class viscosity is not difficult to assess (and see Prodanovich 2020 on histories of Indigenous surfing in San Diego). As Simon Leung observes, speaking about a Huntington Beach art installation he created entitled *Surf Vietnam* (about a surfing-under-gunfire scene in *Apocalypse Now*), "When you look at surfing historically, you get a story of colonialism, the real-estate development of southern California, the rise of the military industries (fiberglass and foam, what surfboards are made of, came from military technology), not to mention the beautiful, golden California child" (Sturken and Leung 2005, 141).

There is also the inside story, how bodysurfers think of waves. One of the few books on the subject, *The Art of Body Surfing*, suggests that a desire for "being the wave" has been in circulation for a while: "This is a feeling that can't be explained, but you know when it's happening. You're moving; you're a part of the wave" (Gardner 1972, 32). That same book also weaves in a narrative from an earlier historical sensibility, one that has the sea as an adversary: "To the surfing purist, body surfing will always be the supreme test of man's age-old struggle to conquer his most ruthless, dangerous, and implacable enemy—the sea. This is because the body surfer challenges the sea at its most violent moment—the thunderous breaker—and he does it without artificial help or assistance" (2). That story—and its gendered cast—has origins in Euro-American beach going. As Alain Corbin records in *The Lure of the Sea: The Discovery of the Seaside in the Western World 1750–1840* ([1988] 1995, 73), early British bathing, aimed at pressing bourgeois people into healthful encounters with the sea, assumed waves were oceanic others: "Bathers and physicians both agreed that the sea should offer three major qualities: it should be cold, or at least cool, salty and turbulent. Pleasure came from the whipping of the waves. . . . Bathing among the waves was part of the aesthetics of the sublime: it involved facing the violent water, but without risk, enjoying the pretense that one could be swept under, and being struck by the full force of the waves, but without losing one's footing." Bourgeois men and women—who approached waves with the aid of working-class helpers called "bathers"—were trained to experience waves in ways proper to their sex. Men thought it manly to face the waves "alone"; with bathers standing at some distance, men would pretend "to be crushed" by waves (76). For women, "The emotion of sea bathing arose from sudden immersion. . . . The 'bathers' would plunge female patients into the water just as the wave broke, taking care to hold their heads down so as to increase the impression of suffocation. . . . [A] Dr Bertrand indicated the manner in which 'the wave must be received' on various parts of the body" (73–74).

Nowadays, after more than a century of swimming lessons for middle-class people, a "one with the sea" story for bodysurfers dominates. It tracks changes in masculinity—real surfer men no longer have to be steely and stoic, having adopted laid-back affects long ago. It also bespeaks a turn-of-the-millennium environmentalist sensibility (Evers 2009). *Deep in the Wave: A Surfing Guide to the Soul* rings the changes: "Bodysurfing is a great way for a beginning surfer to get to know the feeling and flow of the surf because you are deep in the wave, not on it. . . . Every real surfer wants to be one with the wave, not just to ride the wave" (Woznik, with Aronica 2012, 30). The champion bodysurfer Mark Cunningham says, "You know, it's just very pure and simple and it was as tight with nature as I could possibly get."[2] An online entry in the *Surfer's Journal* genders the sea as a mother/heterosexual lover: "The ocean forever trades us bumps and scrapes for only about 30 seconds of pure unfiltered joy. . . . But isn't that part of the reason we bodysurf? . . . To be told by Mother Ocean that we are small, that we mean nothing, that I, Mother Ocean, may be dangerous at times, but fuck, am I ever going to take you dancing!" (Brown 2011).

I ask Jeff what has changed in bodysurfing since he took it up. There are fewer of us, he says. He blames boogie boarding, a genre of surfing that sees people riding belly-down on hydrodynamic foam boards. It has become too easy.[3] There are growing numbers from other water sports, too, so the style is changing; today Jeff sees athletes—water polo people, junior lifeguards—rather than "kids like us, who just fell into it." Some techniques have changed; the both-arms-back, shoulders-hunched move, which Jeff and I favored, has vanished. Back in 1972, it had been enshrined in Gardner's how-to book: "Maneuver to that critical point where the wave is breaking, but try to be right on the edge of the break, then take off exactly the same as before, with a stroke, a kick, hunching the shoulders to get the center of gravity forward so that you are starting *down*" (Gardner 1972, 43).

Another participant, a sunburned white man in his forties, has a different view. "Bodysurfing is having a renaissance" precisely *because* people are mixing it with other sports, discovering it as "a dimension of surfing"—an account I also hear from a twenty-something, self-identified surfer girl straight out of Comer's book, who tells me how she thinks about the difference between surfing and bodysurfing. As a surfer, one is "conquering the wave." As a body surfer, she says, "you have the wave flow over your body. It's a whole different frame of mind."

People are eager to talk about the comparison. One of my age-mates, Tom, says he likes bodysurfing because "the energy catches you. I try to get in a dolphin mentality. You're out there at the source. You're not going against it; you're going with it. You're part of it." His friend Dave goes further: "Waves kind of talk to you sometimes. . . . Bodysurfing is like a painting and you're part of the painting. You're out there making your own designs." As Tom and Dave ready for their turn in the waves, they say they hope people aren't "all aggro" (meaning "all aggravated," too intense) out there. There's a relaxed slowness to this approach that is about age, but also about *when* one came of age. What was once a casual activity has become competitive. There is the viscosity of race and class, too—with the new, largely white, privileged water polo-playing types injecting focused speed. Dave (Latino) turns to Tom (African American) and says, with a studied wryness, "Let's win one for the minorities."

Techniques transform. Marcel Mauss ([1934] 1973, 70–71) wrote about morphing modes of bodily comportment: "My generation did not swim as the present generation does . . . we have seen the breast-stroke with the head out of the water replaced by the different sorts of crawl. Moreover, the habit of swallowing water and spitting it out again has gone. In my day swimmers thought of themselves as a kind of steam-boat. It was stupid, but in fact I still do this: I cannot get rid of my technique."

There also exist geographical differences. I fall into conversation with two men from O'ahu, and they tell me this is their first time competing in California. They, too, note the presence of athletically minded people.

One remarks that California attracts people (water polo players get another mention) who "train in controlled water. Their bodies move differently." I am reminded of "Local Motions: Surfing and the Politics of Wave Sliding," in which Eric Ishiwata suggests that *he'e nalu*—Hawaiian for "wave sliding" (board surfing)—takes waves to be, above all, cultural artifacts, even cultural patrimony. "The surfbreaks of Hawai'i," writes Ishiwata (2002, 264), are places where "locals carve out microresistances" to tourist encroachments, countering the commoditization of surf leisure by moving into, occupying, and (re)territorializing Hawaiian waters with distinct habits of individual and aggregate movement (see also Walker 2011). Hawaiian stories of bodysurfing make the large waves of the archipelago a plot point and often script the size of waves into narratives about boys testing their manhood against the sea. Residents of Hawai'i also told me their bodysurfing often emerged from working as lifeguards in the big surf. Lifeguarding in Hawai'i is usually about saving tourists who underestimate the ocean. Bodysurfing lifeguards stand for life not only because of their vocation but also because, in contrast to tourists encountering the sea as a force of death, lifeguards' capacity to "be the wave" positions them as intimates with the sea.

Being "one with the wave" is buoyed by being able to catch it, and nowadays swim fins are crucial to that intimacy. But it was not until after World War II that swim fins became available. The US Naval Special Warfare Command's Underwater Demolition Team was the first to standardize fins, using a design created by the American yachting Olympian Owen Churchill, who claimed he took inspiration from Tahitian swimmers wearing woven leaves on their feet but who also licensed plans patented in 1933 by French naval officer Louis de Corlieu. It was only in the early 1950s that fins were commercialized and became popular. Before, "in the pre-swim fin days—roughly prior to World War II—the art of body surfing was severely limited by the inability of the surfer, no matter how skilled, to generate enough quick speed to control the wave. Thus, in the 1920s and the 1930s, it was 'straight off' or 'over the falls,' with a reckless disregard of consequences" (Gardner 1972, 6). Fins are a material element in bodysurfing phenomenotechnique. In 1934, Gaston Bachelard (1984, 13) wrote, "Instruments are nothing but theories materialized. The phenomena they produce bear the stamp of theory throughout" (see also Idhe 1991). The fusion of fins with surfers' bodies concretizes a folk theory of waves that has waves as entities with biographies of their own, biographies with which bodysurfers can fuse.

It is time for me to get in the water. I stand at the start of my heat with five other middle-aged men at the foot of the pier. Most of us have Voit

Duck Feet swim fins—a generational marker (and a pointer to Voit's 1953 retooling for recreational markets of a product created for the Navy). The starting horn blasts. I think: always dive under waves, avoid the surge, and get out to sea. I stroke ahead. It is going well. Then it is not. "The last phase of wave riding is disorder. The broken wave—the adjective allegorizes itself—froths confusion" (Mentz 2019b, 441). Pushed to the sand below, I recall that I used to enjoy underwater takeoffs, eyes open. I lurch up. I head back out to get *something*. I remember Jeff's advice: go left! I get a wave. If I were thinking in words, I might teleport back to my teenage vocabulary, declaring myself *stoked*. By now, though, a rip current has dragged me northward, out of contest bounds. But, hey, isn't bodysurfing all about being alone? The worst thing one can do is cut someone else off. Bodysurfing is about navigating watery social space so one can have that communion-with-the-wave moment. If that is what it is, maybe I am winning. I leave the water. I was not really the wave.

Only once back on the beach do I think I am lucky to have made it. I come upon emergency medical technicians, crouched over a swimmer, to whom they are giving oxygen. Jeff tells me this happens: "We had a guy die out here a few years ago."

What is it like to be a wave? It is to participate in an experience of the *sublime*—"That quality in nature or art which inspires awe, reverence, or other high emotion," notes the *Oxford English Dictionary*, and "encompasses an element of terror." For big-wave surfers, that sublime is massive and terrifying. For bodysurfers, it is an intimate sublime. But it is an experience not only personal, but also political. So if, as the surfing theorist Douglas Booth (1999, 44) writes, "Deep in each surfer's subconscious is an interpretation of an ocean wave," that intuition is also shaped by the thickness of history, by the politics of who learns to surf how, when, where, and with whom.

Set Two

Second Wave

Radio Ocean

Ocean waves travel at a range of frequencies and, like radio waves, can be sorted along a spectrum from short to long, whitecaps to swells. While *radio-*, the combining form of the Latin *radius* (beam), gave its name in 1913 to the wireless travel of electromagnetic waves carrying sound, its sense as "ray," "ray-like," or "by means of radiant energy" might as easily, in an alternative history, have come to encompass ocean waves, which, like radio waves, radiate energy. Stretching the frequency band beyond the 3 kilohertz to 300 gigahertz of the conventional radio spectrum to include everything from 10 hertz sea-surface ripples to those 10^{-5} hertz waves of the daily ocean tides, we might have come to think of the ocean as a radio, broadcasting waves.

Maybe we already can. Ocean scientists tell us that waves propagate across the ocean at a range of speeds, relaying information about storms close and distant. When waves break, the energy they carry crash shifts from aquatic into sonic—plunging, surging, spilling—like electromagnetic waves arriving at a radio receiver, demodulating to become audio. Whether we can decode the messages sent by Radio Ocean may be another story, though, at least in 2017, when I first wrote these thoughts—when Hurricanes

Harvey, Irma, Jose, and Maria sent punishing waves crashing into fragile Atlantic livelihoods—the communiqués of the waves were not difficult to decipher.[1] If the ocean is a radio, the program now transmitted to many parts of the world seems to be one that tells of onrushing climate change.

Radio Ocean has many precursors in poetry, literature, music, sound art, and science that seek to ventriloquize, mimic, measure, or record waves speaking, singing, sounding, noising. The waves of this sea of language, song, and number oscillate, telling at one moment of eternal returns (waves arriving, over and over), at another of irreversible change (waves washing away the past). They are tidalectic. Tidalectics, as the poet Kamau Brathwaite (1999) theorizes it, places dialectics (thesis, antithesis, synthesis) into flux, calls and responses ever switching places. Radio Ocean's tidalectics transmit messages from pasts and presents colonial (waves that take shape in ports carved by imperial travel, waves that variously wash over or reveal the sunken agony of the Middle Passage), military and industrial (waves that carry legacies of Cold War nuclear radiation), and transnational (waves that carry tourists, exiles, and refugees across surf alternately enticing, promising, and deadly). Radio Ocean carries dispatches, too, from futures arriving with the Anthropoceanic (Brugidou and Fabien 2018)—waves that move with the winds of human-made climate change, waves that come to life in a newly melted Arctic Ocean. How these messages are received depends on where listeners are—at the shore, on a boat, under the waves, at the bottom of the sea; at which frequencies and on which channels they listen; and on whether they are hearing through ears, cochlear implants, hydrophones, pressure sensors, the internet, and so on.

Before ocean waves were imagined as possible, like light or radio or sound waves, analytically to array into a spectrum—and before it was possible to listen to breaking waves for acoustic signs of their dissipative energy (Kluzek and Lisimenka 2013)—the meaning of ocean-wave sound often found contemplation in poetry, with waves pressed to speak of loss and longing, distance and death. Listening to the provincial sounds of English-language offerings, one would have heard of "many-voicèd waves" (Percy Bysshe Shelley), the "sullen moaning wave" (Lewis Carroll), "barking waves" (John Milton), "the murmurous noise of waves" (John Keats), "thundering waves" (Charlotte Brontë), and "the ghostly sound of waves rustling like grass in a low wind" (Derek Walcott). Waves speaking, sighing, sounding. The "bleat, the bark, bellow, and roar" of the "waves that beat on heaven's shore" (William Blake) would (and may still) be sounds rolling from the human to the animal to the plant to the mineral to the elemental to the perhaps indecipherable. (Check out the "Ker plotch . . . pish rip plosh . . . Sho, Shoosh, flut, / ravad, tapavada pow, / coof, loof, roof" of Jack Kerouac's "Sea: Sounds of the Pacific Ocean at Big Sur.") Shelly Trower (2019, 20) tracks how in Bram Stoker's short story *The Jewel of Seven Stars* (1903), "Sea sounds cross the borders between the audible and the palpable," as wave sound in that tale preverberates calamity. Beyond poetry, one would have heard (and still can hear) the everyday onomatopoeic (again in the parochial but polyglot English) that have waves whispering, hissing, roaring, sizzling, cracking, and drumming.

What has wave noise meant? The literary theorist John Melillo (2016) has argued that the sound of waves crashing and dissipating has been enrolled into philosophical ruminations on the meaning of noise. Gottfried Wilhelm Leibniz's *New Essays on Human Understanding* ([1765] 1982, 55) contains a passage on ocean-wave noise: "To hear this noise as we do, we must hear the parts which make up this whole, that is the noise of each wave, although each of these little noises makes itself known only when combined confusedly with all the others, and would not be noticed if the wave which made it were by itself." In other words, the sonic phenomenology of the breaking wave is relational, made by a hearing subject who is a live player in making meaningful the noises of waves as "wave sound"—progressive, repetitive, dissipative, tidalectic—at all.

What have wave listeners been listening for?

Maybe music. Start with old-school motifs.

Many composers have mobilized musical *icons*: signs they imagine to have some *likeness* to the way ocean waves sound or appear (think arrangements of notes, chords, rhythms or of the timbre of instruments). Albert Schweitzer (1911, 76) identified a "wave motive" in Bach's cantata *Siehe, ich will viel Fischer aussenden* (No. 88 [1726]):

At other moments, composers work through *indices*: signs they take to have a causal or existential relation to the waves they represent (think field recordings of waves or captures of frequency spectra of a crashing wave). At still other moments, composers have approached the matter through *symbols*: signs that stand for their object by convention (think military march music to represent dynamics of disciplined, steady succession). Often enough, they have done all at once, rendering the distinction between those key semiotic terms—icon, index, and symbol—cluttered, awash in overlaps.

Take the music of sublime reverie: Consider romantic composers' attempts texturally and formally to represent the timbre and relentless arrival of ocean waves. Listen to Felix Mendelssohn's Hebrides Overture, op. 26, of 1830–32; Camille Saint-Saëns's *Le déluge*, op. 45, of 1875; the second movement of Claude Debussy's *La mer*, "Jeux de vagues," of 1905; or the third movement of Maurice Ravel's *Miroirs*, "Une barque sur l'océan," also of 1905. In the evocation of waves in such music, writes the musicologist David B. Knight (2006, 58), "an essential character . . . is of increasing and then decreasing swell. This is conveyed by both a sense of rhythm and increasing and decreasing volume."[2] These are waves to be contemplated in an attitude of rapt intellectual and emotional attention—sometimes imagining oneself above the water, sometimes at the surface, sometimes below.

The music of mathematical and abstract appreciation: take mid- to late twentieth-century electronic music's emphasis on the periodicity—though sometimes also drift—of ocean waves, as in the work of R. Murray Schafer, who, in his String Quartet no. 2, *Waves*, of 1978 offers

> dynamic, undulating wave patterns, the rhythm and structure of which are based on his analysis of wave patterns off both the Pacific and Atlantic coasts of Canada. Incorporated into his music is his discovery that the duration from crest to crest is between six and eleven seconds. To ensure that performers get his meaning, Schafer has written into the musicians' scores the varying wavelengths. As a result, when played, a fascinating musical seascape emerges that reflects an appreciation for the irregular regularity of the sea. (Knight 2006, 58)[3]

Or take "Wave Code #E-1," of 1975, in which the Fluxus composer Takehisa Kosugi repeats the word *wave* into an echo chamber, offering an embodied cyborg enactment of voice, machine, and electricity. By sending the word *wave* through machines that format the sound within the physics of reverb and delay, Kosugi's piece underscores the way the notion of the wave partakes at once of the physically enunciated and the analytically captured.[4] And then there is Tristan Murail's *Le partage des eaux*, of 1995, which tasks a traditional orchestra with playing a score based on "the spectral analysis of a breaking wave," with the result that "strangely coloured and strangely coherent harmonic-timbres" collapse over one another, suggesting ocean waves made of some kind of glittering liquid metal created by melting down strings, woodwinds, and brass.[5]

The electric music of the Black Atlantic: consider the bassy submarine soundscapes of Jamaican dub, created in such sites as Lee "Scratch" Perry's Kingston studio, known as the Black Ark. (Compare with Miami bass, with low-end vibrations that Dave Tompkins [2017] has theorized through the Rossby whistle, the ultra-low sound of the basin of the Caribbean Sea, resonating.) David Toop (1995, 116) suggests that "sonar transmit pulses, reverberations and echoes of underwater echo ranging and bioacoustics" constitute the "nearest approximation to dub" (cf. Henriques 2003). Or listen to the Detroit techno outfit Drexciya, whose sonic works conjure an imagined and unsettling underwater world created and sounded by the mutated descendants of African captives thrown overboard during the Middle Passage, whose alien music haunts a Black Atlantis. Tracks such as "Wavejumper"—"You must face the power of the black wave of Lardossa, before you can become a Drexciyan Wavejumper"—become uncanny messages from waves as both acknowledged and unacknowledged graves.

The music of environmental audit: sound-art recordings of the noises of waves, from the riverine sounds of Annea Lockwood's field recordings (e.g., "A Sound Map of the Hudson River," from 1989, though also, more recently and linguistically, her "Water and Memory," from 2017, which enjoins singers to voice "words for water, for ocean, for waves, from Hindi, Thai, Hungarian, and Hebrew" [Lentjes 2017]) to the hydrophonic recordings of Jana Winderen, including her piece "bára" (one Norse word for "wave"), of 2017, activated by tides at Trieste, Italy.[6] In the trajectory from sublime music to documentary recordings, ocean waves get taken out of the water and, with the rise of electronic composition, treated as abstractions, only to then, with the rise of sound art, be submerged and audited from within the materiality of water. Winderen's work in particular has presented underwater sound as a form for thinking through climate change, as well as the transforming textures of the realms within which aquatic creatures live.

Meanwhile ocean-wave scientists have been listening, too. The 1960s saw key papers asking how noise in the sea can be created by waves (Vigoureux and Hersey [1962] 2005, 489). In 1994, Eric Lamarre and Ken Melville (see chapter 2) audited the sound of waves breaking to determine how they fall apart. Ocean-wave breaking can also be queried for climatological information. Kim Prather (see chapter 2) observes that crashing wave bubbles carry organic materials and gases (Stokes et al. 2013). Which materials rise into the atmosphere can be assayed by tracking the size distribution of bubbles, which can be discerned by listening to them come together, as well as pop. As Timothy G. Leighton (2014) argues, it is possible to engage in "monitoring the transfer of greenhouses gases between atmosphere and ocean using the sounds of breaking ocean waves." It may also be possible to infer changes in water temperature from patterns of whitecap formation in the open sea (Babanin 2011), as well as from patterns of waves beneath the sea surface, which, emerging at the interfaces of density, salinity, and temperature, can be discerned through hydrophonic listening (Barclay et al. 2009). There is also a story to be heard in the lives of underwater creatures, from crustaceans and fish to, of course, cetaceans, hearing a different Radio Ocean than they did decades ago, one filled with industrial and military noise.[7]

To end, let me switch stations, tuning to a composition that brings together poetry, music, and sound about waves and what they might prefigure. The first track of *Paradiso*, a 2017 album by Chino Amobi, opens with the austere and electric voice of Elysia Crampton, the Latinx/Aymara experimental composer, reciting and revising Edgar Allan Poe's "The City in the Sea" over the rumbling roll of crashing storm waves. The waves summon a sinister world, one doubled in Crampton's recitation, which tells of an abandoned city by the sea. We hear that the desolate city sits at the waterline:

> There open temples—open graves
> Are on a level with the waves—

We also learn, though, that the surrounding sea is unsettlingly still:

> For no ripples curl, alas!
> Along that wilderness of glass—
> No swellings hint that winds may be
> Upon a far-off happier sea

But then a wind rises:

> But lo! a stir is in the air!
> The wave! there is a ripple there! . . .
> The waves have now a redder glow—
> The very hours are breathing low—
> And when, amid no earthly moans,
> Down, down that town shall settle hence,
> Hell rising from a thousand thrones
> Shall do it reverence.

The city sinks into a hellish sea beneath the waves. The waves of this Radio Ocean drown it out.

Set Two

Third Wave

Gravitational Waves, Sounded

One of the most arresting wave phenomena of 2016 arrived in the form of a sound, the audio translation of a cosmic vibration from a billion years ago. In a far corner of the Southern Celestial Hemisphere, beyond the dwarf galaxies that make up the Magellanic Clouds, two black holes spiraled into each other (Abbott et al. 2016). Colliding, they generated an undulating train of gravitational waves, oscillations in space-time that arrived at our planet on September 14, 2015. When astronomers at the Laser Interferometer Gravitational-Wave Observatory (LIGO) captured these outer-space vibrations, one of the first things they did was translate them into sound—a sound they made public in February 2016, in time for the centennial of Albert Einstein's theorization of gravitational waves. They called the sound they recorded a "chirp," a sine wave speedily swooping up in frequency, indicating the accelerating coalescence of the two black holes (Adams and Chapman 2016).

But what kind of sound *was* this chirp? How did such a cataclysmic event as the in-spiraling of black holes—which one might imagine as properly represented by the death metal sublime—cash out as something halfway between the sound of a Theremin and the voiceprint of a bird? The answer

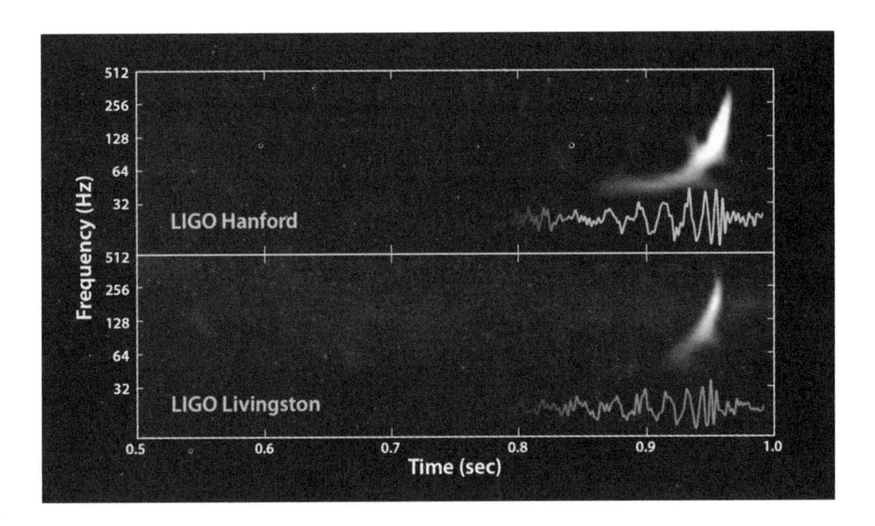

FIGURE 2C.1 A visualization of the chirp pattern of gravitational waves detected by the Laser Interferometer Gravitational-Wave Observatory (LIGO) at its two sites—at Hanford, Washington, and Livingston, Louisiana—on September 14, 2015. Created by LIGO.

is that the chirp was a scientific representation of oscillating phenomena (figure 2c.1), not a recording of sound from space. Although gravitational waves have frequencies that map onto the human auditory range (though at vanishingly low amplitudes, one-one thousandth the diameter of a proton), these waves are not acoustic pressure waves but, rather, wiggles in space-time (Bartusiak 2000).

Still, the sonic analogy was powerful. When Caltech's LIGO lab executive director David Reitze announced the detection on February 11, 2016, at the National Press Club in Washington, DC, he offered an explication that was at once clarifying and poetic: "What LIGO does is it actually takes these vibrations in spacetime, these ripples in spacetime, and it records them on a photo-detector, and you can actually hear them. So, what LIGO has done, it's the first time the universe has spoken to us through gravitational waves. And this is remarkable. Up till now, we've been deaf to gravitational waves, but today we are able to hear them" (NSF 2016). Hearing gravitational waves is the result of a chain of machinic mediations, with LIGO a kind of cosmic cochlear implant, a frequency analyzer that foretells and filters meaningful vibration into a cyborgian instrument of reception. The detector is outfitted with template waveforms, waveform imprints simulated in advance of the arrival of signals. They offer mathematically tuned matches to possible waves that might be generated by cosmic calamities. Such templates—kinds of *theory, animated* (see chapter 3)—anticipate significant vibrations, oscillations that make sense with respect to Einstein's theories of general relativity, which provide an account of how gravitational waves might work. To extend Kodwo Eshun's (1998) concept of sonic fiction, these templates operate as sonic *science* fictions, mathematically precise hypotheses about what it might be possible to "hear" in the cosmos. The chirp of the black hole gravitational wave was sieved out from the background hum of the universe

How was this background itself audited? I found out by talking with physicist colleagues at the Massachusetts Institute of Technology (MIT). These scientists, part of the MIT-Caltech collaboration that is LIGO, have spent a lot of time at the detector, listening.

The LIGO detector, a massive device distributed across two physical sites—in Hanford, Washington, and Livingston, Louisiana—is constantly vibrating, owing to quantum, seismic, and thermal vibration. To separate out a signal in this flux, scientists draw up what they call a "noise budget," a catalog of all of the noises that need to be listened *through*—quantum noise, seismic noise, thermal noise. Once these noises are stabilized, scientists can operate with a background against which to detect a signal. Scientists are listening not against silence but, rather, against a hum that tells them that the detector is "on." As Ragnhild Brøvig-Hanssen and Anne Danielsen (2016) argue in *Digital Signatures*, their book on digital music, each recording medium—wax, tape, vinyl, digital—has its distinctive "silence," which is often, also, a kind of "noise." It is no different with LIGO. But the rumble and fizz that accompanies any detection also suffuses that detection with a contagious reality effect. The chirp—a clean, mathematically streamlined sound, like an electronic whoop—is juxtaposed, emerges from, and therefore acquires the reality of the companion roar and hiss that LIGO is always capturing and creating. Think of listening to a 1950s sci-fi cybernetic soundtrack mixed in with a twenty-first-century soundscape recording of gently droning office noise.

Why was the signal called a *chirp*? The term originates in radar, describing a pulse-compressed signal that shows a sweeping increase (or decrease) in frequency, a sweep engineers in the 1950s likened to the chirp of a bird, bat, or insect. The term was coined in 1951 in a Bell Lab memo titled, "Not with a Bang, but a Chirp," a reference to the final line of T. S. Eliot's poem "The Hollow Men," from 1925: "Not with a bang, but a whimper" (Klauder et al. 1960). By the time LIGO scientists use the term, it has a formal meaning and measure.

But is also operates as a relay between human and nonhuman sound. On the day the chirp was announced, YouTube hosted videos of people making the "chirp" sound—underscoring this sound as a human-inhuman articulation. (Scientists' dogs were also enlisted as chirpers.) At the same time, the *inarticulate*, animal character of the chirp offered by scientists (what the sound studies scholar Alexandra Supper [2015] would call "data karaoke") embodied a sense among these scientists that they were imitating a phenomenon beyond the human: a cosmic wave.

The chirping sound of the cosmos today, then, is at least partially an effect of audio technologies of noise measure and reduction, as well as an articulation of human purposes with cosmic phenomena. It is a sound object for our time. It is recognizable as a wave because of the questions that have been asked about it and because of the ways these queries have been built into technologies of audio observation.

Around 1980, the astrophysicist Rudolf Kippenhahn, director of the Max Planck Institute for Astrophysics in Munich, recollected a talk he gave in 1960. In *Making Noise*, Hillel Schwartz (2011, 827) quotes Kippenhahn as remembering that he

> asked the audience to imagine an instrument capable of transforming all the incoming radiation from space into audible sound. We would hear the constant rushing of the starlight and the radio eruptions of the sun as well as the rushing of the radio waves. . . . [Now, twenty years later], we would hear the heterodyne ticking of the pulsars—the low humming of the Cancer pulsar, for instance. . . . There is not only rushing to be heard in space, there is ticking and drumming, humming, and cracking.

And chirping—a wave sound formatted by structures of anticipation built by the formalisms of physics. We have heard the universe, and it turns out to enfold the sonics and theoretics—the phenomenotechniques—of twentieth- and twenty-first-century technologies of recording, tuning in, and listening out.

Chapter Three

Waves to Order and Disorder

Making and Breaking Scale Models inside and outside the Lab, from Oregon to Japan

BEGIN WITH WAVES IN A BOX. Figure 3.1 pictures an engineer surfing in a wave flume, a long, rectangular container made of concrete that serves as a narrow swimming pool-like channel down which researchers can mechanically generate waves of specified height and period. The photo, with its juxtaposition of surf culture and hydraulic engineering, documents a circa 1974 publicity stunt undertaken to advertise the then-under-construction O. H. Hinsdale Wave Research Laboratory, a site at Oregon State University (OSU) where oceanographers, civil engineers, and others would, over the coming decades, study patterns of waves made to order, waves designed to be set against scale models of shoreside and maritime structures.

The wave flume (360 feet long, twelve feet wide, fifteen feet deep) was the first piece of infrastructure for a facility inaugurated in 1972 by Oregonian coastal engineers invested in building breakwaters at the mouth of the Columbia River, the largest river in the Pacific Northwest of North America. The Large Wave Flume, as the container came to be called, was built to test hydrodynamics relevant to that river-meets-sea purpose; wave modeling here, in early days, was all about learning to read and rewrite

FIGURE 3.1 Surfing in the wave flume at the early O. H. Hinsdale Wave Research Laboratory site, Corvallis, Oregon, ca. 1974.

parts of the Columbia River—a river the environmental historian Richard White (1995) calls an "organic machine"—to ease waterway navigation for the Oregonian logging industry.[1] In 1989, the Hinsdale garnered funds from the US Office of Naval Research to construct a Directional Wave Basin, a 300,000-gallon wave pool (160 feet long, eighty-seven feet wide, four-and-a-half feet at its deep end) that could scale model not just one wave train at a time but whole wave fields. After the Indian Ocean tsunami of 2004, the pool became widely known for an additional simulation purpose—one that had been put in place in 2001, when, after a targeted National Science Foundation grant, it had been renamed the Tsunami Wave Basin (figure 3.2).

The lab tackles a range of problems. Oceanographers come to investigate effects of wave action on shorelines. Civil engineers come to study

FIGURE 3.2 The O. H. Hinsdale Wave Research Laboratory Directional Wave Basin, known from 2001 to 2011 as the Tsunami Wave Basin, 2015. Photograph by the author.

how bridges will weather storms. Oil companies test models of offshore platforms. Wave energy start-ups try out prototype converters.

I visited Hinsdale for a month-long stay in 2015. The lab sits at the edge of the OSU campus in a covered, airplane hangar-size facility, next to the scrubby pastures of the university's Department of Animal and Rangeland Sciences. When I first stopped by, walking past sheep grazing at the border of the building's parking lot, I noted a sign that read "Entering Tsunami Hazard Zone"—a dry joke, since, at fifty miles from the coast, the city of Corvallis is far from such wave danger. Still, as a center of tsunami research the lab has become a key site for working out how scientifically to orient to all of those things: *tsunamis*, *hazards*, and *zones*. In the aftermath of the Tōhoku earthquake off the coast of Japan in 2011, Hinsdale became ever more central for studying tsunamis, those fast, long-wavelength waves generated by sudden displacements of water caused by undersea quakes or landslides. (Tsunami wavelengths can be so long—on the order of five hundred kilometers in some cases—that the ratio between water depth and wavelength can become very small, meaning that, for all their size, these monstrous forms may be characterized as "shallow-water" waves.) The OSU faculty using Hinsdale maintain strong links to colleagues in Japan. One had been a visiting scholar at the Disaster Prevention Research Institute in Kyoto when the tsunami hit and traveled to the surroundings of the devastated Fukushima Daiichi Nuclear Power Plant just afterward, the site of a disaster that unfolded at such scale as to make the very analytic of scale break, setting into disorder in-place calibrations of human lives to the massive forces of the oceanic. Scale effects—mismatches between models and realities—were joined by *scale failures*, dynamics for which ratios and rationality ceased fully to work, where scale became deranged, unable to hold different orders of things in proportion.[2] As Anna Tsing (2012, 505) has observed, "The living world is not amenable to precision-nested scales." Neither, it turns out, is the physics-y world, in which phenomena of all sorts convolve (roll together) to manifest an often overflowing reality.

On one of the first mornings of my fieldwork at Hinsdale, I found the lab's education and outreach coordinator telling a middle-school class about the possibility of a calamitous tsunami arriving at Oregon's shores, a wave that would arrive if there were a fracturing slip in the nearby Cascadia fault. The fault is an underwater subduction zone seventy or so miles off the coast, where the Juan de Fuca, Explorer, and Gorda tectonic plates press against one another. The last Cascadia quake happened in 1700, but with geologists figuring a recurrence interval at about 250 years, there is worry that the Northwest may be due for another, maybe soon. The coordinator showed students a video of a 1:50 scale model of the city of Seaside, Oregon, a color-coded, boxy mock-up of this tourist town constructed on the concrete shore of the tsunami basin. I joined the students in watching the mini city inundated by a twenty-centimeter pulse of water meant to stand for a ten-meter tsunami.[3] Some students laughed, some gasped, and one asked, with a mix of anxiety and defiance, "Are we all going to die?" Everyone was amused to learn that on-foot evacuation plans may be aided by adapting the *Zombies, Run!* fitness app, which tracks a person's movement with a blue dot on a smartphone map, helping them outrun imagined waves of the undead. Uneasy laughter at the scale model indexed something of the unimaginable character of what it might herald.

How do modelers settle on scaling principles, to represent waves and humans, at interacting state, coastal, community, neighborhood, and individual levels? How do they grapple with *similitude*, discerning likenesses, making comparisons across waves? What defines a "wave" in a simulation circumstance in which mechanically produced waves exist both as physical processes in real water *and* as forms, mechanically impressed into liquid, that materialize established theories and models—as, that is, a superimposition of the real and ideal, the actual and virtual? And how do researchers, conjuring lab waves, discern which aspects of their creations are artifacts resulting from their model and which may be transposable to the world?

In making scale models of waves, engineers make scale models of *time*—usually compressing it—to manipulate it as a lab parameter. In generating understudies for real waves, researchers also develop transformed perceptions of space-time, shifting their senses of speed in ways that have more than a little in common with cinematographic special-effects thinking. They learn, too, to think backward: to hypothesize, from wave observations in world and lab, what causal forces might have produced the phenomena they see. In the first part of this chapter, I argue that ordering the space, time, and scales of models is to make claims about which aspects of coastal risks, hazards, and disasters, in Oregon and elsewhere, are significant. And to whom. In the second half, I consider how, in the wake of the 2011 Tōhoku tsunami, the Pacific Northwest and northeastern Japan—and their tsunami pasts and futures—have come into comparison for wave scientists and their publics. Reading wave detritus; Ruth Ozeki's tsunami novel, *A Tale for the Time Being*; and Japanese tsunami science and video evidence through one another, I track how in the making are modes simultaneously of conjoining and pushing apart US and Japanese cases and history. Waves, ordered and disordered, become vehicles for orienting toward the forces—scientific, technical, political—that condition calamity where the breaking sea meets land.

Theory, Animated

When I arrived at Hinsdale in July 2015, Director Pedro Lomónaco was ready with assigned reading. He handed me Robert G. Dean and Robert A. Dalrymple's *Water Wave Mechanics for Engineers and Scientists*, from 1991. This was one foundation text at the Corvallis lab, a book of waves to consult when generating waves in the Large Wave Flume or the Directional/Tsunami Wave Basin. It contains a compendium of *wavemaker theory*: theory not so much about what generates waves in the wild (wind, earthquakes) as about how to reverse-engineer, from desired outcomes, paddle- and piston-made laboratory waves that imitate, that look similar to, real-world waves.

Laboratories are sites where what scientists call nature is imitated, isolated, partitioned into processes of interest, modulated into extremes, and summoned forth in technically constructed form (Knorr Cetina 1995; Kohler 2002). The logic of lab nature is realized through simplifications that take abstracted parts to stand for complex wholes (e.g., isolated DNA to stand for inheritance, subatomic particle collisions to stand for nuclear reactions in stars) and that record the interactions of such parts through inscription, traces that index findings about the world (Latour and Woolgar 1986). Lomónaco affirmed to me that "most of the time we are trying to reproduce something natural—or our *representation* of nature. Sometimes we make sinusoidal waves—which are not necessarily realistic, not something one typically sees on the ocean, but which can be very good representations of our mathematical understanding of waves." He went on, saying that lab waves "are always a simplification of nature. We have to simplify three-dimensionalities. And in the lab, turbulence and viscosity introduce forces that are not always as relevant as gravitational and inertial forces—and so in many cases, we don't focus on those." He continued: "You cannot actually do tsunamis. Not *real* tsunamis. You can try to reproduce some *pieces* of the big picture. The bore, or the wave itself, or forces, or just the beginning of those forces. So we simplify the system and then generate the tsunami—a wave that resembles a *moment* of the tsunami or just some fractions in the *forces* of the tsunami." Despite worries in the 1990s that computer simulations would replace physical wave tanks, facilities such as Hinsdale continue to have utility, especially since the interactions of waves with various materials they might hit—rocks, trees, sand, concrete—are difficult to capture in computer models. If wave basins have been imagined as "laboratories" to the ocean's "field," they have also become sites of "field testing" for computer models.[4] What count as "lab" and "field" are not prespecified but the result of how paradigms, parameters, and values of replicability are scripted, arranged, and prioritized.

Lomónaco had just in 2015 taken up the directorship of Hinsdale after twenty years leading a similar facility in Spain; this had followed education in his native Mexico and in the Netherlands, at Delft (where he worked with many people I interviewed for chapter 1). He was eager to talk and had already become a practiced ambassador for the facility. Almost as soon as we sat down the first time, I heard a loud crash just down from his second-floor office: the rushing collapse of a wave in the wave flume. The *whoomph* of the wave took me by surprise. (I learned that the flume generally runs five-foot-tall waves at twelve feet per second—quite fast.) Hearing a crashing wave indoors was weird, and I immediately wanted to know what chain of human, machine, and watery agency had created it. Lomónaco took me to the control room. There, a research associate at a computer was generating impulses using a Windows interface that could permit him to choose such features as wave height and speed.[5] The wave-inaugurating software I was seeing in action was coded in a numerical computing environment called MATLAB and was written to interface with wave paddles.

Over the next few weeks, scientists told me about how they conjured the lab waves they wanted. A striking poster in the Hinsdale hallway helped me comprehend the process (plate 10). The image, read left to right, morphs Katsushika Hokusai's famous woodblock print "Under the Wave off Kanagawa," from 1829, into a digitized, scientific wave, suggesting a wave in the world being formally dissected and rematerialized in a computer.[6] In this transposition, the fishers in the boats vanish (the people who are "under the wave"), leaving a bitmapped nature. At Hinsdale, there's a next step of transformation, of course: mathematized waves are rematerialized in water, which then makes them "real" again, now as technical rather than unpredictable things.

Software that animates wavemaker theory is a switching point where models of wave action are set into operational motion. Such models have a history: they enfold established, community-certified knowledge about wave dynamics. The knowledge is different from the sort Karin Amimoto Ingersoll examines in her *Waves of Knowing: A Seascape Epistemology* (2016), which documents how Kānaka Maoli (Native Hawaiians) come to know waves through oral history, chants, and surfing—a genre of social epistemology thicker than this lab species, which emphasizes calculative views of waves from outside or above. At Hinsdale, waves are rendered as abstractions around which engineers might work. Any wave in the tank is a convolved artificial-natural hybrid, an ideal wave form made manifest in the material of a real wave.

I met up with Solomon Yim, a structural engineer who has taught at OSU since the late 1980s. He summarized the question before wave modelers seeking to imitate real-world phenomena: "How do you infer from a known wave pattern what it was that initiated it?" This was, he told me, an *inverse problem*, taking a known outcome and trying to reverse engineer an answer to what may have led to it.

What goes into modeling a wave? Yim reminded me that waves were characterized by spectra, the frequencies of which they were made up. Making artificial waves means imparting to water a suite of energies that re-create a desired spectrum.[7] Such spectra, Yim emphasized, are mathematical descriptions and as such are conceptual tools that change as ocean science communities fine-tune them. The theories installed in wavemaker algorithms, he said, were in many instances only up to date as far as the

1980s, even as wave theory has moved on. I asked Yim: "Does that mean that if engineers today are working with wave spectra that were well theorized in the 1980s, that they are creating *1980s waves in 2015 water*?" Were they still playing Madonna songs in an age of Beyoncé? His answer was an emphatic yes. I thought of 1980s vinyl recordings transcoded into 2010s digital streaming files. Matters started to snap into focus: wave machines animate equations, even theories—inherited disciplinary knowledge. What then happens is that the physics of the water and the tank itself become the *media* for theory, *theory, animated*—and in ways that, scientists hope, do not interfere or produce artifacts added on top of the target waves. Artificial waves are chimeras, thought experiments made water. If, as Nietzsche once wrote, in one of his rare typewritten personal letters, "our writing tools are also working on our thoughts" (quoted in Kittler [1986] 1999, 200), here is an example of how mathematical writing in/through water participates in making new ideas about waves. Yim gave me a more down-to-earth reading: "Mother Nature still governs the lab, too, right?" An artificial model of a real wave is still itself a real, physical, wave.

Still, any given model may be deficient. Make the analogy, he instructed me, of waves to animals and consider that the task of wave modelers might be to replicate the diversity of the animal world; to create models, say, of both cats and dogs. Current wavemaker theory, he suggested, could only make "cats," when we know that there are also "dogs" out there. Some tweaking can be done to make some cats look more like dogs, but in general, he told me, "Wavemaker theory is really only *cat theory*; the linear wave theory is our common shorthair American tabby." But there were, he said, more interesting nonlinear waves—*dogs*—out there. I could not help but think of these domestic animal analogies as pitching artificial waves as entities to be sculpted, carved down to size, by human work (compare chapter 1). Mechanically made waves install the formalisms of the people whose work is reflected in them, a realization one could attach, fancifully, to the lab's press image of one of its open house events (figure 3.3).

Yim arrived at another metaphor. The lab, he said, is engaged in the *3D printing* of waves, in water.[8] When I relayed this metaphor to Lomónaco, he ran with it, telling me, "Right, so when you print a picture with a dot matrix printer, you can tell it's correct, but it's not exactly the same. It's blurry. It's imperfect. But when you have a color laser printer, you will have a very nice picture that is very close to reality. So, yes, that analogy applies." Scientists know that their waves are artifactual and that modeling choices—which pick out which dimensions of waves are significant—then

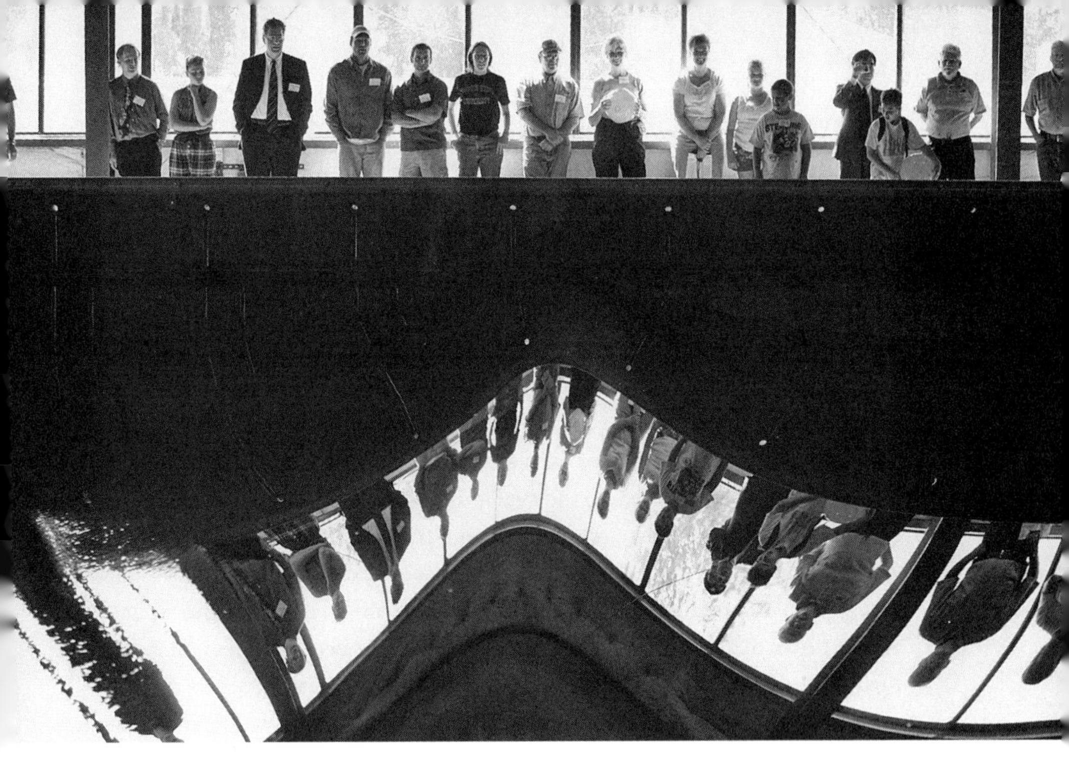

FIGURE 3.3 Attendees at an open house for the Hinsdale lab reflected in a wave traveling down the Directional Wave Basin. Photograph by Sol Neeman.

provide limits on how faithfully they can render lab waves similar to real-world ones. They are producing likeness, not sameness. I began to see waves in wave tanks not just as mathematical texts written in water, but as pictures—even moving pictures.

Waves as Cinema

One day, a video crew arrived to gather footage for a documentary on tsunami modeling. They shuttled lights around, setting themselves up to film some half-meter waves speeding across the Directional Wave Basin. The tank's concrete shore hosted a one-and-a-half-foot-tall model of a building (a "hotel"), giving the incoming waves a scale. The video team zoomed in on the paddles generating waves—paddles, they had been informed, that were set in motion by a person at a computer in a control room perched above the shore side of the basin. As that man looked down through a small window, he reminded me of nothing so much as a movie projectionist peeking out from an overhead booth. The scientists on the lab floor communicated with him via walkie-talkies, asking him to "run some waves."

When a member of the visiting film crew was given control of the walkie-talkie, she offered overlapping jargon: "Roll 'em!" I could not help but think, now, of these artificially generated water waves as three-dimensional (3D) movies; as movies in a material—water—that, like celluloid, tape, or streaming video, has its particular affordances and resistances. This was a coincidence between *real waves* and *reel waves*: not so much now the book of waves I was given at the beginning of my visit, *Water Wave Mechanics for Engineers and Scientists*, but a cinema of waves. A book of waves made into a movie.

The coincidence between the real and the reel had a strange temporal texture. All was not linear, frame-to-frame time here. There is something hidden from the untrained eye about time as it exists in a scale model. When wave scientists make a 1:100 (geometric) scale model of a tsunami, space scales down at a different rate from time. This is so because the molecular structure of water and the force of gravity cannot themselves be miniaturized. (It is also the case that the tank does not use saltwater, since salt wears on the infrastructure. As a result, as one engineer told me, "We have to use all kinds of fancy math to calibrate.") So modeling a ten-meter-tall tsunami by generating a lab wave that is one hundred times smaller (ten centimeters tall) does not yield dynamics that unfold one hundred times faster, as one might think (or, as some might also think, at the same speed). Rather, the temporal unfolding of such a wave is described by the ratio

$$v / \sqrt{gd}$$

where v = the velocity of flow
g = gravitational acceleration
d = depth of water

The ratio, proposed by the nineteenth-century English hydrodynamicist William Froude, suggests that if one wants to watch a real-time film of a 1:100 scale-model tsunami and have it look anything like a real-world event, it needs to be slowed down by ten, not one hundred, times, in order to keep the flow characteristics in the model similar to the real-world target system (accounting in this way for those aspects of water that are not scalable). Hollywood filmmakers know this well and do such calculations whenever they destroy a scale-model city with a wave of real water, whenever they summon what the environmental historian Gregg Mitman (1999), in his analysis of nature films, calls *reel nature*. To watch a video replay of a wave run in the lab, *reel time* needs to be slowed down by the square root of the length scale—the *Froude similitude* (said out loud, these words rhyme)—to give the impression of *real time*, which, recursively, is available only through such reel time.[9]

Similitude, in medieval European traditions an aesthetic that permitted correspondences to be drawn between, say, celestial and earthly bodies (think astrology), is nowadays mere disenchanted measure. If making waves and filming them retains any feeling, to quote the anthropologist of cinema Anand Pandian (2015, 8), of "participating in the creative process and potential of a larger universe beyond the human," that *beyondness* is meant to be wrestled back into sensorial access by lab and media similitude—though, as I discuss in connection with experiences of tsunamis, such *beyondness* can reassert itself, arriving in the register of the traumatic, sometimes unspeakable, terror-sublime.

Viewers of artificial waves, of time-adjusted video of such waves, and of footage of real-world tsunamis often gather their sensibilities about how to compare these phenomena though experiences across screens—cinematic, computer-y, and documentary. They develop what the anthropologist Cristina Grasseni (2004) calls "skilled vision," a hybrid viewing that permits them to see in physical objects those features that might be modulated by technical action. (Grasseni's examples are from the biological world, having to do with how breeders learn to see different potentials in animals they domesticate.) The doubled vision into which I was being initiated as I began to see a slice of the world through a mathematical formula made me see these waves differently. These scientists around me now struck me as science-fiction aliens able to see in two temporalities at the same time. Lomónaco said: "You get used to that. Because everything is relative, you get used to having a mental picture going. Things are happening at a different pace than one expects. So when you see a wave in the lab that is moving fast—and that might be rocking model boats at speed—you adjust your frame of reference, understanding that everything you see should be happening in slow motion." Some of this is about speed as a sensation, "not sheer velocity as such but the impression of shifting from one velocity into another" (Pandian 2015, 222)—where such impressions, for wave scientists, require not just tuning for "exaggerated speed and slow-motion torpor" (Pandian 2015, 226), but, back to the equation, knowing how to move through exponential, not linear, time.

Waves to Order

When the lab is not running waves for television documentaries, what sorts of waves does it model? The facility is designed to run hurricane and tsunami waves for many kinds of coastal and ocean study and has done experiments examining wave effects on almost anything that might go into the water: breakwaters, submarine pipelines, elevated platforms. Lomónaco told me that, after early work on structures guarding the mouth of the Columbia River against Pacific waves, the lab saw a number of projects on "biofouling," the accretion of unwanted organisms—algae, animals—on wet marine structures such as piers. The question was how wave action on such conglomerations might put measurable pressures on marine infrastructures. Then, in the first decade and a half of the century, tsunamis came into focus, and some 90 percent of lab work was on these sorts of waves (see Briggs et al. 2010). When the dedicated tsunami grant ran out, around 2014, a new public rhetoric of relevance appeared: "multiple hazards." The tank was now pressed to account for waves not just from tsunamis, but also from hurricanes and winds. *Resilience* became the watchword, Lomónaco said, with hurricanes Katrina and Sandy as reference points. More recently, another set of interests has emerged that has turned to the lab to explore *wave energy*, the mechanical extraction of power from the rise and fall of ocean waves. Waves are ordered by their historical period, approached, at one moment, as dangers, and at another, as promising energy infrastructure. In this way, too, the lab is like a film studio, with clients contracting the facility to produce movies to order for audiences whose orientations and preferences change over time.

Getting model lab waves to behave takes negotiation. Clients might order a certain periodicity—a wave every eight seconds, for example—or a reliable height. Or they might want a more general picture. Lomónaco drew a distinction for me:

> If you have a single wave, you can analyze all the details of the shape of that particular wave. But if you have a sequence of one thousand waves, you cannot study all the details. You have to represent those with a single parameter, such as significant wave height, to average what's going on. I might approach one case by focusing on the shape of individual waves and another by focusing on the spectrum and the probability distribution of waves.

It is not always easy to meet clients' desires. The lab facilities have idiosyncratic parameters, and it takes time to set up experiments. Dan Cox, the former director, told me, "People come to the lab with certain demands, or they want a certain spectrum, or they want a certain frequency, or a certain kind of set of profiles for wave things, and then you come back to them and say, 'Yes, we can do that,' or 'We can *kind of* do that.'" Some clients rush in after having done models on computers and then expect the in-water world to behave the same way. They may assume waves will break at a particular point because the mathematics gives them that expectation. As Cox told me,

> So if they do a lot of numerical work first, they kind of synthesize a certain sea state. And then we'll generate waves, and just naturally they'll break, and they'll come out a little bit small. And then the client is very unhappy that the waves are too small. And you have to explain to them, "You're never gonna get 'em," right? Just because the numerical models and textbooks tell you you're going to have a wave break when it reaches .78 of the stable height . . . it's not reality. Because waves will break sooner than that.

Lab engineers may argue with clients about the model they are after, too. Scale is not a given but a discussion. As one engineer at OSU told me,

> In a recent proposal, I was reproducing a fraction of a port in 1:50 scale, which is about the limits of what you can do with scaling when you're studying the stability of breakwaters. Sometimes you can go 1:60, but it depends. And the client says, "Can you do this in 1:40?" And then I take a look at it and say, "I think I can, but there is a trade-off. I cannot reproduce all of the breakwaters you want to reproduce; it's gonna be smaller. And you will have some other effects."

The wave basins also have properties to be taken into account. As Cox told me,

> A whole other part of the problem is that the laboratory itself introduces artificial walls. There's re-reflection and limitations on what the wave makers can do. These are all effects of being in a laboratory. The most notorious one is the current. In the ocean, currents are just going to go along the coast or out to sea, but we're in a box, and those currents are just coming back in. And if you keep running things for a long time, it's just a soup of currents.

Getting a wave experiment set up requires physical and social engineering—thinking inside the box—from the scale of the lab to matters of concern in the world.

Scaling at Seaside

The creation of the Hinsdale laboratory's scale model of the coastal Oregon town of Seaside, a hamlet of about seven thousand people south of the mouth of the Columbia River, was motivated in part by the Indian Ocean tsunami of 2004. Cox, the lab's director when that wave hit, told me how the disaster had prompted modeling efforts for Oregon. He and colleagues had the idea of modeling a generic coastal city, using a black-and-gray checkerboard with blocks of different heights. When that effort did not get funded, they added color coding to call out residences, commercial sites, parking lots. And when *that* did not get funded, they modeled a specific place to make the local relevance of their science laser clear. They selected Seaside because, with its flat shore and downtown area, it was dramatically vulnerable to a tsunami coming from the Cascadia subduction zone, with a possible wave arriving just twenty minutes after a seismic event. Some Seaside residents worried the town would lose tourist revenue with the public attention that might come from this simulation project: some tsunami warning signs were taken down in an effort to protect the town's image as a safe harbor. Eventually, though, most residents embraced their status, with some even making light of their newly described locale with restaurant names such as Tsunami Sandwich Company and Tsunami Bar and Grill.

One of the papers Cox and his colleagues published (in early 2011, before the Japan tsunami) about the Seaside model, "Optical Measurements of Tsunami Inundation through an Urban Waterfront Modeled in a Large-scale Laboratory Basin," illustrates how settling on scale models entails making judgments about social as well as scientific values. The paper spells out reasons for choosing Seaside:

> First, the constructed environment (seawall, hotel, residential and commercial buildings) were typical of coastal communities at risk to tsunamis with populations concentrated within the first 200 m[eters] of shoreline. Second, the bathymetry at Seaside was fairly easy to construct, with uniform, shore-parallel contours and a relatively flat spit on which the buildings were constructed. Third, the US G[eological]

S[urvey] report gave reasonable guidance for the expected tsunami height triggered by a Cascadia Subduction Zone event. (Rueben et al. 2011, 230)

The choice of scale brings some things into focus and not others. As the media theorist Joanna Zylinska (2014, 32) suggests, "Seeing things across different scales is more than an attempt to represent the universe: it actively produces entities and relations." The model was built at 1:50 scale, a degree of resolution that, unlike previous 1:200 scale models, permitted researchers to grapple with such things as the effect of building shapes and densities—here, "eight large hotels and five large commercial buildings," as well as "residential structures"—on the flow of a tsunami across land while leaving out "vegetation, small-scale roughness, debris or sediments." They learned, Cox told me, about how large buildings at the shore affect smaller buildings farther away: "At first, we thought that the large buildings would provide sheltering from the wave, but it turned out that some areas further back were subjected to flow convergence—it seemed that the forces could *increase* because of larger buildings." Spaced close together, this might focus wave energy.

Different aspects of wave behavior had to be weighed. "The biggest scale effect is the wavelength," Cox said. "Tsunamis inundate the coast for several minutes, and this continuous inundation is hard to reproduce in the lab, so we focus on the leading edge." Arguing that a single, solitary pulse from the lab's wavemaker might create a wave that was too neat and might underestimate durations of inundation, Cox and company sought to create a bit of noise, or error, around the "solitary wave algorithm" used to generate the model tsunami. Such error had to be distinguished from "short term variations due to building vibration or wind response and long term variations due to thermal expansion or building creep" (Rueben et al. 2011, 232).

Getting information out of the model required technologies of digital video, which had to be treated to bring into visibility those aspects of concern to people who might need to run away from a wave. Revaluing pixels in the video permitted researchers to see patterns of turbulence and wake areas behind hotels—something Cox told me he saw evidence of when he looked at the aftermath of the Japan tsunami in 2011. Running the experiment over and over (136 times) generated a spread of likely wave fronts.[10] They could be looped into evacuation planning. Visits to Seaside "to look at building stock, to talk to the city engineer and some of the building owners to get a sense of the structural integrity of the buildings" provided knowledge to be fed back into the physical model.

One of Cox's colleagues, the OSU civil engineer Harry Yeh, reflected on the back-and-forth between such models and the world:

> Tsunamis interact with large scales of bathymetry and topography, so scale effects are a crucial issue for studying tsunami dynamics. Tsunamis are rare, and so we do not have a sufficient empirical "feeling" for the phenomenon. Prior to the 2004 Great Indian Ocean event, video footage of real tsunamis was almost nonexistent. So we attempt to understand tsunami behaviors in the controlled laboratory environment. The large tsunami basin in Hinsdale is often used to simulate tsunamis, but there are still potentially substantial scale effects we can't avoid.

Just one month after the work on the 1:50 model of Seaside was published in *Coastal Engineering*, the Tōhoku earthquake off the coast of Japan generated the tsunami that devastated Miyagi, Iwate, and Fukushima prefectures on March 11, 2011. Cox was in Japan and visited the Tōhoku region:

> I was not prepared for how much of an impact it had on me when I went there for the first time. I had been at the museum at Hiroshima the previous week, but still I would say that the most immediate thing was how different photographs are from reality. We think photographs are kind of like a bit of reality, but it's totally different. You know, once you're there, you're looking around, you're looking up, and it's like, "I can't believe this happened." I was not prepared for what I saw. I was a textbook wave guy, not really prepared for that first field visit.

The 2011 tsunami in Japan now hovered as a cautionary tale, an event revisited again and again as both similar to and different from what might happen in the next decades in cities such as Seaside along North America's northwestern coast. Cox pointed to differences:

> Thinking back to the Japan tsunami of 2011, there were a lot of regions that were affected that had different coastlines. Some of the most heavily impacted areas were along Japan's ria coast, which is quite different from what we have here in the Pacific Northwest. The ria coasts were formed from drowned river valleys and look more like fjords, with deep, narrow embayments that funneled a lot of water into the towns. The Sendai plain was wide and fairly flat, so the local tsunami run-up was a lot lower, and the tsunami inundated that area for several kilometers. Even in the "same" event, local conditions made a huge difference. So

for the Pacific Northwest, we would also need to really look at each area carefully.

Yeh pointed to similarities between eastern Japan and the Pacific Northwest: "The source of the 2011 tsunami was approximately 350 kilometers long and 150 kilometers wide. Tsunami-affected coasts are larger than the source because tsunamis propagate. A potential Cascadia event could be similar, with large affected coastal areas, from British Columbia to Northern California. This is quite different from hurricane or earthquake hazards, which are more local." Thinking at different scales, in other words, makes similarities and differences slide in and out of view. Scale making is similitude and difference making. (On how comparisons among hydrological models of Cambodia's and Vietnam's Mekong River Basin generate contests over which scales are best for which purpose, compare Jensen 2020.) Such scale making can fail, as effects from one scale spill over into or swamp another.

Similitudes across the Sea

In July 2015, in the middle of my Oregon fieldwork, the *New Yorker* published an article about the prospect of a catastrophic earthquake visiting the Cascadia subduction zone sometime in the coming fifty years, a quake that would launch a tsunami toward the coast, a wave that would strike the shore just twenty minutes after an initiating undersea rupture (Schulz 2015). Scientific knowledge of the tsunami-generating power of the Cascadia fault, the article reported, turned out to be recent, dating back only to the end of the twentieth century (Atwater et al. 2005; B. Henderson 2014). In the 1980s, geologists, applying theories of plate tectonics, found evidence in the Pacific Northwest of sudden coastal subsidence in "ghost forests," stands of trees abruptly lifted down, underwater, and then layered with sand that was streaming in from the sea. Tree-ring dating, which showed a quick drown for cedar and spruce trees, indicated an early eighteenth-century event—an earthquake followed by a tsunami. That, in turn, resonated with Indigenous oral traditions—notably, from Makah tribal sources and Huu-ay-aht First Nation narratives—that had preserved accounts of a quake and flood likely to have occurred between 1640 and 1740.[11] These clues then pointed to a possible trigger for a contemporaneous event recorded farther away, across the Pacific, in Japan, in the year 1700—a tsunami that, because of its unknown origin, for centuries had been known in Japan as the "orphan tsunami." Historians studied Edo Period records of Japanese coastal towns—settlements that reported unexpected, unheralded flooding in 1700—and, doing the inverse problem, fixed an exact time and place for the genesis of the wave: 21:00 Pacific Time, January 26, 1700, in the Cascadia subduction zone. With this datum, geologists were able to make sense of a long chronology of Cascadia quakes, which, they argued, appeared to have a recurrence period of 243 years.

When the *New Yorker* story broke, local and national press contacted Hinsdale. Lomónaco, Cox, and others explained how to understand statistics and models and how to heed evacuation instructions. The lab had been integrating the story of a possible Cascadia tsunami event into its community outreach, as I had seen. The story reverberated across the coast, putting into circulation a sense that this could happen again, and soon. For some, it recalled a tsunami in 1964 that had been generated by an earthquake off the coast of Alaska, a wave that had reached down to seaboard to kill 122 people, including five in Oregon. I headed to the coast to see what official and everyday apparitions of tsunami futures I might discern.

The "Entering Tsunami Hazard Zone" street sign I had seen in the Hinsdale parking lot appeared frequently, now not as a joke but as an official warning that gave me a sense of being inside a double-exposed hologram, the present and a possible future superimposed. I spent the night at a motel in Seaside—a couple of people I knew at OSU said they would never risk that—picking up the town's evacuation map and making a point of simulating an escape on foot with my anthropologist spouse, Heather Paxson, and our kid, Rufus, at age nine ready for an unsettling game of pretend. It was not clear whether we made it in time to high ground. Back in town I bought a book titled *The Next Tsunami* and noted that the bookseller had a rolling backpack "go bag" ready just behind her cash register. The drive along the coast phased in and out of hazard zones. On beaches there were warnings about smaller waves—"sneaker waves," about which I had learned from the OSU coastal engineer Tuba Öskan-Haller, whose research had been featured many times on the local news, reporting on her work at the Hinsdale tank simulating these unexpected kinds of waves. Outside Yachats was a memorial created for two men who had drowned in 2011 because they were taken unaware by a sneaker wave. The memorial, a sculpture in the shape of a crashing wave, was at once sentinel and warning, its legend reading, "The Ocean Is a Treacherous Wonder."

Tsunamis remained the big story. The *New Yorker* piece amplified public attention to tsunamis in Oregon that had been on the rise since the 2011

FIGURE 3.4 Dock fragment from Misawa port, Aomori, Japan, at Hatfield Marine Science Center, Newport, Oregon. Photograph by the author.

tsunami in Japan, which had had echo effects on the Pacific Northwest's coast.

In June 2012, a 105,000 pound, sixty-six-foot-long chunk of debris from the 2011 tsunami washed onto the shore at Agate, Oregon—a concrete fragment of a fisheries dock ripped from the port of Misawa, Japan, and ferried by currents across the Pacific. A seven-by-seven-by-seven-foot segment of this dock is on display outside OSU's Hatfield Marine Science Center, in coastal Newport (figure 3.4). Despite the cartoon wave graffiti on one of its sides, this was a sign of death and disaster, a gravestone and warning. Think back to the wave flume at Corvallis and imagine an abrupt rescaling, an inverted similitude; if the 1974 photo of the engineer surfing in a cement block showed a wave dominated and domesticated, this dock block indexed a gargantuan wave outside the box of concrete control. Explanatory signage adjacent to the dock struggled to name the emotions the structure might have prompted when it arrived: "The dock generated many feelings: sympathy for the lives lost in Japan, curiosity about its sudden appearance, and surprise at the number of organisms still living on the dock." Attention to the organisms on the dock, described on the sign as "invasive species" ("Scientists counted over 118 different Japanese species on this one piece of debris alone"), detoured the story into one about marine biodiversity out of place and into a tale about the *distance*—in space and species difference—between the United States and Japan.[12]

The dock fragment was made more disorienting by its juxtaposition, in the science center's parking lot, with a large yellow prototype of a wave energy machine, a device featuring a moving float eleven feet in diameter (designed to move up and down with waves) ringed around a twenty-three-foot-tall spar buoy (designed with an interior coil to transduce up and down motion into magnetic and then electrical energy). Explanatory copy about *that* artifact explained that electricity might be generated by sustainable energy created by the "wave nursery" that was the near-shore sea. Here were two material totems that worked as portents of two different sorts of ocean waves: death-dealing versus life-nurturing.

Returning attention to the Misawa dock fragment, I saw that one of its sides held a plaque that indexed the terrible story of Tōhoku:

> This monument is dedicated to those who lost their lives
> in Japan's Tohoku Earthquake and Tsunami
> on March 11, 2011.
> It honors the multitude of lives saved
> by the preparedness efforts of the
> Japanese government and its people.
> May this dock's transoceanic journey remind us
> of the great power of the ocean to shape our lives,
> binding us to our natural world, and to each other.
> Dedicated by the Community of Newport, Oregon
> March 2013

The "power of the ocean" narrative, of course, stepped to one side of the more dreadful tale of the Fukushima Daiichi nuclear disaster, of radiation and, indeed, glossed over what many observers identified as the inadequate response of the Japanese government and the continued efforts by its nuclear industry to protect its investments.[13]

Perhaps this is not surprising, given the imagined readership of the plaque, far from Japanese politics. Still, it is possible to read the dock as a geographically displaced version of a common kind of Japanese monument, the *tsunamihi*. Megan Good (2016, 140) explains that the word *tsunamihi* is

> a term derived from a combination of the word "*tsunami*," meaning a very large ocean wave caused by an underwater earthquake or volcanic eruption, and "*hi*," meaning a stone monument with an inscription. *Tsunamihi* are large stone tablets or elongated rocks ranging from three to even ten feet tall, set into the ground and featuring inscriptions. When the recent Great East Japan Earthquake and Tsunami struck on March 11, 2011, some residents turned to these monuments for guidance on where to find safe ground, and recalled the messages inscribed on their surface which contain warnings from their ancestors of the dangers of earthquakes and their ensuing tsunami. The earliest known *tsunamihi* date as far back as the 14th century.

Good notes that *tsunamihi* from different historical periods index shifting senses of the meanings of these waves; a *tsunamihi* in Tokushima Prefecture from the year 1380, marking an event that took place in 1361, holds an inscription that points to then dominant Buddhist beliefs that victims of disaster have only their own ill behavior to blame; a *tsunamihi* in the same area from the year 1605 incorporates a Shinto shrine, marking a competing belief that disasters result from divine negligence. After an event in 1933, declared by scientists to be definitively seismological in character, *tsunamihi* begin to contain safety instructions.[14] These emerged alongside *ireihi*—cenotaphs that record loss of life—marking a division between warning and mourning.[15]

What sort of *tsunamihi* is the Newport block? Good reports that, after 2011, many survivors advocated for the preservation of tsunami ruins as more effective warnings than *tsunamihi*. Perhaps the dock block is ruin as *tsunamihi*—post-tsunami disaster heritage (Littlejohn 2021). But who is its message for? Most obviously, it is for Oregonians and people who visit this part of the Pacific Northwest coast—though that message, supported by signage that points to the dock's origins, acknowledges neither the ancestral nor ongoing presence on Oregon land of Indigenous people (posed on the sign as inhabitants of long vanished "Indian villages") or the location of the block on the territory of the Tillamook, Siletz, and Yikina peoples. That erasure paves the way for simple US-Japan comparisons.

The message of the dock prefigures a coming disaster in Oregon. It sits as a monolithic visitor from the future, a prophecy of coming natural disaster or of the coming home to roost of sociotechnical shortsightedness (see Dupuy [2005] 2015). An apprehension of the dock as foreshadowing a tsunami future haunted descriptions by Oregonian newspapers when the fragment landed. The anthropologist E. Summerson Carr and her colleague Brooke Fisher catalog these framings, for precisely this dock block, in "Interscaling Awe, De-escalating Disaster" (2016), whose title summarizes what they argue has been the rhetorical effect of dominant readings of the object. They recall that the *Oregonian* newspaper pronounced, when the fragment arrived, "The mass of concrete was bigger than anything anyone dared imagine, a harbinger, it seemed, of our worst fears."[16] Carr and Fisher examine how commentators shifted the scale of this block in their accounts. Press coverage was "characterized by an almost compulsive interscaling that moves briskly between the dock's heaviness, its height, the distance it traveled, and the enormity of the natural disaster that sent it on its way, as well as the quantity and diversity of 'non-native' marine species attached to it" (134). Carr and Fisher observe, too, that those people who were engaged in this scale making "figured the dock as a fragment of ecologies, histories, and futures too disastrous and overwhelming to otherwise imagine" (134–35). In many ways, the dock was a communiqué from a traumatized Japan—a message from the recent past about a possible future to which Oregon's coast might be subjected. "Scaling," write Carr and Fischer, "is a practice that can—among other things—spawn a sense of intimacy and an ethic of interrelatedness" (136). When I saw the object, it was not clear to me how, whether, and for whom this dock awakened intimacy and interrelatedness, though especially after the *New Yorker* story it did seem to invite questions about how residents of northwestern North America and northeastern Japan are and are not kin to one another, living under Pacific threat of either shared or mirrored wave disaster.

Accounts of Japan in mid-twentieth-century American and European social theory often posed the country as a mirror version—similar but with key traits reversed—of "the West." Seeking to critique American individualism and self-interest, the sociologist Robert Bellah held Japan up as home to an alternative modernity in which neo-Confucian and Buddhist tradition anchored a secular, industrial, capitalist society consistent with communitarian networks of social obligation. As the anthropologist Amy Borovoy (2016, 469, 489) argues, however, "holding up Japan as a mirror" operated for Bellah, as in much mid-century Euro-American Japan studies, by treating the country as an "'empty box,' divorced from historical and social context." The empty Misawa dock box, similarly, cannot straightforwardly foreshadow, invert, or mirror Oregonian time and space.

Still, the Japan disaster had prompted cross-Pacific comparison. One scientist at OSU told me that, for him, Fukushima was a reminder of the disastrous folly of never realized (but researched, by the US Nuclear Regulatory Commission) plans in the 1980s to site nuclear power plants on artificial islands off Oregon, enterprises that would have called on Hinsdale to do wave modeling for the coastal armoring of such plants. For Cox, the tsunami was no question a cautionary tale for Oregon, one that redoubled his commitment to preparedness planning.

In the aftermath of the *New Yorker* article, though, there were also publics that refused any connection, any lesson, either from American seismology or Japanese experience, folding their reflections into more local, US-centered narratives. Some offered anti-expertise, right-wing survivalist pronouncements. The comments section on "*The New Yorker* Earthquake Article Unleashes Tsunami of Social Media: 8 Takeaways" on *Oregon Live*, featured examples:

> Deltaforce: Liberal scare tactics from liberal scientific intelligentsia (*sic*) and liberal media designed to increase taxes just like the global warming scare tactics.
>
> Dead Duck: Only Hope is to crown Obama King. . . . [B]ow down and ask for his blessing. All seriousness aside, preparedness is the key. Knowledgable (*sic*) people have been storing up a little more food and water . . . preparing for the possibility of no utilities . . . power . . . water . . . heat . . . coordinating with family, neighbors, church, civic groups etc. Security plans . . . communications etc.
>
> Azgolf02: I'm FAR from being an "expert," but I know well enough to know these models aren't exact replicas of Seaside and Cannon Beach. It doesn't duplicate the exact slope of the sea floor, let alone the elevations of the land. Sorry, it makes for a fun video, but it is FAR from scientific![17]

These opinions operate in a register apart from one that would see the US West Coast and Japan subject to common large-scale processes in geology or world history—or, indeed, as modern nation-states with entangled investments in infrastructure. They also articulate a long-standing frontier ethos among some residents of Oregon, an ethos often coupled with apocalyptic antigovernment rhetoric, as well as with traditions of white supremacist separatism that date back to the state's "Black Exclusion Law" of 1844 (repealed only in 1926), a history that conditions the overwhelming whiteness of the state. These visions refuse any similitude with Japan—and evidence, in their survivalist extreme, a resistance to locating individual, first-person experience within such scientifically spelled-out time frames as 250 years, the rough period of past and projected Cascadia quakes.

There are, of course, crucial dissimilarities between northwestern Pacific and Japanese cases. For wave scientists, they became visible through comparison. Drawing on his work as a civil engineer on both sides of the Pacific, Yeh told me:

> The major differences between Japanese coasts and Pacific Northwest coasts are the populations and coastal structures. Japanese coasts were meant to be protected by seawalls from tsunamis [though the wave of March 11, 2011, overtopped them], while there is no such protection along the Pacific Northwest coast. Another difference is bathymetry and topography. The source of the 2011 tsunami was along the more than ten thousand-meter-deep Japan trench, and the coastline it hit is a jagged ria coast. No such distinct trench exists off the Pacific Northwest—though there are similarities with the source of the 2004 Great Indian Tsunami—and the coastline is "more or less" straight.

But to return to Carr and Fischer's suggestion that inter-scaling can "spawn a sense of intimacy and an ethic of interrelatedness," in what sorts of transoceanic intimacies are residents of Oregon and Japan bound up, post-2011? The scale of that question—which residents? when? in which settings?—makes it challenging to answer. And, as the anthropologist Ann Stoler (2001, 862) suggests, analysts would do well to "historicize the *politics of comparison*, tracing the changing stakes for polities and their bureaucratic apparatus" (see also Choy 2011). Some comparisons circulated around terrors of contamination, a fear of partaking in what Kath Weston (2017) has called "the unwanted intimacy of radiation exposure in Japan." Fear of that intimacy found a paranoid example in some American news media's misreading of vibrant color-coded maps made by the National Oceanic and Atmospheric Administration of wave heights issuing from the undersea Japan trench (plate 11). They were taken, instead, as tracking the travel outward from Japan of radiation, probably because maximum wave amplitudes were coded in a highly saturated fire red, imparting a kind of symbolic heat to the image. The *Ventura County Reporter*'s cover of June 7, 2012, using this image—a mash-up of map and graph—pressed the claim that Fukushima was sending well-defined streams of radiation across the sea, as though Japan were sending nuclear tentacles out into the ocean.[18] Eiji Oguma (2013, 240) has suggested that, "for readers in Europe and the

United States, Japan's nuclear disaster is referred to in no small part as 'Fukushima' because it rhymes with 'Hiroshima.'" Rhyming offers a kind of similitude that both activates and damps historical meaning.

A more historical story would highlight how Japanese nuclear infrastructure came into being in large part because of efforts promoted, as early as 1953, by the US Atomic Energy Commission (AEC) to rebuild Japan in line with President Dwight Eisenhower's Atoms for Peace program, with a representative of the AEC at the time arguing that, "while the memory of Hiroshima and Nagasaki remains so vivid, construction of [a nuclear power plant] in a country like Japan would be a dramatic and Christian gesture which could lift all of us far above the recollection of the carnage of those cities," perversely transmuting Japan, a victim of a nuclear attack, into a beneficiary of it (quoted in Beauregard 2015, 103; see also Muto 2013). There is also a story to be told about transnational civil engineering projects that build seawalls around the globe, a story of global concrete markets—and note that the town of Misawa, where the dock came from, is also host to a US Air Force base. There's a story to be told about Japan-US history, which saw the internment of Japanese American citizens during World War II in parts of Oregon, a history that I cannot help but hear in mentions of "invasive species" traveling from Japan on the Misawa dock.

Let me try additional paths toward bringing Pacific Northwest and Japanese tsunami histories and futures into a common frame—and toward thinking about the relations among simulation, forecasting, hindcasting, and unimaginability that such a frame suggests. After taking up a work of fiction about the transoceanic connections, in the wake of the tsunami, between Japan and the Pacific Northwest, I move to a history of Japanese tsunami science, guided by a Japanese OSU scientist trained in this tradition. I then return to the theme of waves and film, discussing the social media and scientific life of survivor videos of the tsunami.

Ruth Ozeki's novel *A Tale for the Time Being* (2013) revolves around a Japanese Canadian novelist named Ruth—a fictionalized version of Ozeki who phases in and out of similitude with the author—who discovers, washed up on the shore of her island home off the Pacific coast of Canada, the diary of a young Japanese American teenager; the teenager had recently relocated to Japan with her father, a software engineer who had failed to succeed in Silicon Valley. The journal was written, it seems, in the months before the tsunami, though the teenager, Nao, also announces that by the time someone reads it, she will have taken her own life. Diving into the diary,

Ruth is riveted, disoriented. Is the girl still alive? Does the Japanese North Americanness that Ruth shares with the girl, even across generations, offer, to borrow Carr and Fisher's words, a "sense of intimacy and an ethic of interrelatedness"? The tsunami hovers as a presence throughout *A Tale*. The tsunami "is the means by which Nao's diary *might* have been delivered to Ruth across the Pacific, while the nuclear meltdown *might* have brought unspeakable harm to Nao's family, though neither of these details are ever confirmed" (Gullander-Drolet 2018, 307; see also Glassie 2020).

Nao describes herself as a "time being." She says, addressing her unknown reader, that, really, everyone is a "time being," not just moving through but *made of* time. And in the book, time is far from linear. The girl may be alive or dead. Japan, Ruth ruminates, may be her real home, or it may not. The long shadow of Hiroshima may have passed, or it may not. Consider the anthropologist Ryo Morimoto's (2014) ethnographic work among people in Japan dedicated to retrieving and cleaning photographs damaged by the tsunami, work conducted in the indeterminacy of not knowing whether the people in the ruined images are living or dead, work in which photo-cleaning volunteers often conjure hopeful biographies for their subjects. In Ozeki's transoceanic story, the tsunami becomes an oscillating relay, an unreliable messenger across Pacific time, a force that at once carries and confuses time beings.[19] In this sea, people themselves may be waves—as Nao's great-grandmother Jiko, an anarchist feminist, avers at one point in Nao's diary:

> "Surfer, wave, same thing."
>
> I don't know why I bother. "That's just stupid," I said. "A surfer's a person. A wave is a wave. How can they be the same?"
>
> Jiko looked out across the ocean to where the water met the sky. "A wave is born from deep conditions of the ocean," she said. "A person is born from deep conditions of the world. A person pokes up from the world and rolls along like a wave, until it is time to sink down again. Up, down. Person, wave." (Ozeki 2013, 194)[20]

There is a dizziness in scale and similitude here—how to place the tsunami, the radiation, in time both historical and biographical. The name that eventually attached to the event, 3.11, stood both for the date (March 11) and for the triple disaster (earthquake, tsunami, meltdown). As a made-for-media time stamp, 3.11 also echoed the figure of 9/11 in the United States, which, with its companion vocabulary of "ground zero," bizarrely called back to (and eroded the historical singularity of) Hiroshima—that earlier, and nuclear, ground zero. Even with such time loops in view, I want to refuse any too fast treatment of the horrific events of 3.11 and after as material for psychocultural theory or for drawing or refusing similitudes (or, for that matter, assimilating the 2011 tsunami into a rhetoric that militarizes disaster [see Tierney et al. 2006]). Still, a certain speed in theorizing 3.11 may have its proper place in wave science, which may find it crucial to work quickly to grapple with making worlds, coastal and otherwise, tenable and livable.[21]

Tsunami Model Circulations

The word *tsunami* first appeared around 1612 (Cartwright and Nakamura 2009). Though *tsunami* as "harbor wave" may have been coined by fishers focused on wave effects in their local harbors (which may have appeared as tides—hence, the sometime translation as "tidal wave"), tsunamis came to be defined as fast, long-wavelength waves generated by sudden displacements of water caused by such forces as undersea earthquakes and landslides. The term began to move from Japanese into the world's physical oceanographic lexicon in 1886 when the writer Eliza Ruhamah Scidmore introduced it in *National Geographic Magazine*. Over the years, tsunamis came to name—and symbolically suggest—overwhelming and sudden oceanic forces. If wind waves, as Rachel Carson (1951, 115) once had it, are "crying a warning" of nearing storms, tsunamis bellow disastrous futures with near-immediate speed.

Japanese tsunami researchers point to 1896 as the opening moment of their field, when a tsunami that killed 22,000 people was traced definitively to an originating undersea earthquake (Shuto and Fujima 2009) (see plate 12). Earlier tsunamis had been recorded over the centuries, in bureaucratic documents and in cautionary storybooks for children, but this was the first for which tide records showed a long wave period, consistent with fault motion.

Tsunamis in 1933, 1946, 1960, 1983, and 1993 prompted not only the relocation of houses to higher ground, the building of dikes and seawalls, the creation of evacuation protocols, the making of buffer zones, but also a good deal of scientific work led by such figures as the oceanographer Kinjuro Kajiura (1963), who fused the study of historical records with mathematical modeling. More recently, the oceanographer Nobuo Shuto has compiled a store of historical—newspaper and photographic—sources, seeking to bring them under a scientific optic and treating them as data that suggest inverse problems to be solved.[22]

Harry Yeh has spent a career studying tsunamis. Born in Japan in 1950 to a Japanese mother and a Chinese father, he finished a bachelor's degree in economics at Keio University in 1972 and, after moving to the United States, a bachelor's degree in agricultural engineering at Washington State University. After completing a master's in civil engineering from the same school, he took up a job as a hydraulic engineer at Bechtel Engineering. He told me, "I like the fundamental approach, findings and explanations based upon real physics, findings that can be applied. And if I can make a simple quantitative model, that gives me pleasure. But I want to feel like the things I'm doing have some impact, too." After earning a doctorate at Berkeley in hydrodynamics in 1983, he turned his attention to tsunamis, conducting his first field research in 1992, in the aftermath of a large wave in Nicaragua. Subsequent work took him to Indonesia, Papua New Guinea, Peru, and, in 2011, back to Japan. Yeh's understanding of tsunamis emerges from a network of comparison (see also Santiago-Fandiño et al. 2015).

Having talked with Yeh in Oregon in 2015, I caught up with him again in 2019, during a sabbatical year he was spending at the University of Tokyo. We chatted over Skype. With his permission, I recorded the conversation using transcription software—which, strangely, as an artificial intelligence-enabled product, kept rendering *tsunami* as *zombie*. (When *wave front only* was transcoded as *Trump army*, I wondered whether the web chatter on which the software was trained was delivering a premonition.) We discussed a paper on the 2011 Tōhoku tsunami on which Yeh was the lead author. It reported on in-the-field measurements—with global positioning system (GPS) buoys and sea-bottom pressure transducers—of the tsunami as it traveled to the coast of Japan, measurements that showed "a small leading depression wave, followed by a gradual increase, and then a rapid rise to reach a 6.7 m wave height" (Yeh et al. 2011, 133). The gradual rise, Yeh and his colleagues wrote, "could have confused some people and delayed evacuation, especially because a similar gradual rise in sea level had been experienced in the same region one year earlier, when the 2010 Maule Chilean tsunami reached the Japanese coast" (138). Once the tsunami hit the shore, it was also subject to dispersion, meaning that "the tsunami could not maintain its solitary wave identity." It became a chaotic fleet of churning waves. Time—for lab tsunamis an abstract metric that can be compressed—here uncoils into real tsunami time, a time, especially as experienced by people on the ground, that does not flow separately from the wave but is convolved, con-fused with it.

Yeh told me that the Indian Ocean tsunami had marked a major transformation in tsunami science: here, for the first time, were stores of video records of tsunami events. Yeh already knew, from his site visits, that "each tsunami is different," but now it was possible to see footage of different patterns of flooding depending on topography. He approached these videos with a mixture of trepidation and interest. He recalled his visit to Nicaragua: "I had never seen such suffering. It really gave meaning to the work I was doing. . . . Everyone who works with tsunamis ought to visit these sites—it's a *big* difference. No matter how high-resolution the video, it is different in reality." In the aftermath of 3.11, many Japanese officials—particularly those in the nuclear industry—had been taken to task for saying the event had been *sōteigai*, "beyond expectations" (Kimura 2016). Critics went after this claim, pointing to incompetent safety planning and incomplete understandings of how seawalls might work not just to block seawater but to amplify its energy before overtopping. Yeh suggested that, after 3.11, any simulation that did not account for *that* feature of seawalls—that did not imagine them in its 3D movie of a scale-model tsunami—would be inadequate.

Thinking back to the tsunami basin at Hinsdale as a kind of movie theater, a site where moving 3D animations of wave theories unfolded as a kind of cinema, and prompted by Yeh's mention of tsunami video, I revisited video of the 2011 tsunami that had been posted on YouTube, video I had never been sure how or even whether to approach. I turned to a passage from Ozeki's *A Tale for the Time Being*:

> In the days following the earthquake and tsunami, Ruth sat in front of her computer screen, trawling the Internet for news of friends and family. Within days, she received confirmation that the people she knew were safe, but she couldn't stop watching. The images pouring in from Japan mesmerized her. Every few hours, another horrifying piece of footage would break, and she would play it over and over, studying the wave as it surged over the tops of seawalls, carrying ships down city streets. . . .
>
> Most of the footage was shot by panicked people on their mobile phones from hillsides or the roofs of tall buildings, so there was a haphazard quality to the images, as if the photographers didn't quite realize what they were filming, but they knew it was critical, and so they turned on their phones and held them up to the oncoming wave. Sometimes an image would suddenly blur and distort as the photographer fled to

higher ground. Sometimes, in the corners and the edges of the frame, tiny cars and people were caught fleeing from the oncoming wall of black water. (Ozeki 2013, 112–13)

Such videos present a different genre of waves as movies from those I saw in the Hinsdale lab.[23] They offer embodied, first-person vantage points, often pointing back to a cameraperson caught between standing still to document and bolting to escape. The rawness of the video transmits a sense of "being there," a presence that does not narrate, but offers mobile, shifting experience. The videos are disorienting to watch since to let them unspool is to witness an awful unfolding of an event that is already over, a virtualized image "of a present that peels away from itself, producing in the moment, now, a memory already" (Lippit 2015, 14). Their existence points to likelihoods that the videographers survived. But their existence also testifies to the reality that many did not make it. In Saeki Kazumi's 2012 short story "Hiyoriyama," set just after 3.11, one character speaks to the view from the ground:

> "You know, you probably think that a tsunami comes down on you from above, don't you?"
>
> I nodded, recalling a scene from a surfing film, where the crest of the wave, looking like a shark's mouth full of sharp teeth, was engulfing a person from above.
>
> "That's not what happens. It comes right after you while you're running away from it, licking your heels, then pulls the ground beneath your feet away. You get knocked off balance, you fall backwards, then the next surge catches you." (Quoted in DiNitto 2014, 352)

Nearly twenty thousand people died and some 2,500 went missing in the 2011 tsunami. In *Ghosts of the Tsunami*, Richard Lloyd Parry (2017, 106) reports on survivors describing "sightings of ghostly strangers, friends and neighbors, and dead loved ones. They reported hauntings at home, at work, in offices and public places, on the beaches, and in the ruined towns." This is not about scale but about how the signifying practices of everyday life break—become disordered—making interpretation grapple at once with an absence and excess of representation. Images of the tsunami become chronicles of mourning, in which evidence is woven into "a tangled temporality in which various unrepresentable 'losses' are repeatedly reawakened, or relived and (re)integrated into the present" (Hayashi 2015, 173).[24] The break becomes the wake.

If viewers in retrospect now see plot and story in the grainy videos that survived, videographers at the time did not, even as their on-the-run attempts to offer close-ups, pans, and landscapes were also formed by visual conventions built into their phones and, likely, their own filmic intuitions. In aggregate, the multiplication and sharing on social media of these videos—an unplanned crowdsourcing—has by now been leveraged into narrative, sometimes activating translocal intimacy, imagined community, and cosmopolitan empathy (as with tweets hashtagged #prayforjapan [Wilensky 2012]). If "the photographer in citizen images emerges as an embodied collectivity, as a figure inviting herself to be imagined as 'anyone'" (Pantti 2013, 210), such photographers have also been—in eventually curated compilations, many of which excised images of dead bodies—scripted into a Japanese national narrative of collective tragedy (and even, unstably, resilience). In many cases, especially when brought into journalistic frames, the videos became anonymous, "*captured by a camera phone*" (Pantti 2013, 210). Together, they became an archive of loss and alarm. Whether their warning was one about coastal risk or the politics of nuclear irresponsibility depended on whether those later video compilations separated the "natural" disaster of the tsunami from the "human-made" disaster of the nuclear meltdown, a move that depoliticized the infrastructural catastrophe that was Fukushima.[25]

And whether such videos produce intimate community or activate trauma, denial, dissociation, or victim blaming depends on how they are accessed, aggregated, and framed and by whom. Anthony McCosker (2013), writing about disaster media in the wake of the Indian Ocean tsunami in 2004 and Hurricane Katrina in 2005, argues that the way footage of these disasters was selected by dominant news outlets was shaped by moral judgments about what kinds of nations and people might be "deserving" and "undeserving" of pity and empathy, judgments cross-hatched by the politics of international aid, and, in the case of Katrina, white racist victim-blaming amnesia about how predominantly Black neighborhoods had been persistently disenfranchised in coastal infrastructure planning. The "similitude" of social worlds in disaster videos to the communities of those outside who are watching is not given but created. Japan's status as a hypermodern technological power makes it a potent double for those who would worry about the American West Coast. Bonnie Henderson, in *The Next Tsunami* (2014), the book I purchased in Seaside, follows geologists in Oregon who look to 3.11 as a wake-up call about a possible Cascadia quake.

She writes about an Oregonian geologist watching online video while sitting in his home in Seaside:

> He and the rest of the world had seen the shaky images tourists caught of the Sumatra tsunami back in 2004, but here was a tsunami in a developed country where it seemed everyone had a smart phone capable of video. . . . It was astonishing and horrifying: houses exploding, houses floating. . . . Fishing boats sailing down what had been city streets, boats and minivans pouring over seawalls that had become nothing more than grey waterfalls. . . . Horrifying for those caught in the maelstrom there. And edifying—or should be, Tom hoped—for anyone paying attention in the Pacific Northwest. Cities built upon a flat coastal plain, like Seaside. . . . A fault zone right offshore, where two tectonic plates collided. Just like Seaside. (B. Henderson 2014, 280)

The snapping into similitude of Japanese and Pacific Northwest cases is brokered by a parallel between local undersea fault lines, but also by an analogy between "developed" countries, with similar markers of technological infrastructure—seawalls, minivans, smartphones.

Videos of 3.11 have become a heterogeneous archive: a repository for national memory, a store for documentarians, a reservoir for calls for Japanese citizens to come together, a cache of images for protest. In the year after 3.11, phone videos even became sources for scientists to reconstruct some of the tsunami's hydrodynamics (Fritz et al. 2012). I learned from both Cox and Yeh that the footage of two eyewitness video files, for example—shot from different rooftops in the port of Kesennuma—had become data for wave science, permitting the reconstruction of tsunami flow velocity fields, a mode of using media for forensic wave science, for solving a real-world inverse problem. Reading the past to forestall an imagined future. Here, the circulation of video promised something in addition to digital intimacy and shared trauma: future preparedness. As David Novak (2013, 17) writes, "Circulation is not just movement and exchange, but performance and process." What was being performed here was the promise of a scientific ordering of chaos, an urgent grappling with how to theorize massive destruction as quickly as possible, to turn experience into data. As the historian of Japan Richard Samuels (2013, x) writes, "What had been a discussion about the *slow* devastation wrought by debilitating demographic and economic forces now became a pressing set of demands about how to respond to *sudden* devastation." Those demands had a future-facing cast in the work of scientists such as Yeh.

What does that future look like? As the anthropologist Michael Fisch (2021) has documented, post-3.11 construction in Japan has seen the making of ever-more massive seawalls, the construction of an extreme infrastructure delivering a "fortress-ification" (*yōsaika*) not keyed to the scale of the day-to-day:

> The newly constructed seawalls are concrete behemoths. Dominating the shoreline, they rise as high as 15 meters in some places, with a base stretch as wide as 80 meters across. . . . One of the most common reactions one hears from residents about the new seawalls is that they are massive, overwhelming, and alienating. . . . Indeed, standing behind the wall, it is hard to image that a vast ocean lies only a few meters away. . . . The air is dead still and silent, with neither ocean wind nor the sound of waves able to penetrate the two-meter-thick concrete.[26]

The massive order of these seawalls is tuned to scales of temporal reasoning out of alignment with the tempo of everyday life, "in sync rather with global finance, actuary practices, and Japan's cement corporations."[27] They are materialized derangements of scale, calibrated only to the most dramatic possible wave. As the anthropologist Shuhei Kimura (2016) reports, however, there is local resistance to such walls; there are calls to lessen their heights and to create boundary markers—stands, perhaps, of cherry trees, doubling as memorials—that would leave the ocean visible to fishers and others; that would keep wave memory alive rather than attempting to obliterate it. At stake in work to protect coastlines is how to navigate between the concrete and abstract, how to think outside the boxes of received scales.

Set Three

First Wave

Massive Movie Waves

In "Wave Theory," published in a 1996 issue of the movie magazine *Sight and Sound,* David Pirie suggested that Hollywood genres rise and fall in waves. Adapting theories of crowd psychology from the accountant Ralph Nelson Elliott, who in 1938 proposed a model of investor behavior called the "Elliott Wave Principle," Pirie observed that the genre of the disaster movie, which had fallen into a lull since its 1970s crest, was coming back as film companies in the 1990s financed a new cycle of disaster movies that banked on new special effects techniques and a surge of millennial anxiety.

I here spin off from Pirie's claim in a literal-minded way, looking at changing representations of waves themselves in disaster and science fiction movies, mostly, though not exclusively, in American and European settings, reading them for what they tell us about mainstream ideological accounts of human-ocean relations. Movies featuring colossal waves have realized such forms through a range of effects, from the use of scale models to photographic trickery to computer-generated imagery. Movie waves are what the film theorist Kristen Whissel (2014, 6) would call an "effects emblem": "a cinematic visual effect that operates as a site of intense

FIGURE 3A.1 Sequence from Étienne Jules Marey, *La Vague* (1891).

signification and gives stunning (and sometimes) allegorical expression to a film's key themes, anxieties, and conceptual obsessions." The dramatic roles filmic waves have fulfilled have been multiple. Towering waves in film have operated as emblems of (1) the elemental power of inhuman, arbitrary forces; (2) the return of the social-environmental repressed, with waves moral messengers paying humanity back for sins against orders of nature or social justice; and (3) the fantastic power of cinematic media themselves.

Étienne Jules Marey, the nineteenth-century physiologist and originator of "chronophotography," who is famous for his twelve-frames-per-second studies of the movements of animals, made *La Vague* in 1891 (figure 3a.1), an under-a-minute film of a wave crashing into rocks in the Bay of Naples. Marey's purpose was documentary, extending experiments realized by Eadweard Muybridge in Palo Alto in the 1870s, in which Muybridge famously generated stills taken in quick succession of the movement of horses. After Marey and Muybridge, it became imaginable to slice time up into framed visualizations and to then suture these together in linear space—laying them out along segmented spans of nitrate, celluloid, and other film stock. It became possible to watch things over and over again; to slow, pause, reverse (see Crary 2001, 140).

As the philosopher Henri Bergson ([1934] 1946, 4) noted, in chronophotography "the terms that designate time are borrowed from the language of space." The result, for Bergson ([1907] 1911, 332), was that, "instead of attaching ourselves to the inner becoming of things, we place ourselves outside them in order to recompose their becoming artificially." The wave is portioned into instants and reassembled in the space-time of the viewer, who comes to see the wave-movement-image as at once an empirical record and a lively phenomenon in itself.[1]

In *Cinema 1: The Movement-Image* ([1983] 1986) and *Cinema 2: The Time-Image* ([1985] 1989), Gilles Deleuze draws on Bergson, and, reading films from the twentieth century, complicates this view, arguing that with film movement comes to be inscribed *within* the still. Cinema waves, I suggest, are movement images *and* time images par excellence, for their pitched present always points to their immediate past (their swelling arrival) and their imminent future (their about-to-breakness). More, the time they gather up is not simply linear, unfolding time; nor is it only the subjective time of a viewer. It is also a time of fantasy, of the "unthought, the unsummonable, the inexplicable, the undecidable, the incommensurable" (Deleuze [1985] 1989, 214). Techniques of computer-generated imagery (CGI) extend the parameters of fantasy imaging (and have jumped away from the physicality of film) but also largely hew to what Stephen Prince (1996) calls "perceptual realism," even as they have amplified the possibility of creating impossible points of view, human and nonhuman, as well as camera-eye zooms across sweeps of scales.

Consider a canonical disaster movie, *Tidal Wave*, from 1975, featuring an out-of-scale catastrophe confronted by people struggling against or submitting to their fate (see Casper 2011; Keane 2006). Its poster freezes a movement-image that speaks of momentum, threat, and danger (figure 3a.2). *Tidal Wave* belongs to a lineage of films that pose waves as massive unthinking intensities that test human hardiness or that demand that people band together across social divisions thrown into disarray by disaster.[2] Close on the heels of *Tidal Wave* came *Meteor*, in 1979. In both films, the challenge for filmmakers was to create a large wave that looked realistic, and in each case, the challenge was met with a combination of splashings of water onto actors, scale models of inundated cityscapes, and double exposures that superimposed actors against onrushing waves. The use of real water in scaled-down city models came widely to be regarded as looking fake (see Newman 1977). It turns out that the molecular structure of water and the force of gravity are not themselves scalable. Trying to correct for this by slowing down the action, as Hollywood films have, cannot work well since not all the processes in play scale at the same rate. What becomes jarring about wave images such as those from *Meteor* is not only their material and spatial clumsiness, but also a sense that human time and wave time

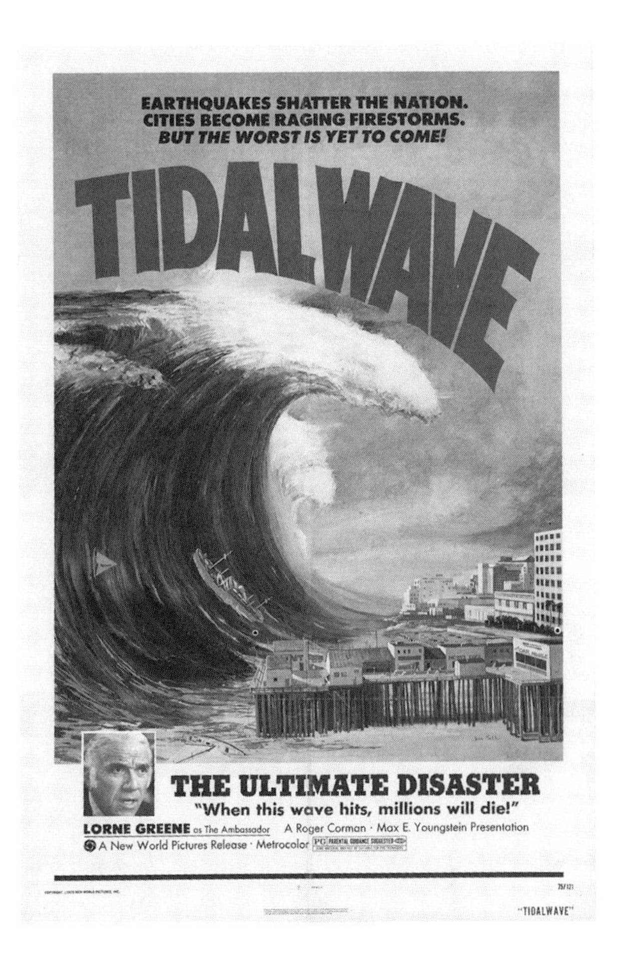

FIGURE 3A.2 Poster for Andrew Meyer, dir., *Tidal Wave* (New World Pictures, 1975).

are not calibrated. For a large wave to be convincing, it must exist, with humans, in the same filmic and narrative time.

Wrangling with scale in depicting monstrous ocean waves has been aided by computer graphics, which permit filmmakers to fuse filmic and digital elements (but see Davis 2014). Filmmakers can avail themselves of, say, physics simulations that render water surfaces with mathematical exactitude, but they may also choose to leave some properties of material water behind, to indulge their imaginations of what a wave *could* look like (see Sobchack 2000). The virtual can be layered onto/into the actual (see Hansen 2004).

Take as an example of the unreasonably immense wave the mammoth comet-impact-generated wave that strikes New York City in *Deep Impact* (1998). In one widely reproduced still from the movie (figure 3a.3), the wave is represented as at once a brute fact, a terrifying and unyielding character,

FIGURE 3A.3 Still from Mimi Leder, dir., *Deep Impact* (Paramount Pictures, 1998).

and a force of action, joining what Deleuze would call the perception-, affect-, and action-image. The effect is fantastical, though inserted into a regime of perceptual realism, with the computer-animated wave given lighting, coloring, and timing coordinates that assimilate it to the scene in which it appears. In the face of this onrushing force, humans in the film mostly run away or, in moments of resolve or surrender, submit to its immensity. These scenes govern the speed of the wave, since the wave must move in sync with narrative elements—disbelief on people's faces, tender goodbyes, last-minute heroics.

What to make of the gargantuan wave of the movie *Interstellar* (2014), which appears when astronauts arrive on an ocean planet and realize that that world hosts periodic four thousand-foot waves? The wave in the film, as the effects supervisor Paul Franklin has reported, was

> a combination of very, very detailed, rigorous simulation of what water should actually do using physics simulations to work out how do the waves splash and sweep the spacecraft away, and how does the water surge over the hull of the spacecraft. And that was a quite tricky thing to do, was to work out what would be the surface details that would tell you this thing is 4,000 feet high. And of course, this all then had to be integrated seamlessly into the photography that we'd done on the location.[3]

Franklin concluded: "We actually treated the waves as though they were animated characters" (quoted in Hawkes 2014).

In the disaster film *San Andreas* (2015), which features an enormous wave coming to wipe out California, cinema waves deliver yet another trick. When the wave arrives at San Francisco's Chinatown with a large cruise ship on its crest, it plunges the "camera" underwater (see Brinkema 2014; see also Jones 2007; Pierson 2015).[4] Though there is a fantastical perceptual realism at play, it is clear that we have left the indexical promise of old-fashioned film. These cannot be images that are "direct and isomorphic transcriptions of a moment that existed concurrently with the camera" (Hadjioannou 2008; see also Mangolte 2003). Digital cinema no longer offers a *record* of time; instead, it offers a *simulation* of time. The fluid transition across air and underwater is an illusion, one from the age of digital cinema, "defined by spatial and temporal continuity and by a rejection of the cut" (Brown 2013, 9). More: "the relationship between animation and film has been reversed following the advent of digital technology, such that if animation was once a subset of cinema, live action analog film is now the subset" (Brown 2013, 10, paraphrasing Manovich 2001, 302).

Filmic re-creations of real, historical tsunamis, as in *Hereafter* (2010) and *The Impossible* (2012), both of which tell tales of people surviving the Indian Ocean tsunami of 2004 and seek a certain verisimilitude, retain an accounting of waves as inhuman energies but script their tales of human resilience by placing individual persons at the center of their stories. In each of these Euro-American productions, these are not just any people: they are white European families caught up in a disaster that, even as it is arbitrary, is coded in its outlines as that which more usually befalls the struggling thousands of the third world. European families, their lot thrown in with local bodies, survive only by luck and wits meant to be coded as universal, even as their narrative work is accomplished by staging them as bodies out of their usual place and time.[5]

Bringing humans and gargantuan waves into the same time frame offers explicitly humanist—rationalist, emotional, ethical—narrative opportunities, and a good portion of the waves of disaster film are explicit moral devices; waves become characters that carry messages to humanity. Often, spectacular destruction is visited on cities, sorts of hyperculture inundated and undone by wave hypernature.

Waves as a narrative climax for stories of human foolishness also appear at geopolitical scale. A canonical movie is *The Day after Tomorrow* (2004), which features a wave created by climate change that hits New York. The wave in the film is a mix of the realistic and fantastical, an impossible sublime. Viewers are pressed to see their todays and tomorrows as moments of suspension before a fully realized, climate change–induced flooding Anthropocene (cf. Fay 2018). The wave of *The Day after Tomorrow* is *anthropoceanic* (see Brugidou and Fabien 2018). It has the sins of humanity inside of it—and not just any humanity, but the humanity of the global industrialized North, as, at the end of the film, the president of the United States confesses that climate change has come about because of overconsumption from overdeveloped countries, which must now, humbled, turn to the (less hard hit) Global South for support (see Keane 2006; Rust 2013). The wave fits in the category of "effects emblems . . . deployed to give a spectacular representation to scenarios and events that strongly imply, or threaten to bring about, 'the End'": "the end of freedom, the end of a civilization, the end of an era, of even the end of human time altogether" (Whissel 2014, 18, 60). The wave of *The Day after Tomorrow* is also a force that is ground clearing, the beginning of a new era ("The sea can wash away all evils of humankind," said Euripides [see Patton 2007]), though a narcissistically Northern one, since, in the real world, low-lying lands of the Southern Hemisphere are now perhaps most dramatically feeling the waves of rising sea level.

Resonances with Noah's flood are thick. We might read the form of the disaster in *San Andreas*, in which cruise ships flip upward, container ships upend bridges—as shorthand for human transgression, the ocean punishing sea-borne globalization. The becoming vertical of the social world is key. Whissel (2014, 30) writes: "As an effects emblem, verticality creates an interpretative framework for . . . rising and falling bodies and matter in these films that goes well beyond their reality effect: by mapping complex struggles for power onto the laws of physics, verticality can make historical change a matter of inertia or inevitability." She continues, "Gravity . . . acts as a historical corrective" (32), as it does in *Titanic*, when the order of things—rich at the top; poor at the bottom—becomes subject to the leveling force of physical law. If the world of film gives us narratives that are "the recognition of a temporal-sequential or 'horizontal' connection across shots or scenes" (Ivakhiv 2013, 89), the verticality created by large waves signals a world turned sideways.

Such moral messaging has precedents. Take *The Last Wave* (1977), a film by Peter Weir about a white lawyer contracted to defend four Aboriginal men in Australian court (figure 3a.4). As the lawyer is drawn into the world of his clients, he is haunted by visions of—and, perhaps, sees, in

FIGURE 3A.4 Poster for Peter Weir, dir., *The Last Wave* (World Northal, 1977).

a crossover to the Dreamtime—a tremendous wave coming to wipe out Australia. The wave symbolizes not only the power of nature to destroy human enterprise, but also the disaster of colonial dispossession and the return to the colonial power of the repressed.

Cinematic play with time has motivated reflexive representations of waves for a while. The Ur-text is Jean Epstein's film *Le Tempestaire* (1947). In this drama, a folkloric figure—a *tempestaire* (storm tamer)—is able to control a gale arriving at a coastal village in Brittany. We see this old man cradling a crystal ball that holds an image of a bay tossed by waves (figures 3a.5a–b). He blows at/into the ball, and the film cuts to a scene of waves slowing and, eventually, reversing (see Keller and Paul 2012). The old man is, of course, cinema itself—viewing from far, from up close, speeding up, slowing down, reversing.

FIGURES 3A.5A AND 3A.5B Stills from Jean Epstein, dir., *Le Tempestaire* (France Illustration, 1947).

This effect has a digital descendant in the film *The Abyss* (1989). In it, extraterrestrial visitors to Earth capable of sculpting water (and whose spaceship lands and lurks underwater) manipulate large ocean waves to threaten humanity. The aliens eventually call the waves off, unrolling them backward, like the *tempestaire*. Like their wavy ancestor in Marey's *La Vague*, they run forward and back because of the phenomenotechnique of

FIGURE 3A.6 Still from Ron Clements and John Musker, dirs., *Moana* (Walt Disney Motion Pictures, 2016).

frame-by-frame film, even as the waves themselves are the result of computer animation. These waves emblematize the rise of computer graphics, coming to sweep the techniques of older cinema away, along with, perhaps, "the human" as the only author of film. In *The Abyss*, humans are viewed by aliens as children in want of instruction.

This brings me to children's films, in which the potential disastrousness of waves can be stayed. Consider Hayao Miyazaki's *Ponyo*, the story of a fish who becomes human. In it, waves phase in and out of being living whale- and fishlike things and being active water. They are torrential natural forces that might live alongside humans, if only humans can tune in to the oceanic in the way the fish child of the sea Ponyo does.

In Disney's *Moana* (2016), waves—as animate arms/faces of the ocean—are friends with the child protagonist, hovering over her, cradling her, giving her high fives (figure 3a.6). Waves in *Moana* defy gravity and suspend their own crashing. Moana's youthful wonder places the wave outside of time, history, a placement supported by Moana's representation within a Disneyfied, romanticized South Pacific fantasy space.[6] *Moana*'s wave promises a tuning back into the sea, a kinder, more harmonious anthropocean, anthropomorphized. Think of *Moana*'s wave alongside Kelly Slater's Surf Ranch in Lemoore, California, where artificial waves are generated in an inland pool so that surfers can surf a "perfect" wave each time (Minsberg 2017).[7] Like *Moana*, the wave pool can be read as a container of childlike optimism, a swerve away from the all-too-human damage that now saturates the sea, a place to play a game of waves.

Video games provide a companion pedigree for computer-generated waves (Alarcon 2019)—from Nintendo's *Wave Race 64* to today's open world survival game *Subnautica*, which animates a wave description first formulated in the early nineteenth century (Gerstner 1804)—and link forward to waves made by artificial intelligence (AI) waves. The computer artist Francisco Alarcon has proposed a machine-learning program that can generate digital waves—animations whose realist verisimilitude might improve as the program is trained on movies of waves (Alarcon and Helmreich 2020). Alarcon observes that, as such an AI network "learns," it would, like many pattern-recognition algorithms, use more electricity. One might employ the magnitude of that energy usage as input to grow the height of virtual waves, creating a work that could become an allegorical enactment of the electro-anthropogenic driving of increased wave height in world oceans. The fact that Alarcon does not plan actually to create this piece does not mean its logics are not already enacted by digital effects companies. Artificial intelligence waves, coming soon to a screen near you, leave in their wake the breaking news, the carbon imprint, of unspooling disaster, the reeling horizon of oceanic sight and sound now.

Set Three
Second Wave

Hokusai Now

Katsushika Hokusai's woodblock print *Under the Wave off Kanagawa* (1829) is the world's most iconic portrait of an ocean wave. It has been reproduced, quoted, and repurposed over the past two centuries in a widening circle of representations of the unruly, powerful sea. Today's reimaginings of this storied Japanese image often comment on the damaged state of the ocean. In the early 2010s, some re-versions referred directly to the Tōhoku tsunami and Fukushima nuclear plant disaster of 2011.[1] Artistic and graphic adaptations have come increasingly to operate as a synecdoche for anxieties about catastrophic climate change and ocean pollution, acidification, and plastification. Such reimaginings of the Great Wave do so, significantly, by drawing attention to the materials of which such critical artworks are nowadays frequently made: plastic, trash, and other sea-borne detritus, the flotsam and jetsam of a sea damaged by the thalasso-geo-historical practices of (some) humans.

In *Art in the Anthropocene: Encounters among Aesthetics, Politics, Environments, and Epistemologies*, Heather Davis and Etienne Turpin (2015, 3) argue that "the Anthropocene is primarily a sensorial phenomenon: the experience of living in an increasingly diminished and toxic world." Contemporary revisions of Hokusai's wave seek to be sensorially unsettling precisely by bringing into visibility that which might otherwise be beyond everyday view (e.g., islands of swirling trash in the middle of the Pacific). Such works, often incorporating garbage as a medium, call into question the generalizing and ahistorical character of the Anthropocene as a "charismatic mega-concept" meant to reframe an entire geological epoch (6). Ocean trash, after all, is not so much about the long history of *Homo sapiens* as it is about a set of particular social practices that have produced this waste: commodity production and consumption. These swirls of refuse are less a sign of the Anthropocene than of the Capitalocene or Plasticene (Haram et al. 2020). Or perhaps they are a sign of what Jussi Parikka (2014) calls the Anthr*obscene*, the toxic material accompaniment of computational, tablet, and smartphone media culture, which, far from ushering the contemporary world into a "paperless" ecotopian sublime, fills the world with poisons consequent on producing, consuming, and discarding the devices that permit (some) people to conjure images of the globe at all.

The art historian Christine Guth's *Hokusai's Great Wave: Biography of a Global Icon* (2015) is the essential account of the worldwide travel of the wave image. Guth examines the mutating social life of Hokusai's print, from its production in late Edo Period Japan to its emergence as a French, Euro-American, and then global signifier of all things Japanese; from its work in Japanese nationalism to its play in capitalism; from its status as shorthand for international high art to its manifestation as a lifestyle brand; and from its early conscription into classical Western aesthetics of the sublime to its increasing association with looming environmental disaster. Artists today are rematerializing Hokusai's wave to call attention to the beleaguered life of the sea in the age of a toxic modernity. The wave break is a symbol for a broken ocean.

European and American assessments of "Under the Wave" often tag the image as quintessentially Japanese, an example of the "floating world" style of the Edo Period, a piece that represents waved nature as *yin* and the confident (if imperiled) fishermen in their boat as *yang*. Locating the print within wider political economic matrices, however, tells a less insular, less Orientalist, story. The first receptions of Hokusai's wave keyed to a transitional moment in Japanese history. Where once waves symbolized the divine agency of the sea as a substance protecting the perimeter of Japan, in the early nineteenth century, as Russia, the United States, and England angled for access to Japanese ports, the wave came to suggest forces of foreign incursion and influence. The print itself, a view of Japan from outside (Mount Fuji marks the place), was caught within these forces, quoting what would have been recognizable in its time as Dutch pictorial practice: a horizontal composition, a prominent featuring of peasant fishermen, and a demand that the picture be read left to right, rather than up-down and right-left.

The blueness of the wave crystallized translocal connection, too; prints were made using not the usual indigo but a newly fashionable synthetic dye called Berlin Blue, imported to Japan from Germany via China. This blue, Guth (2015, 29) suggests, "materialized the relation between Japan and the world beyond its shores, making the medium part of the message. Hokusai's 'Under the Wave off Kanagawa' thus participated in a discourse in which the beholder did not simply imagine China and Europe but experienced them bodily." This wave sat symbolically, visuo-chemically, in a moment of nineteenth-century globalization.

Hokusai's wave then, in turn, traveled transnationally. Its aesthetic made its way into Europe, where Claude Monet, who became a fervent collector of Hokusai's prints, created his painting *The Green Wave* (1865) in homage. Monet, like many others in the tradition of wave painting, was fascinated by how the materiality of different sorts of paint, properly deployed, might summon the materiality of water waves. That concern with materiality has been recently amplified in a moment when the substance of the sea is itself in jeopardy. Building on the arguments of the art historian Irmgard Emmelhainz (2015), who has offered that the Anthropocene is too overwhelming to image whole and that therefore "the Anthropocene era implies not a new image of the world, but the transformation of the world into images," such images now increasingly draw material *stuff* into their ambit.

This transport of Hokusai's wave into a conversation about viscerality sets up the life that Hokusai's wave has in today's catastrophe-conscious world. In the newest rescriptings of the Great Wave, there is less of the contemplative and more of the calamitous. The wave now jeopardizes not only the small fishing boat in the image but also a wider translocal polity. It operates as a harbinger of danger to/for the viewer, who is addressed as a person caught up in a world in which catastrophe is globalized. This leverages earlier habits of viewing toward new purposes; Guth (2015, 42) writes, "The vantage point adopted in 'Under the Wave off Kanagawa' erases the boundaries between subject and object, transforming the viewer into a participant in this watery drama." The participant becomes the environmentally freaked-out citizen.

A version that appeared in a sustainable design magazine is diagnostic; it shows a biospheric city floating on a turbulent sea, about to be swamped by a wave.[2] Mount Fuji is replaced by a fragile urban utopia cradled in a globe-like bubble carried on the foam of a modernity turning against itself.[3] As Guth (2015, 137) suggests, "The elimination of Mount Fuji, a feature of many of its commodified articulations, has helped the image migrate by unmooring it from Japan." The substitution of a globe for Fuji makes it speak to planetary matters.

Many usages zoom in on the trashy texture of today's ocean and do so by using trash from the sea itself. In such representations, the *medium* becomes crucial to the message. Media, of course, have long been central to representations of the ocean in motion. Painters such as J. M. W. Turner and Winslow Homer are famous for attempts to render the sea with oils and watercolors, with various canvases and watercolor papers and techniques such as blotting, scraping, and rewetting (see Athens, Ruud, and Tedeschi 2017).[4] But those techniques and media—paint, paper—have not generally been understood as *participating* in the substance of sea; they are one-step-removed representations. Contemporary wave art, particularly that which comments on the increasing load of garbage in the sea, now seeks to incorporate the material detritus of large-scale consumption. As Guth (2015, 207) suggests, Hokusai's wave is easily adapted to a "growing focus on the real world operations of nature—volcanic eruptions, typhoons, floods, and earthquakes," perhaps particularly because so many "have occurred in the Pacific region" (and see Jones et al. 2022 on disaster memes that riff on Hokusai's image).

And, of course, more than nature: the detritus of culture—trash.[5] Bonnie Monteleone, of the Plastic Marine Debris Lab at the University of North Carolina, Wilmington, "took the plastics she had gathered in each ocean and modeled them on the well-known woodblock print, 'The Great Wave,' by Japanese artist Hokusai done around 1830. She wanted to demonstrate how altered the oceans are now, compared with less than 200 years ago. . . . Monteleone wants the art exhibit to go from coast to coast, much like the AIDS Memorial Quilt that went around the country" (Hance 2014). If the Berlin Blue of Hokusai's 1829 woodcut materially indexed an earlier moment of globalization, this 2011 version (see plate 13), made of trash, concretizes the garbage globalization of our day. If imperial Europe used paint to summon subjective impressions of a sublime sea, meditative and malignant, work now employs the debris of the sea, underscoring the entanglement of representations of the sea with the very substance of oceans themselves (see Davis 2022, 3, on artist Tejal Shah's installation "Between the Waves," which features people/performers that "lie in the sand, with waves passing over them, entangled with all kinds of debris, including Styrofoam and plastic-coated wires"). If, as the anticolonial geographer Max Liboiron (2021) has argued, *pollution is colonialism*—that is, that generating trash and putting it *elsewhere* depends on territorial dispossession—trash waves can be read as signs of how colonial enterprise has irretrievably spilled and smeared into the sullied sea.

Guth (2015, 7) writes that waves are "resonant signifiers of both the deterritorializing effects of globalization and the hybridity resulting from the mixing of their waters." No surprise that Hokusai's iconic image, then, appeared immediately in journalistic pieces about COVID-19 and its waves, its foam sometimes rendered as a lather of red coronaviruses, its suspended arc a warning of surges to come. And with biomedical waste nowadays washing up on the shores of the pandemic-age ocean (Probyn 2021), the newest waves of trash come in face-mask blue.

Set Three
Third Wave

Blood, Waves

The narrator of Italo Calvino's "Blood, Sea" ([1967] 1969), Qfwfq, is a volume of blood—possibly a cell, a drop, the quantity encapsulated in a single human body, or maybe all blood everywhere, ever. Qfwfq's biography stretches back to a time "when life had not yet emerged from the oceans" (39), and Qfwfq spends the span of the narrative doing two things: remembering a past of flowing freely within a salty sea and making sense of a present-day circumstance in which Qfwfq finds itself swirling through the veins of a passenger in a Volkswagen automobile that is speeding along a snaky road in northern Italy. Calvino's story pulls the reader into a tale of primal and ongoing communion between the substances of blood and seawater, tracking what the author, in an italicized scene setting, names the "primordial wave" traveling from Earth's oceanic past into today's human bloodstreams. A portion of this splashes at the end of "Blood, Sea" onto the metal of Qfwfq's Volkswagen container after it swerves to avoid an oncoming Jaguar and crashes.

The religious studies scholar Gil Anidjar (2011) has dissected Calvino's "Blood, Sea" for the way it veers around the politics of blood. Far from unifying all things—let alone all humanity—blood historically has been called into service to create differences and inequalities: by bloodlines, races, sexes, sexualities, health chances, and more. From Patristic Christian fears of menstrual blood to Iberian Catholic notions of the purity of blood lineage; the one-drop rules and blood-quanta policing of African American and Native American identities; the semiotics of sexuality that suffuse HIV-infected plasma; and cross-species experiments in xenotransfusion, the flow of blood has been less like an unbounded journey on an open sea than like a passage through striated rivers and channels.[1]

How, then, do we conjure with Calvino's "primordial wave," his unifying formal abstraction arcing from the ancient to the contemporary? How do we historicize this wave? The history of blood waves might start with William Harvey's call in 1628 to recognize that blood circulates through the body, a call that made human veins look a lot like the channels of exchange being set up in Harvey's day to make money circulate from one part of the political economic body to another. By 1698, the English mercantilist Charles Davenant would describe trade and money as "like blood and serum, which though different juices, yet run through the veins mingled together" (Davenant [1698] 1771, 350). Blood as formal flow, as rise and fall and rise, becomes a managerial abstraction.

These days, cardiologists continue to characterize a variety of waves that travel through human blood. Mayer waves, for example, are low-frequency blood-pressure oscillations in the arteries, speedings or slowings of which may point to conditions of hyper- or hypotension (Julien 2006). Pressure waves of other sorts move through the bloodstream (with three primary orientations: radial, axial, and circumferential), and their velocity and amplitudes can be indicative of cardiovascular health. Drinking coffee turns out to increase the velocity of pressure waves, leading to an increase in "wave reflection"—that is, waves bouncing off arterial walls and returning to, say, the heart's left ventricle, which can suffer an increased workload as a result (see Karatzis et al. 2005). Blood waves sit inside bodies. Far from being primordial, they form inside biographies, histories, and environments.

The same can be said for another sort of wave associated with the movement of blood: the wave described by the beating of the heart, often visualized in the electrocardiogram (EKG), which registers changes over time in the heart's electrical potential and permits the monitoring of individual heart health. Consider one arena within which these representations circulate: networks of implantable cardioverter defibrillators (ICDs), devices implanted in heart patients to monitor and manage their heartbeats. The EKG wave profiles from these devices can be transmitted wirelessly to hospitals' computers, where they can be accessed remotely to track patients' health. Once EKGs arrive at hospitals' websites, automated algorithms and on-call medical technicians sort through them for concerning waveform tracings. Behind the abstraction of a web page full of EKG profiles are individual biographies, structured by the politics of who has heart problems, who has insurance, who gets an ICD, and more. (In 2007, the *Journal of the American Medical Association* reported that "black women were 44% less likely to get an ICD than were white men; white women were 38% less likely to get an ICD than white men, and black men were 27% less likely to get an ICD than white men."[2]) Far from tracking the "primordial wave" moving from the ancient earthly ocean into today's human bodies, these collections of EKG waves track a bloodscape of difference, of health disparities and inequalities, perhaps markers of public health crises and slow, structural violence. Like data about wave heights and profiles gathered by ocean buoys around the world—data transmitted to shoreside computers to put together a picture of global ocean weather—this information about cardiac waves maps out a sea of difference, an ocean of blood burbling inside people and populations with different life chances.

The crash that carries Calvino's Qfwfq from the salty insides of a human body to spilled crimson over the metal of a crumpled Volkswagen is, as Calvino writes at the story's end, "a number in the statistics of accidents over the weekend" (51). Qfwfq, then, can be imagined as less of a singular figure than as a number, a data point in a wave of bloody car accidents, a wave consequent on such historical forms as speed limits, seat belts, blood-alcohol levels, and more. Reckoning with *blood sea* as a substance requires us to account for the forms that blood waves take, forms less primordial than historical and more social than primal.

Chapter Four

World Wide Waves, *In Silico*

Computer Memory, Ocean Memory, and Version Control in the Global Data Stack

HOUSED IN THE CURVING ARC of a silvery white four-story building, lined with thin bands of windows running along the length of its façade, is the Center for Weather and Climate Prediction, an arm of the US National Oceanic and Atmospheric Administration (NOAA). The center's windows are framed by exterior sun shelves that deflect direct sun but also bounce light deep inside the building, where weather forecasters do their everyday work. When I entered in May 2015 to talk to NOAA's ocean wave modelers, I was signed in by a security officer who asked me to provide the serial number of my laptop, recalling to me that the headquarters of the National Security Agency were just a half-hour away from this future-facing, spaceship-like construction, built in 2012 in one of Maryland's newest office parks outside Washington, DC.

I was met by Hendrik Tolman, director of the Environmental Modeling Center. As he walked me across an outdoor dining terrace, he pointed me to a four-story waterfall fed by rain from the building's green roof. Arriving at an atrium where two bending wings of offices met, Tolman told me the space was aimed at encouraging clusters of scientists, meteorologists, and data managers to cross paths. The architectural firm that designed the building, HOK, declared in a press release that "the building form is organic, with 'waves' of space."[1] The lines of convergence, Tolman confirmed, were meant to evoke hurricane "feeder bands"—though he joked that, to him, trained as a civil engineer, it was disorienting to be in a place where "there are no right angles. It's difficult to figure out where to put furniture!"

Tolman is from the Netherlands and holds a doctorate from the Delft University of Technology, where he developed expertise in wave modeling. He brought me to the fourth floor: the Ocean Prediction Center, an open-plan office space divided by low cubicle walls that marked out areas dedicated to different oceans. I was guided into a section labeled "Atlantic High Seas Forecast," where I met a meteorologist sitting at an L-shaped desk that hosted a sweep of seven computer screens. He showed me a digital map of the eastern seaboard of the United States, an electric-blue chart of seas divided into irregular lattices. On the next computer over sat a satellite map of the same area, over which was overlaid a grid of green dots, scaled to offer markers every ten miles.[2] Significant wave height (Hs) values pop up at the dots when the meteorologist clicks on them. Some values correspond to reports from buoys; others are forecast values generated by a computer model that feeds in information about the speed and direction of winds known to prevail at each spot. The screen delivered a layering of the measured and modeled, a palimpsest of the real and simulated, a juxtaposition of the present and the projected future.[3] "We're able to mix together a lot of things," the meteorologist remarked, telling me one could query each dot for further information. All the displays were undergirded by layers of programming and simulation protocols—a *stack*, in computer-science jargon—about which working meteorologists need know very little; it is their expertise in weather, aided by an evocative graphic user interface, that is called for here.

Just past desks with names such as "Pacific High Seas Forecast" and "Tropical Desk" were banks of computer screens displaying not weather but real-time flowcharts of the progress of computer subroutines, as well as data on the health of the supercomputers on which weather models run. Wave data, it turns out, are not "in the cloud"—distributed across remote servers—but hosted on a dedicated bank of supercomputers in nearby Virginia, keeping computational power closely coordinated with NOAA's center of climate calculation.

In this chapter, I examine the computational programming and simulation package behind NOAA's global wave forecast, WAVEWATCH III (figure 4.1).

I draw on conversations with WAVEWATCH's inventors, contributors, and maintainers, as well as on a weeklong summer-school course on wave modeling in which I enrolled, along with twenty-three other students and professionals in oceanography and coastal engineering. I argue that WAVEWATCH's operation requires its users to reckon with multiple orientations to *time*: the real time of waves at sea; the backward-looking, or retrodictive, time required to generate predictions; the sped-up time of simulation; the reshuffled time of computer processing; the time of distributed collaboration on and authorization of wave-describing computer code; and the operational time of providing forecasts useful to shipping, hurricane wave prediction, beach going, and more. Zigzagging through these kinds of time are—sometimes loosely, sometimes tightly—the career times of wave modelers, times that trace out individually and institutionally shaped arcs of motivation, as well as, in many cases, trajectories of transnational migration and mobility. Wave modelers are an international group, and many of NOAA's US-based researchers grew up and were educated elsewhere, mostly in Europe, Latin America, and Asia.

What do such interweaving, back-and-forth, staggered timelines mean for how ocean waves—as empirical entities, as measured phenomena, as mathematically modeled forms, as computationally processed records of quantities—manifest in WAVEWATCH? I have argued throughout this book that scientifically represented waves often carry the imprint of the media through which they are apprehended. Here, I first claim something analogous: that computationally modeled waves become *informational forms* that take the shape of those mathematical measures meant to stand in for physical processes (ideal-typical energies, frequencies, spectra). In becoming informational, these digital waves become subject to processes of *reformalization*, mathematical and computational rephrasings that obey the demands of processing and prediction time. I next argue that that which remains *in excess* of such representations—visions of waves as scary, pleasurable, numinous, though also, more empirically, as material forms with many dynamics still only partially captured into representation (e.g., their interaction with mud and sand; their sometime manifestation as "rogues")—is part of what makes wave modeling compelling (emotionally, personally, professionally) for practitioners. James Hamilton-Paterson (2007, 7), in his melancholy paean to the ocean, *Seven-Tenths*, writes that, for many of us, "'The sea' survives elsewhere in piecemeal images, scattered pictures which never link up. They include beach scenes from summer holidays, an aunt being seasick, storms, sunken treasure, pirates, monsters

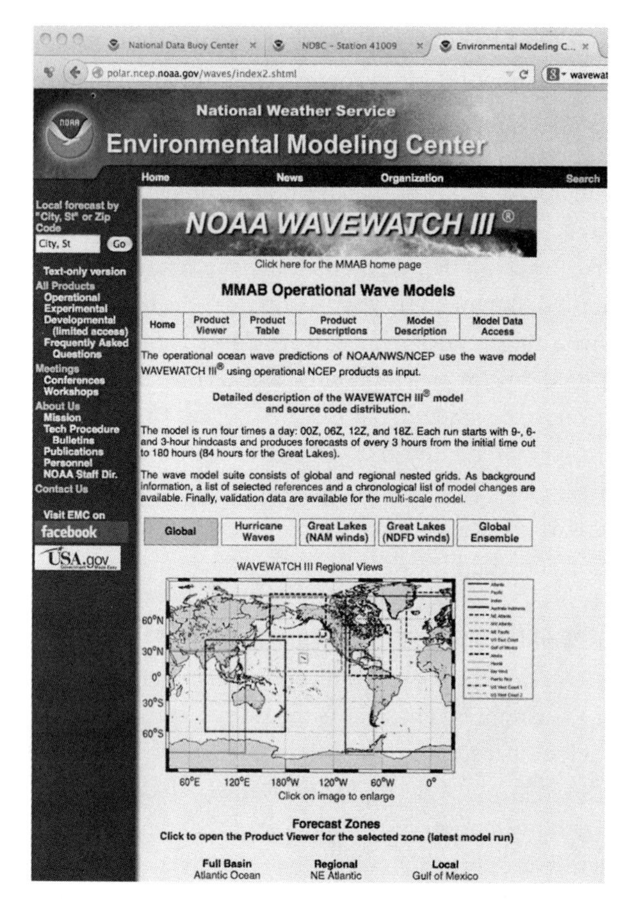

FIGURE 4.1 WAVEWATCH web portal as of September 30, 2020.

of the deep." For wave modelers, piecemeal images of waves enter as anecdotal pointers to their childhood, adolescence, early careers—often in settings other than the United States. Waves imagined, recalled, kept alive in memory may even be tokens that speak to their political awakenings to the uneven distribution of wave science and wave effects in the world. If, as the feminist tenet has it, the personal is political, it may be fair to say that, for many of these scientists and engineers, the personal is the oceanographic is the political.

Now used in an Earth-spanning web of about two thousand organizations, WAVEWATCH is the world's most widely used simulation and prediction platform for ocean waves. It is maintained at the Center for Weather and Climate Prediction by a handful of experts who work for the National Centers for Environmental Prediction, under NOAA's National Weather Service

(NWS), and who vet and accept suggestions from a community located in a patchwork of sites around the globe. It is an open-source program written in Fortran 90, a 1991 descendant of FORTRAN, a language created in the late 1950s to accommodate high-speed numerical analysis and that still operates as a lingua franca for scientific computation. The WAVEWATCH team works to keep program "version control," cataloging, managing, and harmonizing revisions of the source code to keep the identity of the software steady.

The program had modest beginnings. It started as Tolman's thesis project at Delft (see Tolman 1989). He crafted it as an improvement on an existing model called WAM (for Wave Model), based on the work of a team in the Netherlands, Germany, Italy, and the United States (see chapter 1; see also Komen et al. 1994). After moving to the United States—with his fiancée, who started her career working at Johns Hopkins University, where she also completed a nursing degree—Tolman further developed WAVEWATCH at the National Aeronautics and Space Administration (NASA) and eventually, when he began working at NOAA, with funds from the NWS and the Office of Naval Research. It has grown enormously—so much so that WAVEWATCH is no longer considered so much a wave model as a wave model *framework*. One oceanographer told me it was like "a Swiss Army knife that is a good tool for many different applications."

Tolman created the backronym WAVEWATCH to stand for WAVE height, WATer depth, and Current Hindcasting. With the framework's increasingly rich graphic interfaces, which permit globewide views that can be zoomed in and out, the word *watch* also tags its monitoring function.[4] It stands watch.[5] The word *hindcasting* refers to a mode of generating predictions by *looking back*—for waves, by testing a model against previously measured wave data to make prognostications about the kinds of waves that will appear in the future. As Tolman told me, "I don't *need* any observations of waves to get a good wave forecast, except for the fact that I need observations to develop the model, to tune it, and to validate it, and I have to monitor it later." As Paul Edwards (2010) argues in *A Vast Machine*, his authoritative history of computational climate modeling, data are always recorded with respect to an anticipated model, and models always assume data will arrive in prespecified formats. For wave models, these data include such measures as significant wave height, speed, and more.

The play between measured and reconstructed past, simulated present, and projected future points to the constant shuffling and calibration of time orientations that WAVEWATCH modelers and users must undertake to *get their work done*: to deliver timely accounts of world waves and to do so on machines—computers—that themselves continue to be associated with the futuristic. As the software studies scholar Wendy Chun (2011, 9) writes, "Computers as future depend on computers as memory machines, on digital data as archives that are always there. This future depends on programmable visions that extrapolate the future—or, more precisely, a future—based on the past." Think of WAVEWATCH as layering informatic memories of an ocean past, with these memories nested inside a worldwide "stack," a "vertically interrelated and interdependent series of hardware configurations and software protocols that make high-level media computation and networking possible" (Lewis 2019, 219). This stack, suggests the media scholar Benjamin Bratton (2015), is now planetwide, a meta-infrastructure of grids, clouds, apps, and internets through which many genres of governance increasingly pass, an infrastructure (made, also, of rare earth minerals, extractivist economies, digital waste) that, as Jennifer Gabrys (2016) reveals in *Program Earth: Environmental Sensing Technology and the Making of a Computational Planet*, includes the oceans (not least as an environment for submarine cables [Starosielski 2015]). If the ocean holds memories of its past in its centuries-long circulations of cool and warm water, its decadal cycles of nutrient flows, and its weeks-lasting current patterns, then many social memories of waves now partially exist, in scientifically mediated form, in the stack (see also Bowker 2005).[6]

This chapter is about the format of those memories, as well as about how they intercalate with human memories, individual and institutional, of what waves do and are. By chapter's end, when I recount the summer school's demo simulation of 2005's Hurricane Katrina, it should be clear that memories are always enmeshed in historically specific concerns. Stabilizing wave memories, in consciousness and code, is political work. Achieving version control over which waves are significant when and where and to whom requires zigzagging between the particular and the general, repeatedly revisiting past data to set its narration on a path that can better predict or prepare possible futures. Like characters in Dexter Palmer's time-travel novel *Version Control* (2017), in which protagonists live through permutations of the same events over and over again, nudged by the chance outcomes of algorithmic automation, which tweaks everything from family tragedy to the contours of everyday racism and US presidential politics, wave scientists are ever working to both steady and improve how phenomena they care about are ordered, sorted, and remembered in the silicon second nature that is WAVEWATCH.

Wave Time, Remediated and Reformalized

In July 2017, stepping into a windowless classroom with fluorescent lights in the Atmospheric and Oceanic Science building on the University of Maryland, College Park, campus, I joined the WAVEWATCH summer school's twenty-three other students as they located their seats. Some were oceanography graduate students working on dissertations. A couple worked for the Army Corps of Engineers. One was an employee of a company delivering wave forecasts to surfers. Seven were international, having come to the summer school from Canada, China, Mexico, Portugal, and South Korea. Of those based at US institutions, many were nationals of elsewhere: Bangladesh, Iraq, Pakistan, Turkey, and, again, China and South Korea. About one-third were women.

Tolman opened the proceedings with a history of WAVEWATCH. We might find it odd, he offered, that the NWS runs a summer school. The NWS's mission, he reminded us, is "to save life and property," not to teach. But instruction has become central to keeping WAVEWATCH growing. A few years back, he said, NOAA had one of the top-ten fastest computers in the world, but now there are so *many* computers—and so many as yet uncharacterized wave dynamics around the world—that, as he put it, "we cannot do it alone. We need to do it with a community—and especially an international community." Since 2010, his group has offered weeklong workshops in Bangkok, Thailand; Dakar, Senegal; Hyderabad, India; Nairobi, Kenya; São Paulo, Brazil; and Taipei, Taiwan—though with funding cuts and with some groups (e.g., Taipei) up and running on their own, he and his colleagues had scaled back to the once-a-year US-based workshop. Running the class on the Maryland campus allowed his team to make sure everyone was running the same operating system; previous workshops had become confusing as people brought in incompatible Linux machines—pointing to the social and technical work required to keep software universal and to enroll people in communities of practice (Mackenzie 2008b).

The aspiration has been to keep WAVEWATCH in international circulation—one reason that it runs on Ubuntu, an open-source software operating system (whose motto is "Linux for Human Beings"). This, Tolman emphasized, could enable users to contribute ideas for the governing code. (Because WAVEWATCH is a US-based program, however, it cannot be licensed to users in Cuba, Iran, North Korea, Sudan, and Syria, countries against which the United States has trade embargos, demonstrating the geopolitical limits of Ubuntu's utopian "human beings" motto.[7]) Computer hardware matters, too. Tolman remarked that these days most hardware was "set up for other things than fluid dynamics. The newest chips are optimized for gaming graphics." Keeping WAVEWATCH "weather ready" requires keeping an eye on software and hardware. The framework is aimed at generating usable, real-world information. While WAVEWATCH can permit scientists to do research, it is mostly an operational model (no untested theories allowed in!), meant to generate data to be accessed by everyone from lifeguards to mariners, ports, ship companies, and oil platforms—people planning short- and medium-range futures. In this way, it constitutes a loose kind of wave polity.

Tolman introduced members of NOAA's "waves group" who led us through sessions on wave physics, on creating "grids" within ocean maps, on modeling wave fields using the spectral format, and on validating models using real-world data. Toward the end of the workshop, we were led through a simulation of the wave fields associated with Hurricane Katrina, the category 5 hurricane that devastated New Orleans in 2005 and killed more than 1,800 people. Katrina is widely regarded as a massive failure of US civil infrastructure and governance, whose damage was exacerbated by ongoing structural anti-Black racism—a fact that the NWS's warning of August 28, 2005, which predicted "a most powerful hurricane with unprecedented strength" that would leave "most of the area uninhabitable for weeks," could not itself overcome. The NWS recalls the Katrina warning based on its modeling as one of its strongest moments; the responses to that warning by the Federal Emergency Management Agency and New Orleans Police Department, however, were catastrophically inadequate.

I had met many of the waves team during my visit to NOAA in 2015 and knew they were an international group—part of the reason earlier international workshops had been able to get going, with scientists making use of contacts in home countries. Only one member of the NOAA waves team, at the time of the workshop, had been born in the United States. When Tolman, who is Dutch (or, actually, as he corrected me, Friesian), was promoted to director of the Environmental Modeling Section, the wave physicist Arun Chawla, who grew up in India, took over as leader of the team. Tolman and Chawla, who became US citizens some time ago, work as federal employees, whereas other members of the waves team—from such places as Greece, Iran, and Venezuela—were international visa or green card holders in the employ of scientific and technical support agencies that contract with NOAA.

It felt like a different moment from when I visited NOAA two years earlier. By 2017, the election of Donald Trump had amplified a range of anti-science, anti-environmentalist, and anti-immigrant forces. The arrival of the administration had sent a chill through my own home institution, the Massachusetts Institute of Technology, where some areas of scientific expertise were targeted by funding cuts and a number of international students had been caught up in Trump's travel ban. It was destabilizing, intensifying experiences of precarity among internationally mobile researchers (see also Davies 2021). One scientist had told me: "I think if I was a student now, I would not come to the United States. And I see that in a lot my friends—most of us have left our country. But a lot of them are going to Europe now. They're not coming to the United States. It's bad." In that moment of what people were calling the Trump era, time and history seemed out of joint. The liberal-conservative give and take that was one ideological assumption animating American politics as usual had lost any veneer of version control, any historically accreting source of common, if contested, code.

WAVEWATCH is a model of the worlds' ocean surfaces, particularly as they are modulated by the wind that generates waves. In the model, the ocean is divided into sections called "grids" (Cartesian or spherical), each of which, in the open ocean, stands for about fifty square kilometers, or, for waters closer to shore, seven or so square kilometers. (There are some models—such as SWAN [Simulating Waves near Shore], also developed at Delft—that go smaller.) As Tolman told me, "When we talk about the coastal side of things, that's where what you might call the anthropological aspects actually start to drive the research and development work we're doing. A wave is on the scale of a person." Though they were not much covered at this workshop, there are models aimed at predicting rip currents at beaches. I could have used better information on rip currents in my past, when my father and I almost drowned, conveyer-belted out to sea almost immediately after moving from the East to the West coast, having transposed intuitions from Massachusetts beaches to quite different Southern California places and finally returning to shore only when were able to catch outsized breakers to ride in.

Think of the grids in WAVEWATCH as standing in the computer model for "fetches" of ocean water over which virtual winds blow, generating reliable ranges of frequencies, or wave spectra. Those spectra, coded in Fortran 90, are based in part on measures from the Joint North Sea Wave Project (JONSWAP), that effort from the late 1960s (described in chapter 1) to characterize the range of wave frequencies created by specific wind profiles (and that turned the North Sea into the model ocean for WAVEWATCH, which, Tolman tells me, has been assimilating spectra that characterize other places, including the Pacific, for a while). WAVEWATCH also depends on a "physics" installed in the program, by which is meant a set of equations that specify wind input, wave dissipation profiles (waves' loss of energy due to breaking, turbulence), and the effects of wave-wave interaction. If one is "reading waves" through WAVEWATCH, it would be good to remember that, as Chun (2011, 5) puts it, "computer reading is a writing elsewhere"—that is to say, beforehand.

When the class started in earnest, Chawla gave us a crash course in wind-wave physics. It was a refresher on what I had come to think of as the *anatomy* and *population dynamics* of waves—a lesson, first, in the hydrodynamics that lead to and characterize the growth, propagation, and breaking of *individual* waves, and then, second, in how to understand groups and fields of waves in aggregate, in terms of a *wave spectrum*. The waves modeled by WAVEWATCH, Chawla clarified, do not include tsunamis or tides; they include only wind waves, "the kind," Tolman interjected, "that make you seasick." Chawla provided an idealized cartoon of a segment of seawater to help us along, showing a side view of waves traveling, left to right, toward a beach (figure 4.2). Their undulating crests and troughs and wavelengths were labeled, and cutaway/inset views of the "orbits" of water particles beneath the surface showed circular tracks below deep water while revealing flattening ellipses as shore-heading waves came into shallow water and "felt" the bottom, heralding their arrival at the "surf zone."

Chawla told us that this description mapped out a *four-dimensional problem*: how to understand the travel of waves across a *length*, *width*, and *depth* of water over *time*. He reminded us that hydrodynamicists had a ready—and old!—set of mathematical tools to tackle the problem. The next slide flipped us into that firmly mathematical idiom (similar to figure I.2 from this book's introduction): now we saw not waves on a cartoon ocean but one wave (trough and crest) pinned sideways against the page, its height held on an up-and-down axis, its length along a horizontal one. The sine wave representation was already a bit of an idealization since most ocean waves are pointy, trochoidal. The diagram was sectioned up with differential equations pointing to things such as mass and momentum balance, tagged to the names Laplace and Bernoulli. Chawla observed that these equations offer simplifications of one of the basic equations in fluid dynamics—the

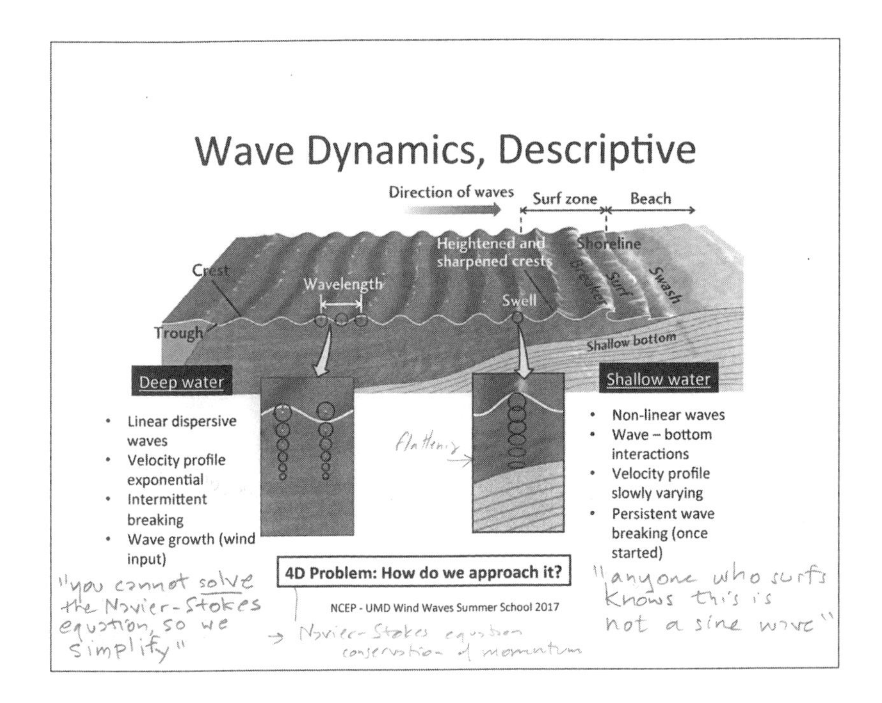

FIGURE 4.2 "Wave Dynamics, Descriptive" slide from National Centers for Environmental Prediction, University of Maryland, Wind Waves Summer School, 2017. My notes are added in pencil.

Navier-Stokes equation (see chapter 1)—which describes the viscous flow that characterizes dissipating ocean waves.

This was a reminder that modeling ocean waves—in drawn diagrams, mathematical formulas, and computer code—is a process that involves *remediation* and *reformalization*. *Remediation*, argue the media theorists Jay David Bolter and Richard Grusin (1999), is a process through which media refer to and refashion one another—as when films quote and redo conventions of live theater; when digital design borrows customs from the drafting table; or when digital audio production offers interfaces that evoke the unwinding of magnetic tape (making it possible to "rewind" even when there are no spools in sight). Reformalization, meanwhile, argues the historian of science Stephanie Dick (2020), is a process through which descriptions move from one symbolic system to another, as when a book is translated across languages; when genetic information moves from an electrophoretic gel to a database; or, more pertinent here, when a mathematical description (e.g., a differential equation) is re-presented in computer code. In WAVEWATCH, waves are reformalized many times over.[8]

The next talks dove into further reformalizations, moving farther from the visual and diagrammatic. We were presented with a description of "velocity potential" for waves, a description that, we were instructed, "is not physical—it is *mathematical*," indicating not only that the *physics* of physical oceanography was being idealized, but that "waves" were becoming differently abstracted objects as we went along. We saw a stream of cosines, sines, tan*hs*, slicing waves up into horizontal and vertical velocity components. We learned about descriptions of waves hitting ocean currents, which could squeeze or stretch these undulations. Lifting waves into mathematics renders them oddly timeless; as Geoffrey Bowker (2005, 4) observes, in such formulations "all contingency has been removed from the law of nature—it is true over all time and in any place. . . . [T]he past that scientists create can be read as an eternal present." We got some knotty formulas representing energy flux. "We don't like to keep things simple in waves," said Chawla. The equations started coming fast, as he reminded us that waves travel in *groups* that have a different velocity from the individual waves they carry, though that changes as they approach breaking, which, Chawla commented, could be dramatic as group and phase velocity catch up with one another to cause waves to topple over. Showing us a slide of such "wave shoaling," Chawla laughed and declared, "I went surfing *once* in my life and then said to myself 'I like my waves in a computer!'" We got to the next step: the spectral model. We were given a capsule history taking us back to Walter Munk, to the 1970s JONSWAP effort, and more. We got a slide showing a "wind wave modeling genealogy tree." It became clear that WAVEWATCH was built on a catalog of accepted wave knowledge. It was a digital book—even a family tree—of waves.

The basic elements of water wave physics—the input of wind, the dissipation of wave energy over space, and the nonlinear interaction of waves with one another—are so complicated that modelers cannot simulate them in all their detail.[9] "Can we represent wave physics fully in our model?" asked a Brazil-born member of the waves team, Jose-Henrique Alves. "No way!" he said, answering his own question. Yet, he continued, if "reality is a 'little' overwhelming," it *is* "reproducible mathematically," especially if wave modelers use *parameterizations*: equations based on wave theories: "There is *physics* in wave generation and propagation. However what WAVEWATCH does IS NOT PHYSICS; it's a *parameterization* of the physics. We have a *mathematical approximation* of the physics in the model." In some cases, solving such equations would eat too much processing time. Wave researchers therefore create "discrete interaction approximations" for wave-wave interactions, along with other approximations for wind input and dissipation drawn from literature in fluid mechanics. When a wave model starts up, these parameters are "tuned" to make sure they generate realistic values for the wave field for which they are meant to be virtual understudies.[10]

The path of reformalization continued as we were led from the mathematics of fluid dynamics toward discussion of how these equations would be represented in WAVEWATCH. One lecturer told us: "We're dealing here with a numerical model. The model you're running doesn't know about your hopes and aspirations. It doesn't know any physics, anything about the world. A model is an abstraction of reality—and we're adding a numerical abstraction *on top of that* because we cannot analytically solve the equations that describe the world we want to model." The numerical abstraction of WAVEWATCH is not only a practicality for representing waves in general. It is a necessity for doing so *in time*: in the time of computation (practically, not just theoretically) and in the time of weather forecasting—that is, in calculation and social times that need to be calibrated to one another to be *timely*. Chun (2011, 24) writes that programming often "focuses on the movement of data within the machine. The relationship between executable and higher level code is not that of mathematical identity but rather logical equivalence, which can involve a leap of faith. This is clearest in the use of numerical methods to turn integration—a function performed fluidly in analog computers—into a series of simpler, repetitive arithmetical steps."

Our lecturer informed us that using computer time efficiently required "time stepping." He drew on the white board a top view of wave bands of different frequencies arriving at a beach. To model what is happening at each successive moment *t*, he explained, WAVEWATCH needed to calculate the propagation of waves *at different frequencies*, over and over again. So rather than simply "marching the solution forward in time," what was instead called for—since we were dealing with waves with periodicities at different time scales—was constantly to keep *rewinding time* for each frequency. I transcribed the speaker's whiteboard drawing onto the left side of my printed copy of his "fractional step solution" slide, which featured a diagram with a timeline zigzagging from present to future back to past to future back to past (figure 4.3).

We were thick in reformalization now, with the computer model of waves not only unfolding as a time-compressed scale model of the ocean, one

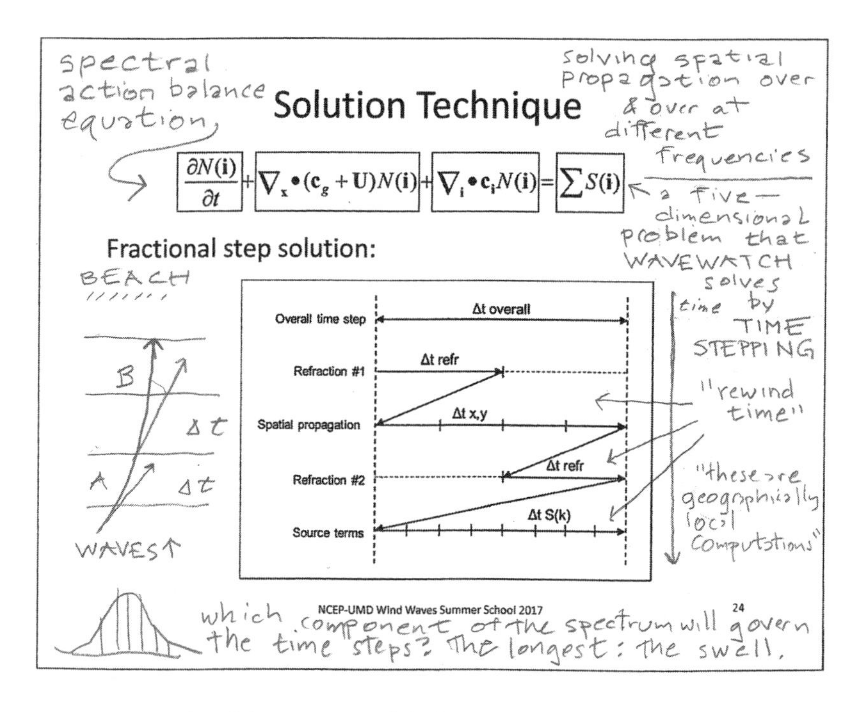

FIGURE 4.3 Fractional time stepping slide from National Centers for Environmental Prediction, University of Maryland, Wind Waves Summer School, 2017. My notes are added in pencil.

that unfurls its results before the actual ocean does, but also, in doing so, operates in staggered time, using back-and-forth time steps that are more about optimizing *computer processing time* than miming *wave time*. In other words, WAVEWATCH was nothing like an old-school animated film, moving frame by frame. If it is true that, for meteorologists looking at wave spectra on the screen, waves might be, as one scientist put it to me, "like cartoon characters" whose stories they watch in sped-up time, for WAVEWATCH, wave time is an *abstraction*, a *measure*, not a *duration*—one that is fractionated to be handled in whatever way makes most sense for computational, institutional, and specific social times and that, calibrated to the synchronized global time now embedded in computers, participates in the very making of *real time* itself.[11] As Adrian Mackenzie (2008a, 52) puts it, writing about the computational signal processing of movies, "The relation between frames is subject to calculated reordering in the interests of accelerated or compressed transport [in computer memory or across the internet]." He underlines a key lesson: "The thinking present in software cannot be reduced to mathematical thought." More, time-rearranging algorithms "challenge cinematic and televisual perception" even as, paradoxically—think of the meteorologist watching the WAVEWATCH user interface like a TV screen—"they participate in making the world *more* cinematic or televisual" (52, emphasis added).

It is not only time that is abstracted in WAVEWATCH, but also space. While WAVEWATCH offers a map divided into grids, the information in those grids does not always need to be processed in anything like geographical sequence (e.g., from offshore to shore). One could take a grid of boxes representing a segment of ocean and process the boxes in whatever order is most efficient. We learned that wave spectra—already an abstraction—might even be mathematically examined by treating their three-dimensional graphs as *virtual physical objects*. We were shown a three-dimensional graph of wave frequency plotted against wave period plotted against wave energy, a graph that partitioned out different "swells," or systems of waves with different frequencies and therefore, probably, origins in different storms at different distances.

"Look," our lecturer continued, "at the spectral *landscape* and think about the topology of the *shape*. Forget that you're thinking about waves and think of it as a mountain landscape. *Now* flip it upside down and imagine it as a valley. And now, imagine that you can send 'numerical rain' into this valley to find its volume—and then use *that* to find the significant wave height across all the wave fields." Mathematicians may find all of these reformalisms, context inversions, and reorientations business as usual, but they underscore the fact that ocean waves are here managed as abstractions—and of many kinds, as researchers flip through diagrams, equations, code. If modelers tried to think through all of these forms simultaneously, they would risk confusion, as the mathematician James Lighthill (1967, 267) confessed to courting when he sought to "summarize my view of the different kinds of wave and their interrelationships in a work of visual art," in which he catalogued waves inside a scheme of fanciful shapes, each "appropriate in some way to the type of wave system concerned" (plate 14). Commenting on this mix of mimetic and conceptual, Lighthill observed, "My whole subject is non-linear," stepping back to remark on where his calligram diagram ultimately landed him: "The lecturer gets worked up to the point where all the different kinds of wave catch each other up and he starts describing them all simultaneously" (293). In WAVEWATCH such simultaneity, when it appears, is managed in part by stacking, not swirling together, abstractions.

Wave Memory in Code and Community

We students learned about WAVEWATCH as an accomplishment, a modeling framework built on a large foundation of wave science. The lessons we received tracked a lineage that by then was familiar to me, from hydrodynamicists such as Sir George Stokes to oceanographers such as Munk and modelers such as Tolman (a list that was pretty much all men, with the exception of Susanne Hasselmann). The more close-to-the-machine lineages of WAVEWATCH itself are visible in the framework's annotated code. In the WAVEWATCH manual, a list of "source terms"—moment-of-creation starting points for wind and wave dissipation in the model—operates as ongoing lineage work, as source terms are named in association with those researchers who first published them.[12] One member of the waves team told me that, "WAVEWATCH has grown and absorbed several different packages and sub-programs and therefore represents an incomparable tool for simulating wave generation, propagation, and coastal effects. It is a complete model covering a broad range of applications and interests."

I caught up with Jessica Meixner, a mathematician who had recently joined the NOAA wave modeling team. She emphasized to me what Tolman said at the opening of the workshop: "WAVEWATCH is a community wave model. We try to be pretty open, and we have about forty active developers all over the world. There's a big group in France, at IFREMER [French Research Institute for Exploitation of the Sea]. There are a couple of people in Australia, Italy. There are some in the Navy." When people in the team want to add a subroutine or module to WAVEWATCH, she told me, they work together to see whether that module adds functionality to the framework. Meixner elaborated: "When one person makes a new development, we will work with them and they'll put it on a branch and we see whether it belongs in the mainline of the code. So that's my job to merge that back into the mainline of the code and make sure it doesn't break anything else. You might say that I am the gatekeeper for WAVEWATCH." Thus, WAVEWATCH can be imagined as something like a wave Wikipedia, with a community of contributors. To join the genealogy, one must speak the code. As the software scholar Simon Yuill (2008, 66) writes, "Code creation is an inherently social act. It involves processes of collaboration, consensus, and conflict resolution, and embodies social processes such as normalization and differentiation." For WAVEWATCH, enacting such "coded conduct" (Dick 2020) means being part of an international scientific community and publishing in peer-reviewed journals, offering theories that can be tested in real-world scenarios once they have been fed into the framework's Fortran 90 format.

Reading through the version history of WAVEWATCH tracks a range of contributions from wave modelers in different parts of the world. Scanning the 2016 manual shows contributors from institutions based in Australia, Canada, Croatia, France, Germany, Italy, the Netherlands, the United Kingdom, and the United States. That variety does not on its face reveal the even larger diversity that might be read from the international character of scientists at any of these sites. The members of the waves team at NOAA, at the time that I met them, included scientists with early training in Brazil, China, Greece, Iran, Mexico, and Venezuela. Tolman said that some of the demographic variety seemed to follow geopolitical transformations, though it was also true that the sample size was small. Twenty-something years previous, he said, when the "iron curtain came down, lots of Russians came into the field—and to the United States. They were often really, really good at math because they had nothing like the computers we had." More recently, he observed, engineers and computer scientists—people who then had to get up to speed on the oceanography—were coming from India and China. Tolman told me that since the team in the Environmental Modeling Center included contract employees, the group hired talented experts from all around the world (many on H1-B visas), making the group very international. Another wave scientist, also from outside the United States, seconded Tolman's account, layering it with an optimistic read of American history: "I think there is generally in science a tradition in the United States of trying to get the brightest minds in the world." Another said to me, about working as part of an international group, "It's fun. We love it. We used to have a map that we would put pins in. Every time we got a new person, 'Where you from? Where did you grow up? OK, we can put a new pin on it.' So we have a grand time, yeah." The demographic profile changes over historical time, and members of the team were keen to talk about it. And the story was complex. As the science studies scholar Sarah Davies (2021, 217) notes in her research on international mobility in science, "Mobility might be simultaneously desired and imposed; permanent and temporary; pleasurable and painful."

Talk of international diversity among the waves team sometimes took the form of researchers suggesting that modelers trained in different places might bring in problems from other geographical areas. Everyone knew that the earliest iterations of WAVEWATCH were based on North Atlantic seas—even on just the North Sea—and, to some extent, the Pacific coast of the United States. Some of this tracked the maritime (and military) history of those sites; some was an effect of increasing population pressure at these places and calls for coastal engineering in connection with harbors. One participant, with training in Australia and alert to some of the Northern Hemisphere–slanted aspects of the model, said, "We want to generalize WAVEWATCH so we [can] actually use it in a place like, say, Australia, where you have some of the larger oil reserves that have exposure to swells from the southern ocean." Australia, he went on, had seen a boost in wave science with the rise of beach and surfing vacations.

The Brazilian coast, I learned from Jose-Henrique Alves, often saw streams of mud flowing into the sea along the southern coast; that mud could attenuate wave energy. This is something a model would miss if it did not treat the mud as a layer of viscous fluid below the waterline, but that had recently been built into WAVEWATCH because of Brazil-based work (see Rogers and Holland 2009). I had traveled to Brazil earlier in 2017 and learned about a particularly horrible instance of river mud arriving in the Atlantic. In November 2015, an iron ore tailings dam owned by the mining company Samarco Mineração burst open, sending sixty-two million cubic meters of muddy waste from the village of Bento Rodrigues, in the municipality of Mariana, Minas Gerais, down the Doce River, killing nineteen people in floods and sending toxic water flowing toward the Atlantic Ocean—a flood described as a "wave of mud" (Creado and Helmreich 2018). The disaster underscored not only the significance of rivers, but also the fact that waves in the world are never abstract. They are always convolved with the material world.

Alves had difficulty pursuing wave science in Brazil because funding from that government was low and often volatile. Oceanographers had tried to work around the fact that the country had just three buoys along its entire coastline by enlisting surfers as "wavewatchers." A project called Sea Sentinels recruited some one hundred surfers and trained them to deliver data in the same format as the Netherlands' Datawell Directional Waverider (see chapter 1)!

There are lots of unknown sorts of waves out there still be to characterized, said Alves, who had hopes, one day, of writing a book titled *The Bizarre Book of Strange Waves*. He passed along some Portuguese-language articles he had written for *O Globo* in the 1990s. One discussed the "pororoca," a wave generated by the coming and going of tides in the Amazon (Alves 1992). In *In Amazonia*, the anthropologist Hugh Raffles (2002, 17–18) quotes a novelistic description from 1925 of this particular recurring wave: "[The Amazon's] current fights with the Atlantic tides and creates a dangerous bore. Three waves which are sometimes as much as twelve feet high rush up the river, sweeping over the low islands with devastating force. . . . This part of the earth is not ready to be the abode of man." Alves was enthusiastic that there were species of waves still waiting to get into WAVEWATCH: "There are waves that are not in there yet because their very basic generation process is not included on the wave model." He told me, too, of the problems that might come from transporting wave intuitions from one part of the world to another. He gave an example:

> In the northeastern part of Brazil, there was a port development project for what was supposed to be a state-of-the-art port. It was developed on the basis of a scale model of the coast made by a Dutch company, who built a physical model in the Netherlands. So they came over to Brazil, and they used a Brazilian apartment building to make this big scale model of the port.
>
> It was in an area that was exposed to trade winds, so the wave climate would be dominated by small-scale waves, high-frequency waves, and so on. But after they built it, the beaches started to suffer significant erosion during the summer, and no one really knew what was going on.
>
> A Brazilian hydraulic company put out a Waverider buoy there, but they couldn't figure out what was happening. My supervisor in Brazil started to have a look at the records and look at the spectra, which were all on paper rolls. Everyone was just looking at the "zero-crossing," which just gives you mean heights and periods. But my supervisor sent me off to look at the rolls, and he told me: "You have to digitize them." It took three months, and then when we ran it through spectral analysis, we saw that there was a very low frequency swell signal over summer months.
>
> And we go, "Hey, look at that. Where does that come from?" We figured it was from two thousand miles away. But where was it coming from? We had to go back to synoptic charts—and this was the '90s, so you had to go to the libraries, sometimes taking buses all over the place—to piece it together.
>
> And we actually found out that these waves were swell systems that were propagating from Greenland all the way down to the coast of Brazil. And they were suspending the sediment close to the beaches, pushing the sediments away from the near-shore area behind the structure of the port and carrying them upstream. So the engineers

who designed the port, they didn't have the knowledge of wave climate they needed.

So today, if you look at a Google Earth picture of Fortaleza, you'll see that they had to build over two hundred kilometers of sand traps along the coast because of that disaster.

Another wave scientist from Brazil told me about growing up on the coast of São Paulo, surfing and sailing and "always trying to predict the waves. As a teenager this became a huge curiosity for me. Sometimes, I didn't know whether I should go to a nightclub or go to bed early because the swell was coming." This man also advocated for more local, oriented knowledge of waves and emphasized that he thought that the quotidian knowledge of people who were not mathematicians might be taken more seriously:

> I finally really fell in love with environmental prediction when I visited small communities of fishermen at some isolated islands in Brazil with a great knowledge of waves, including forecasts.
>
> I ended up studying meteorology, where I started to learn about air-sea processes and wave generation, as well as storm surge prediction. I studied extreme wave modeling, and in my PhD I dedicated a lot of time to improving the simulation of extreme wave hindcasting linked to statistical analyses.
>
> Nowadays, my main motivation is not only to predict good surfing conditions but to better predict extreme winds and waves to save lives, especially those associated with small, vulnerable vessels in South America.
>
> And I believe that ordinary people can also contribute to environmental modeling ideas—that is, including a "human" module into the analysis. Instead of using a one-way forecast, where prediction centers provide information to the users, the inclusion of a user's feedback and knowledge to the physical and mathematical models—two-way forecast—might contribute to better simulations and predictions.
>
> Moreover, as a Catholic person I see the ocean waves as a contemplation of God's creation, looking at the sea from the beach, as well as studying the differential equations from my computer.

Other members of the wave team were interested in sharing wave lore with me, patchwork stories they found interesting or troubling that did not yet fit into the protocols of computational modeling or, in some cases, condensed nostalgia for their countries of origin. A Greek member of the waves team started his talk with a mention of the "three waves" of Plato, the *trikyma,* a trio of waves that some classicists believe were frequent phenomena in the Mediterranean.

Another member of the waves team, who grew up in Iran, was keen to tell me about a four hundred-year-old Persian poem he thought spoke to the existential meaning of waves. He sent me an image of a calligraphed poem by the seventeenth-century poet Saeb Tabrizi in a style called Siah Mashgh (figure 4.4), in which a text (ما زنده به آنیم که آرام نگیریم موجیم که آسودگی ما عدم ماست) is repeated twice. It translates, loosely, as "We are waves; our cessation is our death." He drew my attention to how the form of calligraphy meant to evoke a wave. This was not a call to amplify what WAVEWATCH did but, rather, a pointer to the excesses in waves that kept some of the scientists captivated. Hamilton-Paterson (2007, xii) writes, "The sea's power to haunt sometimes seems as real a force as a wave's kinetic energy." Just so, the more I spoke with people at the workshop, the more I got a sense of wave science as a project suffused by the affects contouring personal, fragmented ocean memory.

As I began to see students in the class as members of the next generation of contributors to WAVEWATCH, I gathered a stack of accounts:

One student living in South Carolina, who had grown up in Iraq and studied at a marine science center in Basra, sought to return to Iraq and be of assistance in modeling storms off the Persian Gulf. He thought WAVEWATCH permitted a politically safer way to model waves than deploying instruments, which could be tricky in the Gulf, where Iraq's recent adversary, Iran, has many territorial claims.

One student based at Texas A&M had started working as a coastal engineer in her hometown of Busan, South Korea, where she had earned a master's degree that prepared her to study port breakwaters and their relations to wave breaking. She grew up fishing with her father and told me that when she was younger, a typhoon had flooded her grandfather's house. The 2011 tsunami in Japan strengthened her resolve to continue in engineering. For her, WAVEWATCH was also a better alternative to measuring waves in the field, which was more expensive.

A student based at George Mason University, who grew up in Pakistan, was interested in how wetlands might be engineered to attenuate coastal flooding, a problem he was looking at in the Chesapeake Bay.

FIGURE 4.4 Poem on waves by Saeb Tabrizi (1592–1676) in the Siyah Mashgh calligraphic style.

A postdoc at the University of Michigan, in the Great Lakes Environmental Research Lab, earned his bachelor's degree in marine science at Ocean University of China, in Qingdao, where he became interested in waves—and saw the subject as a way to bring his taste for math and physics "into practice."

A student at the University of Central Florida grew up in Turkey. She told me, "I have always been in love with ocean waves." She studied civil engineering at the Middle East Technical University in Ankara, Turkey, and, she said, her "interest in waves started in my freshman year at college with winning the title at a floating breakwater design contest." She was interested in wave energy and in "restoring the environment. I wanted to do something that may help with fossil fuel consumption for electricity generation. Then I combined my interest in waves with my desire to help restore the environment."

Another, who grew up in Ensenada, on the Pacific coast of Mexico, now works for a coastal shore assessment company in southern Mexico and has also been involved in modeling the Gulf of Mexico in the wake of the Deepwater Horizon spill.

One midcareer oceanographer working for the US Naval Research Laboratory Marine Meteorology Division in Monterey, California, told me that part of her job entailed maintaining WAVEWATCH for operations—helping calculate navigation paths for ships.

The George Mason University student Ali Mohammad Rezaie, who grew up in Bangladesh and studied water resources engineering at the Bangladesh University of Engineering and Technology, reflected more explicitly on the politics of wave models. He said he was interested in "nature-based solutions to coastal protection" and particularly in the buffering possibilities of wetlands, putting his interest in the context of growing up in Bangladesh. Expressing puzzlement about early Dutch interventions in the landscape there, he remarked:

> So in the coastal regions of Bangladesh we have kind of an annual inundation, almost a perennial inundation in a lot of places, because of riverbed sedimentation, and also because we have a tidal influence. Even if there is no cyclone or anything, the tide goes in and out and floods a lot of areas.
>
> Anyway, in the 1960s some Dutch experts came in and started building polders at lots of different locations on the coast to protect agricultural lands from flooding and saline water intrusion. While it initially boosted agricultural production at places and reduced coastal flooding, at some point people realized that the polders hindered the natural tidal flow, causing sediments to deposit in the riverbed and thus reducing the drainage capacity. As a result, nearby areas had prolonged waterlogging and drainage congestion. With increasing surges in tides—I don't know if it could be an effect of sea level rise—the polders started overtopping, and the water started staying in for a long time. And that had a lot of environmental and cultural impact [see chapter 5].
>
> I started looking at the science. I'm an engineer, you know. And I'm like, "What? These white folks came over from a different world and started doing this? It's neo-colonialism, somehow, as they never asked what people wanted or how they want to solve the problem. Maybe

> there's *some* similarity between the Netherlands and Bangladesh . . . , but not enough; at least not socioeconomically." So the solution should combine conventional engineering and nature and community-oriented approaches that are not only technically feasible but also socioeconomically acceptable.
>
> In Bangladesh, like lots of people, we bury people after they die, and so if your whole village is inundated, flooded all the year, you can't bury your family members. So in 2002, a few people in one village said, "This polder is a curse to us. Let's cut it down and allow the water to come inside properly and to drain naturally." They didn't have any engineering knowledge, any science, or anything. And it worked. It allowed the river to get deeper over time and also the natural flow to come and go.
>
> But the solution itself did not work in the end because of the limitation of institutional management and socioeconomic barriers. And all of that made me interested in management of coastal areas. And in flooding, storm surges—and in waves. I want to know how we can utilize existing resources such as our natural habitats, vegetation, and wetlands to protect people and the coast from flooding and storms.
>
> I would love to apply some of the techniques or knowledge that I am gaining here in the United States and at the WAVEWATCH workshop and to apply it at home sometime.

Think of this story as at least partially about *hacking* the stack of which WAVEWATCH is part, seeking to intervene with other protocols. Reviewing these students' stories from the workshop, I have come to think of them as data from the "ethnographic stack," Héctor Beltrán's (forthcoming) term for the layers of deliberation anthropologists encounter when they talk with programmers and users about what these people see as the active as well as broken links between technical possibilities and social priorities. If, for Bratton (2015), "the stack" is a kind of "accidental mega-structure" that chains together the cascading layers of *earth* (material resources), *cloud* (corporate internet infrastructure), *city* (everyday life), *address* (identification), *interface* (e.g., apps), and *user*, the ethnographic stack asks after how these are inhabited, contested, accommodated, resisted, multiplied, and situated by people and institutions.[13]

Another student, from the University of Aveiro in Portugal, was writing a thesis on "the impact of climate change on port defenses and maritime traffic." In an email he wrote to me, "In order to accomplish my thesis objectives, I need to simulate future hydrodynamic, atmospheric and wave patterns using numerical models, and one of them is WAVEWATCH III." He put his interest in a long national Portuguese context:

> Once the ocean gave us lot of goods. It showed us the world. But now we are getting affected by huge waves, which are destroying property, beaches, and dune systems that protect our coastline. This is getting worse each year. Recent studies show that climate change will increase the mean sea level (from 0.06 to 1.06 meters, depending on the carbon dioxide concentration in the atmosphere), as well as creating more intense storms in the North Atlantic. These two, combined with storm surge, can raise the sea surface, which will result in property damage, as well as cease all the maritime traffic and port operations. Facing this, I chose my thesis theme. So learning WAVEWATCH will allow me to model future wave patterns, contributing to my thesis objectives.

Stories of doing social good through science were recurring. Meixner said that earlier in her life she had been involved in

> modeling oyster larvae and tracers in the Gulf of Mexico, right after the BP oil spill. I've always liked the ocean, but I also decided I hated doing experiments, so I switched to math. Then I met a professor who worked on storm surge, and I liked the fact that you could affect people's lives—and actually work with emergency managers to help when storms hit, to urge people to evacuate. I am not the person who people call at the end of the day, but I *am* involved with the hurricane operational forecast models. I like the fact that you can do good with math.

Many participants described professional interests in waves not only as having emerged from the time of their individual lives, but also as entwined with the time of environments. My conversations with them in the ethnographic stack suggest that to Bratton's *earth, cloud, city, address, interface,* and *user* might be added *ocean* or *water,* a layer that saturates, pressures, churns through all others.

Backing Up an Archive of Waves

That wave forecasting depends on *hindcasting*—looking back—demands that forecasters have access to records of previous wave weathers. One person at NOAA told me about a thirty-year hindcast of wave spectra held by NOAA, hosting data from 1979 to 2009 and available for digital consultation online (https://polar.ncep.noaa.gov/waves/hindcasts). It is something like an archive of modeled waves. Much of it is held on tape, which can serve as a backup when needed. This person explained that

> one of the first versions of the archive was lost during the 2013 government shutdown [see the introduction], when the server started to malfunction and the federal employee who was on duty couldn't do anything about it, because during the shutdown they are only allowed to work on very specific activities. What is now available on the website is what was recoverable from the tape archive, and also from colleagues who had copied the files and were able to offer them back to NOAA once they saw that the website was missing some data.

Tape, of course, has its own vulnerabilities. It can be tricky to keep up with the past. "In operations, the directive is to produce forecasts," this person said. "So, while the Environmental Modeling Center produces the data, there is no directive, or money, to archive it long-term. For our operational data, it can be archived for seven days at a time—it's the most look back you can get." That makes the reanalysis archive precious, important to guard against falling into the "data shadows" (Leonelli et al. 2017). Fortunately, at least in the case of data loss during the shutdown, "some Canadian friends of ours also backed up the reanalysis archive." At the time, I thought of the "data rescue" efforts that had gotten underway in 2017, after Trump took office, which had led many scientists to worry about the durability of climate data. (Many of these efforts were spearheaded by scholars in Canada who had seen climate change data left to decay by the administration of Prime Minister Stephen Harper, from 2006 to 2015.) These were efforts in what Mike Fortun and Kim Fortun (2005) call "care of the data."

In 2019, a Trump government shutdown threatened more data decay. Suru Saha, a meteorologist I met in 2015, was interviewed by the *Washington Post* for a 2019 article titled, "The National Weather Service Is 'Open,' but Your Forecast Is Worse Because of the Shutdown." She is quoted as saying, "Things are going to break, and that really worries me because this is our job. We are supposed to improve our weather forecasts, not deteriorate them. I'm sorry. . . . I'm just really passionate about this. To be sitting at home watching scores go down, it feels terrible. We owe taxpayers the best" (quoted in Fritz 2019). I heard from another source that, during this thirty-five-day shutdown, the Environmental Modeling Center sent everyone home; they were told their mission was "non-essential," even though they provide models that weather prediction centers use.[14] This meant that a lot of contract employees never got paid, so people started to leave these jobs. As version control decays, a narrative about climate pasts vanishes, and waves may fall out of the books to become the eternally cycling, ahistorical forces they were often imagined to be before wave science brought them into a kind of history.

Hurricane Katrina in Ones and Zeroes

The opening three days of the workshop had led us through hands-on tutorials on WAVEWATCH. Having each been assigned our own Linux workstation and asked to download and register for the WAVEWATCH code under an open source license, we had been guided through a series of commands to compile the program. A list of executable programs had appeared on my screen:

```
ww3_grid  (model grid preprocessor)
ww3_strt  (initial conditions preprocessor)
ww3_prep  (input field preparation—wind, current, wlev)
ww3_trck  (wave output along a predefined track)
```

This made manifest that the program was running virtual wind input across a model grid to get wave output. Our test grid was called "Gulf of Nowhere," and, in specifying initial conditions ("the state of the wave field at the first time step"), I had seen that what we would be using to get the model "spinning" would be a wave spectrum:

```
Gaussian in frequency and space, cos type in direction.
fp and spread (Hz), mean direction (degr., oceanographic
convention) and cosine power, Xm and spread (degr. Or m) Ym and
spread (degr. Or m), Hmax (m)
```

More, it would be the JONSWAP spectrum:

```
JONSWAP spectrum with Hasselmann et al. (1980) direct,
distribution alfa, peak freq (Hz), mean direction (degr., oceano-
graphical convention), gamma, SigA, sigB, Xm and spread (degr. Or
m) Ym and spread (degr. or m)
```

As we began to work toward a more specific simulation—of Katrina—we were asked to generate a new grid: one for the Gulf of Mexico. This was done in a piece of software called GRIDGEN, written by Tolman and Chawla in

MATLAB, the programming language so often used to plot functions, run matrices, and write up algorithms. Typing commands into my machine, I called up a few maps of the Gulf of Mexico. One displayed bathymetry, or the shape of the seafloor. Another juxtaposed land with sea. Still another was a map made of 0s and 1s, with 0s the waters of the Gulf and 1s the land.

I looked at our three-ring binder lesson plan:

> In this tutorial exercise we will go through the steps of setting up a realistic, single-grid wave model, run in serial. . . . This model can then be used to explore a wide range of aspects of running WAVEWATCH III. . . . The model [is] set up for Gulf of Mexico during Hurricane Katrina (August 2005) when at least 1,245 people died in the hurricane and subsequent floods, with a total property damage estimated at $108 billion. . . . The simulation is designed to run from 2005/08/28 12Z until 2005/08/30 00Z, involving 36 hours, which should take approximately 15 minutes of computational effort.

Once the "grid" was set up, the next thing to do was to generate virtual winds to blow over the WAVEWATCH sea. One student asked, "Where do the winds come from?" We were told we would be using "reanalysis" information from NOAA—that is, values for wind speeds and directions that NOAA modelers have used to hindcast what winds seemed to be in effect during Katrina. I saw time-updating values scrolling down my screen and was told by an instructor, coming around checking on all of us, "That's the wind blowing over your model!" We had to let those winds "blow" over the model for a while—called "spin up time," which lasted about fifteen minutes. Along the way, we would be using historical data from buoy number 42040 to "validate our analysis." I saw a file titled NDBC42040.txt.

We students operated at our own speeds, stopping to compile, waiting for off-site servers to catch up with us, dealing with unexpected crashes. There was no sense of urgency; we did not have to *deal with* this storm. But speed—or processing time—was on everyone's mind. And that time was experientially sliced into modular parts. As people waited for their winds to "spin up" or for their program to reboot, they texted friends, checked in on dogs on their phones, looked at Twitter.

I had already asked students whether they remembered Hurricane Katrina, twelve years earlier. Many were not in the United States or were too young (or both) for it to be part of their formative memory. But for one participant, Talea Mayo—a professor of civil, environmental, and construction engineering at the University of Central Florida—Katrina marked a key moment in how she got into wave modeling. She told me:

> During my sophomore year in college, I was in Louisiana during Katrina, and I got interested in hurricanes. I could see how it was impacting people—and that it was clearly affecting Black people, people who looked like me—and so I was asking, "How could this be? What can I do?" Everything tends to affect poor communities more, because they just don't have the same resources. I met a professor from New Orleans, and he told me that it's not the storm; it's the *surge*—rising water from low-pressure weather—that destroys property. It's what causes the deaths. Here in the United States, we have forecasting technology that prevents a lot of people from dying in storms, but, as in underdeveloped countries, when storms do kill, it's the largely because of the surge. So in 2008, I went to the University of Texas, Austin, in Computational and Applied Math, and I started working on storm surge modeling. I like working in it because I know there's an application.

Mayo emphasized to me the centrality of questions of race, gender, and justice to her mathematical interest in wave and surge modeling. As we sat next to each other in front of our wave models, Mayo asked a number of questions about what we were seeing. I asked her what she thought about the WAVEWATCH model of Katrina. She said, "Well, I have a question about it. It's hard for me to understand how the model decouples the *surge*. It's modeling gravity waves, but this is not how I look at Katrina. What is a hurricane without surge? The major issue with this storm was the surge, not the wave field." I asked her to say more:

> A lot of times in surge modeling *we turn off the waves* because it takes so much computing power to calculate them. We can afford to ignore the waves since *that's* not what's flooding people. Maybe in Hawaii, but generally not in the Gulf or East Coast.
>
> If you're a coastal engineer building things to survive storms, you might want to know what the regular waves are like. You might want to model all of it. But with surge models, we often turn *off* waves because they are not what we want.
>
> But you have to understand the limitations of your models. So I don't think the problem with WAVEWATCH was that it wasn't accounting for surge. I think that it's not *supposed* to. So you have to know what is the

limitation of this model. This model is going to model waves, but it's not going to model surge, because that's not what it's *for*.

I'm always thinking about limits computationally. You can usually add more details in, but maybe it's not worth the payoff, right? It's going to cost you more to solve your equations. That's kind of what limits storm surge modeling a lot—like, we'll just ignore the waves because if you include them, your model runs so much slower. But it's just that because I specifically study surge, it is odd to look at data that isn't surge. But everything's systemic, right? We want to separate surge from waves like we want to separate sociology from engineering, but in the end it's one system.

So modeling Katrina with WAVEWATCH does not necessarily tell everyone everything they want to know—though it may be one ingredient. Bringing together surge and waves, like bringing together sociology and engineering, will be necessary to track the rise of water. In 1903, in *The Souls of Black Folk*, W. E. B. Du Bois, looking at the question of racial inequality in the United States and in the world, wrote as he looked to the future: "The problem of the twentieth century is the problem of the color-line." Today, I would add—with race an ongoing fracture in the United States—that a key problem of the twenty-first century will be the problem of the water line. And the water line and its effects will emerge as part of a stack—not just of *earth*, *cloud*, *city*, *address*, *interface*, *user*, *water*, and *ocean*, but also of race, region, class, nation, and gender, names for structures of collectivity and experience that are at once sociological and everyday abstractions, as well as thoroughly inhabited, enacted, and unfolding within parameters that format life-and-death consequences.

The Arc of Memory

On the last day of summer school, I headed to Washington, DC, with a couple of coastal engineers from Mexico who were curious to see the sights. After I took them to the Martin Luther King Jr. Memorial, we walked around the outside of the National Museum of African American History and Culture. The building made a contrast with NOAA's sleek Center for Weather and Climate Prediction, which had suggested to me nothing so much as a spaceship ready to ride the weather of the future: NOAA's ark. The African American museum was something else; like the NOAA building, it was up to the architectural minute, but its triangular lines, meant to evoke a Yoruba pot, a New Orleans–style gate, and more, also echoed a ship—a ship like those that brought so many Africans to the Americas in bondage in a centuries-long slave trade that provided a racialized infrastructure for the world capitalism that would eventually play an outsize role in visiting the planet with rising sea levels (Baucom 2020).

Christina Sharpe's *In the Wake: On Blackness and Being* (2016, 17–18) points the reader, as I discussed earlier, to "the wake"—that concept that refers at once to "the path of a ship," to "keeping watch with the dead," to "a consequence of something," and to a coming to consciousness. Sharpe's wake offers a recognition of staggered time, a temporal looping, particularly in her argument, for Black people in the wake of the Middle Passage, that shows how the past is never past, how futures are made by looking backward to go forward. Such operations in reckoning with racial memory are explicitly stacked into the architecture of the National Museum of African American History and Culture. They find uncanny resonance, I think, in wave thinking and programming in such artifacts as WAVEWATCH, which, I have suggested, stacks received, authorized knowledge together—and sometimes in tension—with stories yet to be brought in, versions, subaltern and otherwise, yet to be recognized. The ethnographic "hindcast" that I have sought to deliver in this chapter seeks to backtrack—back stack?—attending to how computer memory may variously enable, embed, and diffract the wakes and breaks of individual memory, cultural memory, and the arc of ocean memory.

Set Four

First Wave

Middle Passages

Hernan del Valle, head of humanitarian affairs for Doctors without Borders, told me about waves. We were outside his office at Harvard University in 2019, where he was visiting for the year, writing about his organization's search-and-rescue efforts in the Mediterranean. He had been working in this sea since 2015, when migrants crossing the sea from its southeastern boundaries, hoping to get to Europe for work or asylum, started fleeing in the thousands from civil war in Syria and Libya; conflict in Afghanistan and Iraq; and poverty, famine, and war in sub-Saharan Africa. Resurgent European isolationism was starting to describe migrants as surging across the sea in "waves," pitching refugees as massed agents of threat, as overwhelming flow, framing them through a xenophobic—and racist—optic that dehumanized them.[1] The wave metaphor performed a rhetorical bait and switch that displaced attention from the waves over which so many of these people had traveled, waves that had become part of the dangerous force field that migrants confronted as they crossed the Mediterranean, often in overcrowded rubber rafts, fishing boats, and dinghies.

Water waves are forms that, with rescue boats struggling to do their work, pattern a deadly, fluid border. In 2015 alone, del Valle said, some 4,054 migrants drowned when their boats sank, a figure that would have been higher if not for interventions from nongovernmental organizations (NGOs) such as Doctors without Borders, which from 2015 to 2018 ran 425 rescue missions, assisting 77,000 people (Abrams 2019).[2] In too many cases, the sea has become a weaponized ecology to which many European countries have outsourced border control (its name, the Mediterranean, poses it as an in-between, what proto-Indo-Europeans considered "in the middle" of known lands). Waves have become graves, a connection made explicit by the Syrian artist Khaled Barakeh in such works as "Multicultural Graveyard" (2015), in which he bears harrowing witness to death by gathering photographs of migrants' bodies on beaches as they are washed over by waves, embalmed by the willful neglect of those who could have rescued them (see Mirzoeff 2015).

That neglect has been at the center of investigations by Forensic Oceanography, a team at the University of London led by the research architects Lorenzo Pezzani and Charles Heller. Since 2011, the team has dedicated itself to reconstructing the tracks of migrant boats crossing the Mediterranean. The team's Left-to-Die Boat project tracked the March–April 2011 travel, just north of Libya, of a boat carrying seventy-two migrants, sixty-one of whom "lost their lives while drifting for fourteen days in the NATO maritime surveillance area" (Heller and Pezzani 2012). As Forensic Oceanography explains, "By going 'against the grain' in our use of surveillance technologies, we were able to reconstruct with precision how events unfolded and demonstrate how different actors operating in the Central Mediterranean Sea used the complex and overlapping jurisdictions at sea to evade their responsibility for rescuing people in distress" (Heller and Pezzani 2012). The sea, Heller and Pezzani say, might at first glance be judged an archive forever undoing itself: "The waters that cover over 70% of the surface area of our planet are constantly stirred by currents and waves that seem to erase any trace of the past, maintaining the sea in a kind of permanent present. In Roland Barthes' words, the sea is a 'non-signifying field' that 'bears no message'" (657). But we are no longer (if ever we were) in Barthes's world. They continue:

> The contemporary ocean is in fact not only traversed by the energy that forms its waves and currents, but by the different electromagnetic waves sent and received by multiple sensing devices that create a new

> sea altogether. Buoys measuring currents, optical and radar satellite imagery, transponders emitting signals used for vessel tracking and migrants' mobile phones are among the many devices that record and read the sea's depth and surface as well as the objects and living organisms that navigate it. By repurposing this technological apparatus of sensing, we have tried to bring the sea to bear witness to how it has been made to kill. (658)

When I met up with Pezzani in Amsterdam, he told me that Forensic Oceanography had been looking at waves, reading them to determine the orientations of prevailing winds and, therefore, the direction of travel of ships and rescue boats. In one case for which Pezzani showed me a wave simulation image, Forensic Oceanography was seeking to defend a German NGO accused by an Italian court—on the basis of an out-of-context photograph—of colluding with Libyan smugglers. Forensic Oceanography, with the help of the Woods Hole Oceanographic Institution oceanographer Richard Limeburner, has been able to reconstruct wave direction.

Back in Cambridge, del Valle told me that understanding waves was something in which mathematicians working on search-and-rescue algorithms had become interested. He sent me a paper, "Complex Network Modeling for Maritime Search and Rescue Operations," written by the computer scientists Alexey Bezgodov and Dmitrii Esin (2014). In it, they describe a simulation method for modeling "floating objects on irregular waves," offered as a way to study "the efficiency of search and rescue operations at sea," including for lifeboats. Individual floating entities are considered an assemblage or network that deforms as wave action separates them.

Here, then, was an up-to-the-minute mode of reading waves. In characterizing such forensic approaches, the literary scholar Daniela Gandorfer (2019, 221–22) suggests that infrastructures and the built environment "function as media since they are storage and inscription devices, while they also interact and affect the very process they record. . . . In this account of reading, subject and object, reader (architect) and text (material interrelations and assemblages) switch back and forth, and thereby constantly affect and influence each other." What this means is that such reading, this seizing of "the means of perception" (Weston 2017), is always and necessarily political and must always take responsibility for its practice, for its role in unfolding historical process.[3]

The scholar of Mediterranean migration Alessandra Di Maio has called the deadly sea at the center of the early 2000s refugee story the "Black

Mediterranean," echoing the "Black Atlantic," Paul Gilroy's name for the territorial connections that have made the African diaspora. *Black* here operates in a register overdetermined by, but also in excess of, racialized Africanity; Di Maio (2013, 42) writes that black "is the color—or rather, non-color—in which all shades merge, that which the sea assumes during the crossings pursued by the million migrants who have 'burnt' it in the past three decades." The death that now visits the Mediterranean echoes earlier terrors of the Black Atlantic's Middle Passage: people on boats captive, sick, dying, crossing a space of dispossession, of unmaking.

Christina Sharpe (2016) draws comparisons between today's migrant deaths in the Mediterranean and the deaths of those kidnapped Africans who perished on crossings of the Atlantic during the days of the triangle trade. One emblem of that deadly passage is J. M. W. Turner's painting *Slavers Throwing Overboard the Dead and Dying, Typhoon Coming On* (1840), a representation of an event that took place in 1781 in which a Liverpool slave ship, the *Zong*, threw overboard—that is, murdered—142 captives to recalibrate its ballast. The vessel was built for 220 and had sailed with more than twice that, then proceeded to claim "property loss" to its insurers. The waves, in Turner's image rendered in smeared copper-red oil paint applied with a rushed brush—what Sharpe (2016, 36) calls a "roiling, livid orpiment"—translate Turner's disgust at slavery, and, particularly its ongoingness in North America. The artist and critical theorist Ayesha Hameed (2014, 713) writes, "The storm is a concatenation of rain water, blustery wind, and crashing waves—a catalyst that puts into motion a chain of events on board a slave ship that refracts the zeitgeist of chattel slavery and maritime insurance." *Slavers*, exhibited at the First World Anti-Slavery Convention in London in 1840, makes physical a repugnance for slavery, in the idiom of the disorienting (the horizon line in the painting is tilted) and sickening terrible, though it is also haunted by an aesthetic of the sublime. It is possible to argue that *Slavers* is an early representation of the Anthropocene on the waves, or, with the depredations of worldwide colonial genocide in view, the Plantationocene.

Turner's *Slavers* approaches its topic in a figurative mode, making the *Zong* stand for all such ships. That generality, along with its shimmering aesthetic, has made some commentators wary. They argue that the people drowning emerge as "exotic and sublime victims" (Dabydeen 1995, 8), making the painting, as "a monument without names" (Wood 2000, 46), an object that trades on images of Black suffering. The tug of war between aesthetics and politics that commentators have mapped—with the nineteenth-century critic John Ruskin arguing that the painting delivers "the greatest union of moral power and poetic vision that British art ever accomplished" and late twentieth-century critic Tobias Döring arguing that "the terrors of the [slave] trade have become transfigurated as aesthetic objects produced for the delectation of spectators" (quoted in M. Frost 2010, 380)—has opened interpretative avenues for later artists, including Sondra Perry in her exhibition "Typhoon Coming On" (2018), which presents full-wall video screens showing digitally animated close-ups of Turner's water, pressing viewers into, rather than away from, its beaten red waves.

Think of Turner's and Perry's works next to wave simulation images and their smooth surfaces. Turner refuses clean lines among waves, bodies, flesh, ship, light, suffusing the image with a bruised terror; for slavers, waves are forces of nature called on to render official murder as the loss of "cargo." Perry queries the line between viewer and work, seeking to upend a distant, outside view and working to immerse the viewer in the specific and general of racialized, oceangoing violence.

Forensic Oceanography, working with hydrodynamic models, seeks to make claims, like Turner, about more than one ship—and, like Perry, about ongoing violence—though this time in the key of the clinical, evidentiary, legally admissible. Waves, made legible through hydrodynamic models, turn into witnesses.

Meditating on the synecdoche Turner offers, in which one ship stands for slaving writ large, Sharpe offers the contrasting, complementary work of M. NourbeSe Philip's 2008 poem *Zong!*, which, working with court case transcripts, seeks to return some singularity to the event. Philip, selectively effacing letters and words from the court case that accepted the murder of people as a matter of maritime ballasting, creates a poem that offers a kind of afterimage-forensis of what happened on and under the waves, a story that, she writes, cannot be told/must be told (facing page).

Philip (2008, 198) writes that the "not-telling of this particular story is in the fragmentation and mutilation of the text."[4] It is a disaggregation and recomposition that, in the quoted passage, suggests a sinking, even a falling orbiting, beneath the waves of bodies at the beginning of what Sharpe (2016, 41) discusses as "residence time," or "the amount of time it takes for a substance to enter the ocean and then leave the ocean." The salt in blood, she writes, has a residence time of 260 million years: "We, Black people, exist in the residence time of the wake" (41). Residence time is present in the breaking of the waves. The Atlantic Ocean, suffused with the afterlives, the blood, of the dead, becomes a Black Atlantic monument.

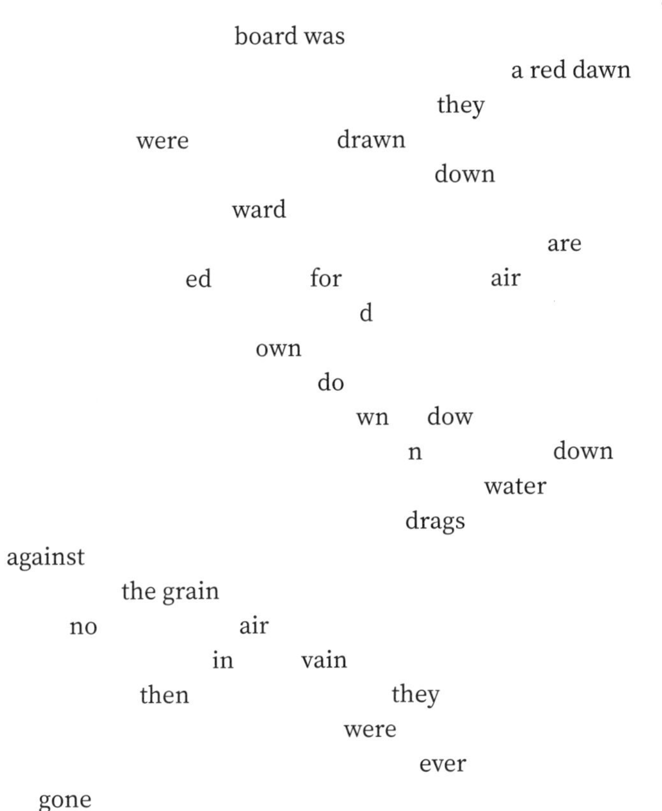

&over
board was
a red dawn
they
were drawn
down
ward
are
ed for air
d
own
do
wn dow
n down
water
drags
against
the grain
no air
in vain
then they
were
ever
gone

Philip is adamant about retrieving some sense of the individual lives lost in the *Zong* massacre (see Shockley 2011, 814). In countering a discourse that would reinforce the legal anonymity of the murdered, Philip assigns these passengers Yoruba names.

That work of centering biographies, similarly for del Valle, can and should be done in discussing the trials of people crossing the Mediterranean. (The story of Alan Kurdî, the Syrian three-year old who drowned, and whose image, dead on a beach, was circulated globally, provides one devastating example.) But this alone is not enough to call structures of politics, economics, and justice to account. Part of what Doctors without Borders and Forensic Oceanography seek to do is present findings in courtrooms in a way that might call law to account in a way it was not at the moment of the *Zong*.

Storytelling, poetry, and sound continue to have their place in projects of memory, intervention, and future making. Tuning her ears to the sounds of sonic science fiction, Hameed analyzes the mythology of the Detroit techno duo Drexciya (see "Set Two, Second Wave: Radio Ocean"), which imagines the descendants of drowned African captives reanimated as founders of a Black Atlantis. She hears in Drexciya's music a call for ongoingness, mixed with a never calmed dread, a nonlinear time: "A key aspect of the Drexciya myth is its temporal proposition: to see time and history as equally in flux as the lapping ocean, to see the afterlife of the middle passage in a futuristic scenario. Its soundtrack is the sound of sonars, of static, of distance" (2014, 714). The residence time that waves write here is one that does not so much underwrite "maintaining the sea in a kind of permanent present" (Heller and Pezzani 2014, 657) as carry the tidalectic echo of seaborne history. Edgar Arceneaux's acrylic and graphite *The Slave Ship Zong* (2011 [plate 15]) explicitly resonates with a Drexciyan mythos: a gigantic black wave looms over disembodied eyes that conjure not the drowned but those watery progeny of science fiction who swim sentinel through the Black Atlantic.

In *elocation (or, exit us)*, the poet Evie Shockley (2006, 18) narrates one woman's mapping of an American city—a city incised by the wounds of train tracks and highways. I cannot help but hear a resonance in the poem of treacherous sea terrains, as Shockley tells us of a territory divided

> like a half-red sea
> permanently parted, the middle she'd
>
> pass through, like the rest, in a wheeling rush,
> afraid the divide would not hold and all
> would drown—city as almighty ambush—
> beneath the crashing waves of human hell.

The migrant Mediterranean may today be a human hell, *a half-red sea*: part promise of Red Sea exodus, part blood-black terror, and part dehumanizing infrared surveillance-scape. It is also a *half-read sea*—a passage still only half-legible to those who would cross it (in flight or in rescue), a space legible in still insufficient "half-tellings" (Philip 2008, 199), a zone crosshatched by the electromagnetic waveforms that Forensic Oceanography seeks still to decode.

In November 2019, at a conference in Germany on Mediterranean migration, I met Amel Alzakout, a Syrian filmmaker who almost drowned in 2015 when a boat she was traveling on capsized just off the coast of Greece, leaving forty-two people out of 316 dead. It was then that I learned about

Purple Sea (Sheen 2020), a documentary she created with Khaled Abdulwahed that made use of footage from a camera she had strapped to her wrist and that, during her hours in the water, recorded the accident and unfurling aftermath (Crucefix and Alzakout 2019; Forensic Architecture 2020). One image from the film, an underwater close-up of a lifejacket set just below the surface of the sea, looks, as the jacket's red-orange is diffracted and extended into fingery glass-brick ripples, like the splash of a red wave.

Set Four

Second Wave

Wave Power

In *Historia Anglorum: The History of the English People*, written in the twelfth century, Henry, Archdeacon of Huntingdon, included a homily about the deeds of King Canute the Great, a monarch who in the eleventh century ruled over England, Denmark, and Norway. Henry ([1133–55] 1996, 367–69) elaborated on a legend in which King Canute attempted to command the sea to cease its tides (figure 4b.1):

> At the height of his ascendancy, he ordered his chair to be placed on the sea-shore as the tide was coming in. Then he said to the rising tide, "You are subject to me, as the land on which I am sitting is mine, and no one has resisted my overlordship with impunity. I command you, therefore, not to rise on to my land, nor presume to wet the clothing or limbs of your master." But the sea came up as usual, and disrespectfully drenched the king's feet and shins.

This story—sometimes known as "Canute and the Waves" (see Lord Raglan 1960)—has been employed to describe the overreaching arrogance of ruling power, particularly when it comes to (under)estimating the forces of large-scale processes, both natural and social. Take as one example

FIGURE 4B.1 Courtiers flattering King Canute's pride, telling him that ocean waves would roll back if he so commanded them. Artist unknown.

the comments of the Louisiana lawyer Stacy Head, who in 2005 blasted the New Orleans City Council's response to Hurricane Katrina—a call "to extend daylight-saving time just for Orleans Parish" (so people would have more "time" to work on repairing their houses)—comparing the council's actions to those of King Canute (Nolan 2009). The Canute story, used this way, points to the folly of seeking to control, in the realm of the political, energies that might rather belong to domains beyond the human or, if human (e.g., enduring social conventions, revolutionary forces), may be beyond full sovereign control. But according to Simon Keynes, a professor of Anglo-Saxon, Norse, and Celtic history at the University of Cambridge, the story is ultimately about Canute's wisdom, for Henry's tale concludes: "So, jumping back, the king cried, 'Let all the world know that the power of kings is empty and worthless, and there is no king worthy of the name save Him by whose will heaven, earth and sea obey eternal laws'" (see Westcott 2011).

This tale is not only about wisdom, but also a medieval king's recognition of God as the master of the earthly realm. Canute's placement of his throne on the beach articulates a theory of human sovereign power that recognizes the limits of that power even as it draws command from an appeal to a higher supernatural authority. Later retellings of the Canute story treat the waves as a symbol for forces of social transformation, and such tellings continue to come in two lineages, activating either the foolish or the wise Canute. The forces in the stories might be tides of immigration, political changes, or, more recently, human-induced climate change or even COVID-19. In such adaptations, the point is to draw attention to the inexorability of processes beyond full social capture.

But ocean waves—to go back to the literal subject—*can*, of course, be captured; they *can* be sometimes made to bend to human power. In winter 2019, I traveled to Denmark (where Canute is known as Kong Knut, the Viking king) to learn about North Sea experiments in the generation of electricity from ocean waves. I arrived at the windswept North Sea fishing town of Hanstholm and met with Christian Nereus Grant, the chairperson of the Danish Wave Energy Association. Grant runs the association out of his house, sited just back from a stunning North Sea shore, a coast marked by permanently windblown trees and, here and there, a small vacation house with surfboards in the driveway. I learned from Grant about efforts to extract power from waves, to bring them under command, either to "harvest" (agricultural metaphor) or to "harness" (working animal metaphor) the energy flux that comes from waves' oscillation and propagation. Grant told me that wave energy, if efficiently extracted, could cover some 15 percent of Denmark's energy needs. Starting ten years ago, at midlife, he began to pin his family's future on that promise. And he came to understand the power of Hanstholm waves viscerally when he was nearly swept out to sea by a ten-meter wave as he worked, during a storm, to secure equipment. In an article in the Danish press about this remarkable episode, he is described as "dreaming of finding a way of extracting electricity from a wave like the one that had nearly killed him" (Frich and Veilmark 2017, my translation using Google Translate).

Grant filled me in on machines proposed over the years to squeeze energy from the motion of waves: the Wave Dragon, the Sea Snake, the SQUID, the Oyster, the Vibristor, the Poseidon, the TETRON, the Kymogen, the Volta WaveFlex—their sea-creaturey and cybernetic names an indicator of the mix of mythic and technological hopes tagged to these devices. Inventors of such machines start by testing scale-model prototypes in wave tanks and then pitch their imagined infrastructural scale-ups to investors. Some seek to anchor prototypes in the waters off Hanstholm, where they are subject to forces of salt, sand, weather, and sea life. Some require small-boat and scuba-diver maintenance.

Back in 2009, a mammoth (but still half-scale) point-absorbing device, made by a company called Wavestar (defunct for a spell but now back in business), stood in offshore waters just outside the Hanstholm harbor, looking like a giant quadrupedal robot staring down at the sea as if scanning for a submerged toy. Grant moved into town from Copenhagen to be the site manager for Wavestar. Now, in addition to chairing the Danish Wave Energy Association, he works for the Danish Wave Energy Center (DanWEC), which seeks to build an at-sea test area outside Hanstholm, an area that might provide sea and land cable infrastructure so wave energy machines can be plugged into the grid, their output measured and certified. (The model is the European Marine Energy Center, based in the Orkney Islands, which has berths for wave and tide energy prototypes.) Grant also captains his own consulting company, OctoMar, which points inventor-entrepreneurs to promising places for wave energy machines; sells them information about North Sea wave heights from Datawell Waverider buoys; and aids them in working with the Danish Maritime Authority, whose permitting process worries about whether wave machines might pollute the sea, make too much noise for marine life, or interfere with fishers. Overlapping property claims and domains crosshatch this wave-energy sea—the water column, the seabed, the shore. The power of waves can be accessed only by navigating a swirl of regulatory frameworks, diving in and out of a mixture of enclosed commons and unsurveyed territory. Relaying wave electricity also requires a network of substations, easement rights of passage, environmental impact statements, and favorable tax and tariff laws for the transmission and purchase of energy. If waves were (or were not) subject to King Canute as a sovereign, nowadays waves are objects within a more distributed, modern, secular polity, woven through with multiple claims.

The anthropologist Brit Ross Winthereik, who put me in contact with Grant, has argued, with Laura Watts, that Hanstholm is an *edge* place (Watts and Winthereik 2018). On the northern shore of Denmark's Jutland Peninsula, Hanstholm is a self-fashioned frontier, a site for experimental futures. Grant more than once told me that the area—which he called "the world's biggest laboratory for testing sustainable green energy"—might become a "Silicon Valley" for wave energy. He gave me a handful of stickers distributed by local boosters trying to brand Hanstholm as the "Cold Hawaii," a name that speaks to seashore fantasies of bracing natural energies.[1] Winthereik (2019, 29) zeroes in on the future-facing discourse of potentiality that shapes wave energy hopes, writing about how a wave energy prototype she saw at a trade show "both demonstrated the capacity of its hydraulic system to smoothly follow the movements of imaginary waves, *and* a green transition that would involve a number of different renewable technologies. With this placement, the wave converter was not just another potential product on display. It was an artefact that made the proposal that in the future, energy consumers will get their energy from a mix of sources, wave energy being one of them."

That future, reports Laura Watts (2018, 68) in *Energy at the End of the World: An Orkney Islands Saga*, her ethnography of tidal and wave energy projects, is most densely occupied in the waters of the Northern Isles of Scotland, where wind blows waves into heights of up to twenty meters. The electricity made by Orkney tidal and wave energy, Watts argues, exists in a space of potential, "in the double sense of electrical voltage and political difference." Such potentials might be positive as well as negative, either catapulting Scotland into the forefront of a new energy world or retaining these northern seas as a periphery in service to the needs of the United Kingdom.

Grant took me to the waterfront to see parts of a wave piston device, sitting in a state of disassembly in a large marine warehouse. He also pointed to the ruins of another machine—a concatenation of plastic yellow cylinders meant to work by floating on the water and flexing, hingelike; this one had a short stay in the water, having been smashed by the ocean pretty much as soon as it was tested. Wave energy machines need to exploit wave energy without being overwhelmed by it. High-energy seas—seas of Sturm und Drang—are desirable, but as Watts (2018, 271) writes about Orkney, "There are always statistical outliers, freak and unpredictable waves, that are dangerous to a generator that is designed to withstand only smaller, anticipated waves."

Think back to that lineage of King Canute that leaves waves to their own devices. If it is near the shore where wave phenomena become most subject to legal attention, whether to do with energy or with coastal engineering, sometimes waves "themselves" become objects for legal protection. Waves may no longer simply be entities to be survived, endured, prevented; much as coral reefs have gone from being experienced as *threatening* (e.g., to ships) to being perceived as *threatened* (Braverman 2018), so have some coastal waves come to be, in the eyes of their advocates, endangered.

Take the case of Save the Waves, an international nonprofit coalition dedicated to preserving the wave profiles of select beaches. The organization, headed by surfers, seeks to protect waves from being modified or from disappearing. They nominate, as part of their advocacy, entities they call "Endangered Waves," writing, "When an epic wave or coastline is under threat from poorly planned development or pollution, we mount campaigns to educate the public and take direct action through our Endangered Waves and Branded Campaigns."[2] Save the Waves, contra King Canute, seeks to create conditions by which waves can continue to operate as they have historically. Among the strategies the organization uses are legal ones. They work with local communities that hire lawyers to take a close look at the legal frameworks behind real estate development projects.

One project to which Save the Waves turned its attention in 2016 was a seawall proposed to protect a golf course in Doughmore, Ireland. It reported on its website: "US President-Elect Donald Trump and his hotel company, Trump International Golf Links (TIGL), seek to build a massively controversial seawall on a public beach to protect his Trump Golf Resort in western Ireland" (Save the Waves 2016). The wall was designed to "run 2.8 kilometers, reach 15 feet tall, and consist of 200,000 tons of rock dumped in a sensitive coastal sand dune system." Doughmore had earlier been designated a "Special Area of Conservation" by the European Union Special Habitats Directive, so there was an existing legal structure within which protection of the beach—and its waves—could work. Save the Waves (2016) reported,

> After a series of winter storms in February 2014, Donald Trump began to illegally dump boulders along the public beach at Doughmore without any permits to protect his golf course. Enraged local authorities quickly intervened and Trump was forced to cease his illegal revetment and is now required to obtain the legal permits. Trump has grown incensed that he needs to comply with the local planning regulations and has threatened to close the golf resort if his permit is not approved. Trump sought special permission from the Irish national government for the wall in March but was rejected in April. The local Clare County Council is now the responsible agency deciding the fate of Doughmore Beach. They have reviewed Trump's permit application and Environmental Impact Statement and have sent a Request for Further Information outlining 51 specific points that they want resolved or clarified.

Here, Trump operates as the overreaching version of King Canute, seeking to control the waves that might compromise a property he owns. The wall is like Canute's throne, placed on the beach to enact a theory of sovereign power over the water. Save the Waves' #NatureTrumpsWalls campaign

FIGURE 4B.2 #NatureTrumps Walls. From Save the Waves Coalition.

(figure 4b.2) collected 100,000 signatures against the proposed wall (Save the Waves 2016).

Whereas Trump famously dismissed the reality of climate change, his organization operated with climate change firmly in its calculations (Rhodan 2016). This is not a shift to the wise and humble version of King Canute. The practice of sovereignty here is cynical. Science and law are rhetorically coproduced at one moment and torn asunder at another.[3] Trump's aspiration to power, part of which he sought to burnish by negating science, operated through attempting to decide when law and science would prevail and when they would not, as when the US Environmental Protection Agency started to scrub climate data from its website in the early days of his administration (Hiltzik 2017; see also Schlangher 2017). King Trump sought not to control waves but to cancel them.

As of March 2020, Save the Waves prevailed. Wiser Canutes won the day, for now. For how long will depend on the turn and churn of social time, ocean tide, and wave power.

Set Four

Third Wave

Wave Theory ~ Social Theory

In *The Democracy Project*, the anthropologist David Graeber (2013, 64) reported on what he saw as "a wave of resistance sweeping the planet." According to Graeber, the wave began in Tunisia in January 2011; moved across the Middle East—Egypt, Libya, Yemen—to manifest as the "Arab Spring"; and traveled on to the Occupy movements that materialized across the United States later that year. Protests in Brazil, Greece, and Turkey in 2013 rolled into view next, all framed as "waves."

Back in 2011, in Tunisia, protests against Zine El Abidine Ben Ali culminated in his removal. An editorial cartoon published on January 25, 2011, in a Pan-African newspaper featured a sunbathing Muammar Gaddafi of Libya relaying news of Tunisia's upheaval to an intravenously fed President Hosni Mubarak of Egypt and represented the revolution as an approaching wave (figure 4c.1). Around the time the cartoon came out, the figure of the wave was organizing a different set of conversations in Tunisia, among scientists planning WAVES 2013, a conference on mathematical aspects of waves in domains ranging from solar seismology to the acoustics of grand pianos, cardiology, and oceanography.

FIGURE 4C.1 Godfrey Mwampembwa (Gado), "Tunisia's Waves of Revolution," *Pambazuka News*, no. 514, January 25, 2011.

I was at the beginning of conceptualizing this book and planned to attend. I couldn't get to Tunis in 2013, but Nabil Gmati, head of the Wave Propagation research team at the École Nationale d'Ingénieurs de Tunis, invited me to a follow-up in 2014 called Inverse Problems, Control and Form Optimization. That conference gave me key concepts to carry forward. "Inverse problems" ask how to retrodict earlier states of a system, inferring, for example, the shape of waves from scattered effects. As I listened to talks with titles such as "A New Approach to Solve the Inverse Scattering Problem for Waves," "Mathematical Modelling of the Electrical Wave in the Heart from Ion-channels to the Body Surface," and "Inverse Spectral Conductivity Problem in a Periodic Quantum Waveguide," I learned not only how wave mathematics and analogies shuttled back and forth across fields, but also about how this work required thinking from ends to means and back.

The figure of the wave haunted Tunis talks on social phenomena, too, from a lecture on the oscillation of human populations after disasters to a presentation on modeling crowds. My conversations with the mostly

Tunisian and French scientists at the gathering kept returning to people's recollections of the Tunisian revolution, which they explicitly described as a wave. Reconstructing events leading up to the Tunisian revolution was itself, I heard, an inverse historical problem. Walking through Tunis—in those days partially soundtracked by the Tunisian DJ collective Waveform—I spotted a poster advertising a play, *Tsunami*, about the revolution's early days. Bids to look back were everywhere.

The Tunis conference left me with a question about how the wave image in social theory operates. I knew that cyclical explanations of social history had a long vintage, traveling back—at least—to the Tunis-born historian Ibn Khaldūn, whose *Muqaddimah* (1377) forwarded a model of dynasties rising in cycles of warfare.

The notion of a wave as a form that might propagate through a social medium arrives in English in the nineteenth century. A scan of the *Oxford English Dictionary* (*OED*) to locate phrases that use *wave* to describe social forces picks out: waves of opinion (1809); waves of population (1852); trade waves (1891); waves of immigration (1893); new wave (1960); waves of feminism (1970s). Two social senses of *wave* are present here. One emphasizes the MATERIALS or BODIES bound up in aggregate through processes of motion. Definition 2c of the noun in the *OED* has *wave* as "a forward movement of a large body of persons (chiefly invaders or immigrants overrunning a country, or soldiers advancing to an attack), or of military vehicles or aircraft, which either recedes and returns after an interval, or is followed after a time by another body repeating the same movement." An example appears in *History of the Anglo-Saxons from the Earliest Period to the Norman Conquest*, in which Sharon Turner (1820, 24) wrote that, though much human migration has proceeded over the sea, "the great waves of population have rolled inland from the east." In this definition, it is the movement of bodies that matters.

A second sense of *wave* (*n.*, *OED* 3b), also from the nineteenth century, has it as "a swelling, onward movement and subsidence (of feeling, thought, opinion, a custom, condition, etc.); a movement (of common sentiment, opinion, excitement) sweeping over a community, and not easily resisted. Also, a sharp increase in the extent or degree of some phenomenon." This meaning names the FORM of a process—and the substances caught up in the wave need not themselves necessarily move; a sentiment may "sweep over a community." An instance comes from the *Panoplist, and Missionary Magazine United* ("On Creeds" 1809, 303), which warns believers against being "cast afloat upon the waves of opinion . . . with no fixed star and no unvarying needle to direct us." Such waves are maritime in metaphorical reference, emerging at a time when, as Alain Corbin (1995) has shown, maritime activities were bringing more Europeans to the sea as a site of work and danger, as well as of leisure and aesthetic production.

The material and formal endure as the primary modes through which social waves are described, though the boundary is always blurred. Indeed, the analytic of the wave of social change wavers between high theory and popular model, between objectivist sociological explanation and vacuous sociobabble, between vanguardist predictions of social revolution and conservative prognoses of political inevitability. The title of this intercalary chapter, "Wave Theory ~ Social Theory," gestures toward such vacillating interpretation. For logicians, the tilde symbol means "is not." For those of a mathematical bent, the tilde denotes "is approximately equivalent to." Right.

In nineteenth-century Europe, techniques of inscription—graphical, numerical, diagrammatic—produced formal claims and hypotheses about rising and falling tendencies in the social body. Statistics helped this happen. Having originated as "state numbers," statistics muster the possibility of representing—and revealing—aggregate social phenomena. The period between 1820 and 1840 in Europe saw the emergence of data collection by bureaucratically minded nation-states and "friendly societies" dedicated to self-insurance for workers, which sought to project average rates of workers' sickness (Hacking 1982). Such record-keeping inscriptions made visible tendencies in the polity, often by graphically indicating the wavelike rise and fall of indicators, which, in turn, secured the notion of a knowable, measurable social body through which processes could propagate.

Alain Desrosières (2002, 8, 10) argues in *The Politics of Large Numbers* that statistics combines "the norms of the scientific world with those of the modern, rational state" and that the "creation of administrative and political spaces of equivalence allows a large number of events to be recorded and summarized according to standard norms." A collection of individuals losing jobs becomes "unemployment"; a number of babies born becomes "the birth rate"—new kinds of objects, *things* accorded the status of phenomena in the world, emerging through what Desrosières calls "the realism of aggregates" (67).

In early public health work, *On the Geographical Distribution of Some Tropical Diseases* (1889), Robert Felkin described "pandemic waves." For Felkin, the wave was not just a line that could be drawn by a record keeper between a run of printed numbers. It could also be inscribed on a map, as if it were tracking a physical force moving across geography. The pro-

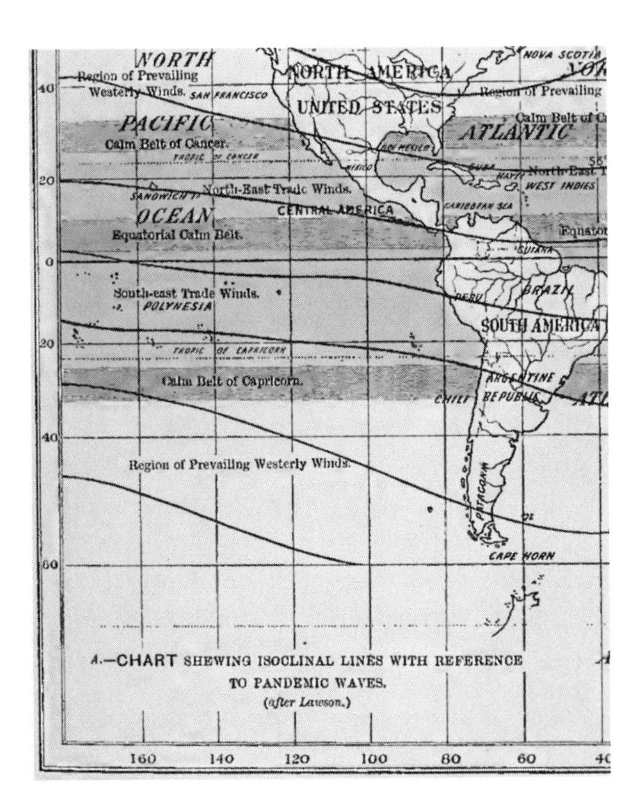

FIGURE 4C.2 Detail from "A Chart Shewing Isoclinal Lines with Reference to Pandemic Waves." From Felkin 1889, map 14.

cess animating disease was, by Felkin's historical moment, beginning to be understood as organic—the germ theory of disease had just been formulated—replacing earlier theories, such as those of Robert Lawson (who coined the term *pandemic wave* in 1861), which had held that such waves followed the flow of the Earth's magnetic flux (figure 4c.2), which Lawson believed could create a "miasma" capable of conveying disease from one place to another.

Data collection would continue to authorize wave lines, as well as claims about waves as material-processual things in the world. Move to Émile Durkheim's study *Suicide* ([1897] 1951, 47), in which he interpreted data on rates of suicide: "The evolution of suicide is composed of undulating movements, distinct and successive, which occur spasmodically, develop for a time, and then stop only to begin again. [O]ne of these waves is seen to have occurred almost throughout Europe in the wake of the events of 1848." Statistical data that present wavelike patterns implied wavelike behavior among people. The pattern pointed to social—and not psychological or individuated—causes.

Waves were in the air—or, better, on paper—in other social sciences. In 1872, the German linguist Johannes Schmidt proposed the *Wellentheorie*, or wave theory of language change, a model he contraposed to the tree model of the neogrammarians, who held that languages branch off from one another in ways that can be described as arborescent patterns. Leonard Bloomfield (1933, 316) summarized the intuition: "Different linguistic changes may spread, like waves, over a speech-area, and each change may be carried out over a part of the area that does not coincide with the part covered by an earlier change." In this model, the medium through which waves travel is spoken language and its phonology, a "thing" materialized in the articulations of language speakers.

Soon came the *heat wave* (1876) and the *crime wave* (1889).

All these waves emerged because of techniques of representation that animated formal claims about tendencies to rise and fall. Nineteenth-century social theory as wave theory delivered the world as a table, a map, a diagram that showed the rise and fall, the undulation, of aggregate phenomena.

As the twentieth century opened, wavy imagery had been established in two genealogies. One had waves as large-scale transformations, often successive, that either dissipate or wash over predecessors. Another tradition had waves as periodic repetitions organizing the world from the atomic scale up.

The modern/modernist moment in Europe and the United States tuned to worries about the relation between reality and representation. Moderns were fascinated by the weird world of contemporary physics: "By the late 1920s waves in motion are all the universe consists in—and they are probably fictitious, 'ondes fictives,' as de Broglie called them, or as Jeans suggests in *The Mysterious Universe*: 'the ethers and their undulations, the waves which form in the universe, are in all probability fictitious . . . they exist in our minds'" (Beer 1996, 295). And modern human minds, in turn, might exist in a field of waves. Virginia Woolf's modernist novel *The Waves* (1931, 278–79) has one character articulate a wavy account of experience: "The sound of the chorus came across the water and I felt leap up that old impulse, which has moved me all my life, to be thrown up and down on the roar of other people's voices, singing the same song; to be tossed up and down on the roar of almost senseless merriment, sentiment, triumph, desire." The physicist Erwin Schrödinger came in 1926 to see the world in the image of his own theory of the quantum wave, writing of "conscious awareness as something emerging in individuals like tips of waves from a deep and common ocean" (quoted in Beer 1996, 315). A moving particle was, he said in 1926, "nothing but a kind of 'crest' on a wave of radiation forming the substratum of the world" (quoted in Joas and Lehner 2009, 344). Woolf and Schrödinger puzzled over human agency and its place in a wavy world; for them, agency was all at once an illusion and an emergent structure.

Such pronouncements about the vibratory phenomenology of experience mostly vanish after the 1920s, reappearing only in the countercultural 1960s. More solemn wave talk emerges in the interim with respect to large-scale economic or sociological processes.

There came to be two modes of conceptualizing waves. The first treated waves as *cycles*—undulations that might either fluctuate in steady states or build. The second treated waves as forces that could *succeed* one another, with waves washing over their predecessors. Cyclical waves emphasized the power of large institutional structures to format social action. Successional waves, which highlighted irreversible historical change, were apprehended as manifestations of social or collective agency.

Consider an example from state planning in a large-scale command economy: the Soviet Union of the 1920s. In 1922, the economist Nikolai Kondratieff forwarded the idea that economic dynamics follow fiftyish-year-long cycles, waves of expansion followed by waves of decline. Kondratieff was instrumental in shaping the five-year plan of 1923 for Soviet agriculture. As he worked for the People's Commissariat of Agriculture, he began to believe that transnational markets were essential to the development of industry; interacting with these markets could put Soviet production in the thick of global waves of exchange.

Economists in the United States, in the wake of the Great Depression, went scouting for accounts of what had happened. In 1938, the accountant Ralph Nelson Elliott wrote *The Wave Principle* (see "Set Three, First Wave: Massive Movie Waves"), arguing that economic periods oscillated between optimistic and pessimistic moments, here building on a tradition of social psychology inaugurated by Gustave Le Bon in *The Crowd: A Study of the Popular Mind* ([1895] 1896), which suggested crowds were prone to unreason, to hyperactive sentimentality. Le Bon had written that "such ideas as are accessible to crowds . . . resemble the volume of the water of a stream slowly pursuing its course . . . [with] transitory ideas . . . like the small waves, for ever changing, which agitate its surface" (48).[1] Elliott's vision of economic waves saw them naturalistically, as predictable.

Different in discipline was the British geneticist R. A. Fisher, whose "The Wave of Advance of Advantageous Genes" (1937) sought a hereditarian, mathematically phrased account of how wavelike changes might unfold in populations. Writing in the *Annals of Eugenics*, Fisher postulated that some populations might, over generational time and geographical space, exhibit "a steadily progressive wave of gene increase due to the local establishment of a favourable mutation." This trend, he argued, could be described using diffusion equations from physics. He offered examples of what such changes over time might look like on a graph. Here Fisher enlisted a visual convention from statistics: the Gaussian (normal or bell) curve. A pivotal illustration in his article suggested that the median and mode of a bell curve might, with "the advance of advantageous genes" slide toward higher (more desirable) values (figure 4c.3). Fisher invited the reader to imagine the bulge of a bell curve moving forward, like a wave (temporalizing what had usually been the more snapshot formalism of the bell curve). The work became an ingredient in new biologically phrased and reductionist racism.

Whether in the hands of conservative or radical social theorists, successive waves of change were—and are—often described as processes that wash away entrenched ways of being. Those descriptions invite political readings. On the one hand, such forces can be read as demanding the dissolution of traditional social ties in favor of those created by modern, reflexive selves that can discern the structures in which they are bound up and act recursively to take advantage of them—"riding" the waves of change.[2] On the other, such waves may herald collective, perhaps revolutionary agency, refusals to think of liberal methodological individualism as the best motor of social change. Accounts of both kinds can prompt worries about selves misled, peoples following false hopes, perhaps toward submission to totalitarianism.

One influential mode of wave thinking came from futurology. Alvin Toffler's *The Third Wave* (1980) posited that "information age" society represented a rising wave of historical development. Toffler's waves were successive, forces of progress, each wave sweeping away the effects of the previous. Toffler (1980, 29) wrote, "One powerful new approach might be called social 'wavefront' analysis. It looks at history as a succession of rolling waves of change and asks where the leading edge of each wave is carrying us. . . . [I]t views each of these not as a discrete one-time event but as a wave of change moving at a certain velocity." Toffler suggested that when waves were powerful enough, prediction of the future—the "wave of the future"—would be straightforward. However, he wrote, "When a society is struck by two or more giant waves of change, and none is yet clearly dominant, the image of the future is fractured" (31). Toffler's theory

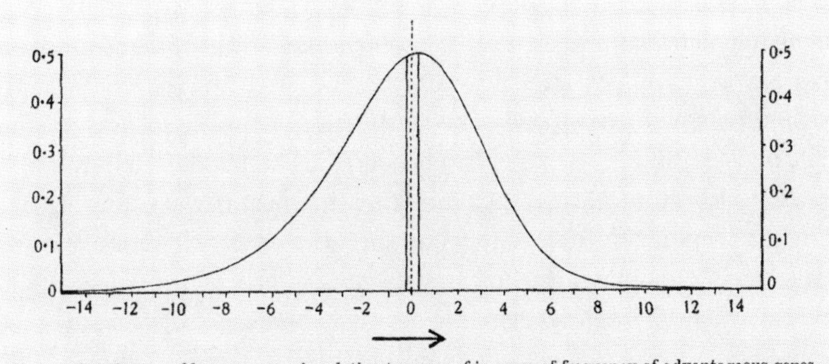

FIGURE 4C.3 The statistician R. A. Fisher (1937, 360) puts a bell curve into forward motion (follow the arrow), turning a statistical distribution into a moving wave.

mixed waves of different kinds, and his theory included “shock waves,” waves that travel faster than a medium can carry them.

At some moments, the “wave” stands for anxiety about the vanishing of individual judgment. The wave, in this guise, appears as a symbol of fascism. Take the Palo Alto high school class that in 1967 was led by a history teacher in a simulation of fascism dubbed “The Third Wave.” The canonical story, often told alongside tales of the Milgram and Stanford Prison experiments—other investigations of the 1960s into social psychologies of obedience—emphasizes how the will of individuals can be submerged under authoritarian rule. Ron Jones, the high school teacher who led the simulation, tells a muddled origin story about the imagery of the wave, one part discipline and obey and one part California surf party: “As the class period was ending and without forethought I created a class salute. It was for class members only. To make the salute you brought your right hand up toward the right shoulder in a curled position. I called it the Third Wave salute because the hand resembled a wave about to top over. The idea for the three came from beach lore that waves travel in chains, the third wave being the last and largest” (Jones 1981, 8).

The wave image is also thickly in play in descriptions of and prescriptions for *progressive* social action—consider the figure of waves of feminism (See “Set One, First Wave: The Genders of Waves”). Social movement theory has long been enamored of the wave. The “insurrectionary wave” as revolutionary form is an early example.[3]

Like their nineteenth-century predecessors, today's social waves continue to be summoned by inscription technologies. In "The Revolutions Were Tweeted: Information Flows during the 2011 Tunisian and Egyptian Revolutions," for example, the media studies scholar Gilad Lotan and his colleagues examine the unfolding of the Tunisian revolution in terms of its "flows" of information, including one significant "wave of retweets." The anthropologist Nick Seaver (2015) has spied in "big data's oceanic imaginary" recurring images of waves "made of blue 1s and 0s"—symbols of fear and excitement about a big data wave, a force that computer scientists, investors, pollsters, and many others must learn to manage. And, as danah boyd and Kate Crawford (2012, 668) caution, calling allied attention to the complexities of inscription, "Too often, Big Data enables the practice of apophenia: seeing patterns where none actually exist, simply because enormous quantities of data can offer connections that radiate in all directions."

By 2020 (following a "Blue Wave" election in the United States in 2018), the spread of the COVID-19 epidemic was narrated as coming in waves. I joined the historian of medicine David Jones in looking to the history of the pandemic wave (Jones and Helmreich 2020). We wrote this:

> Accounts of the wave-like rise and fall of rates of illness and death in populations first appeared in the mid-nineteenth century, with England a key player in developments that saw government officials collect data permitting the graphical tabulation of disease trends over time. During this period the wave image was primarily metaphorical, a heuristic way of talking about patterns in data. Using curving numerical plots, epidemiologists offered analogies between the spread of infection and the travel of waves, sometimes transposing the temporal tracing of epidemic data onto maps of geographical space.
>
> [In] the twentieth century, scientists developed new ways of thinking about epidemic waves, including mathematical models meant not just to describe but also to predict disease propagation. With such models the wave gradually became more than a metaphor and a mere graphical accompaniment to statistical representation; it was now a naturalistic and technical object, one whose causal basis might be discovered and understood.
>
> We live now in a third era, in which the imagery of epidemic waves is widely mobilized in both public health and public culture to describe, prognosticate, and urge—to ask people to band together to "flatten the

curve." As the *New York Times* early on summarized the idea, "Slowing and spreading out the tidal wave of cases will save lives. Flattening the curve keeps society going." The seeming simplicity of the wave image is therefore infused with moral messaging, animating a mix of resolve, fear, and reassurance.

The work with Jones secured a sense I already had: when social transformations are described as waves, we should ask questions about causality, about what mix of form and material are being invoked. What produces such waves, physical and affective? Are waves formed by large-scale historical structures, frameworks that overdetermine social action? Or are they, instead, expressions of collective agencies—networks?—fracturing previously stable structures and finding until now unrealized materializations? Wave talk can make it difficult to see causality. But one can ask when presented with a wave account: Are there legible structuring forces being whisked out of view by the rhetoric of the account? For an election, are matters of gerrymandering, racial redlining, and corporate consolidations of media outlets kept out of dominant accounts? For movements of people, are works of community and institutional organizing—for example, legal advocacy for immigrants' rights, religious efforts to protect asylum seekers—giving form to larger-scale trends? The figure of the wave in social commentary is often invoked when structural, analytic, or causal accounts are either being downplayed, are being strategically funneled through one measure of change (e.g., the basic reproductive number that characterizes viral transmission), or have become difficult to settle on, when they waver—that is, when social theory ~ wave theory.

PLATE 1 Ocean waves on the cover of the Penguin Classics edition of Albert Camus's *The Plague*. Photograph © 2003 Rankin. Reproduced with permission of Rankin/Trunk Archive.

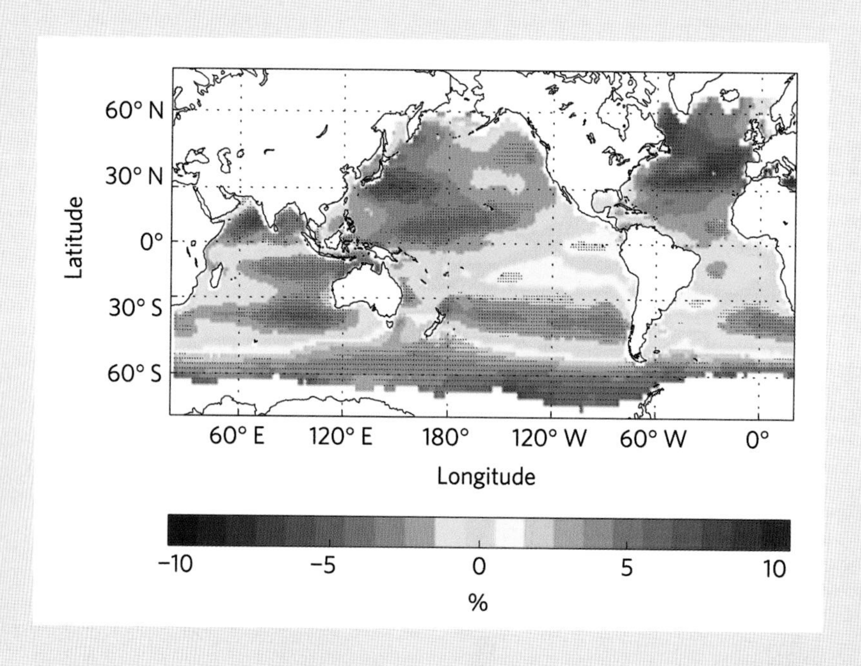

PLATE 2 Projected changes, by percentage, in annual mean significant wave height for a future time slice (ca. 2070–2100) relative to a present climate time slice (ca. 1979–2009). From Hemer et al. 2013, fig. 2b.

PLATE 3 Inaugural wave in Delta Flume, Deltares Research Institute, Delft, Netherlands, October 20, 2015.

PLATE 4 Ivan Aivazovsky, *The Ninth Wave*, 1850, oil on canvas, State Russian Museum, Saint Petersburg.

PLATE 5 Raden Saleh, *A Flood on Java*, ca. 1865–76, lithograph, Royal Netherlands Institute of Southeast Asian and Caribbean Studies, Leiden, Netherlands.

PLATE 6 FLIP upright, at sea, at 33.689°N, 118.989°W, October 2017. Photograph by the author.

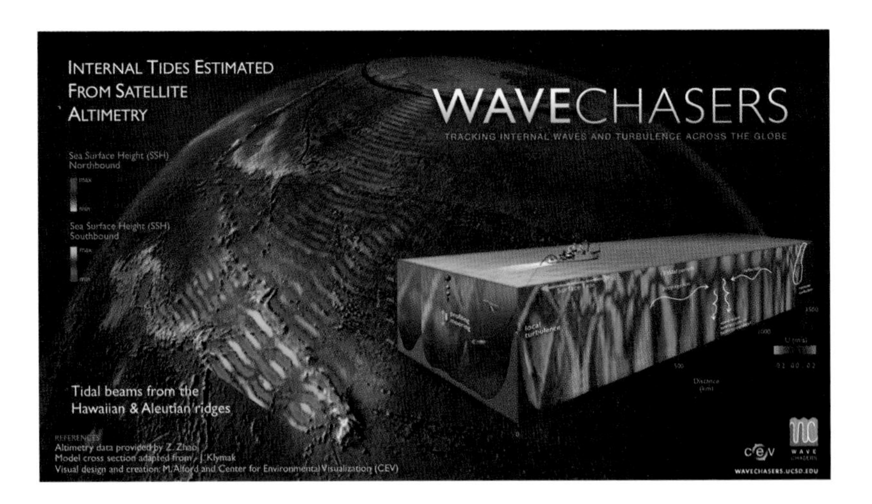

PLATE 7 Internal wave crests, detected from space, travel north (red) and south (blue), away from the Hawaiian and Aleutian ridges. Right: cross-section volume depicting propagation of internal waves. Image design by Matthew Alford, using altimetry data from Zhongxiang Zhao and simulation work by Jody Klymak, in collaboration with the Center for Environmental Visualization, University of Washington, Seattle.

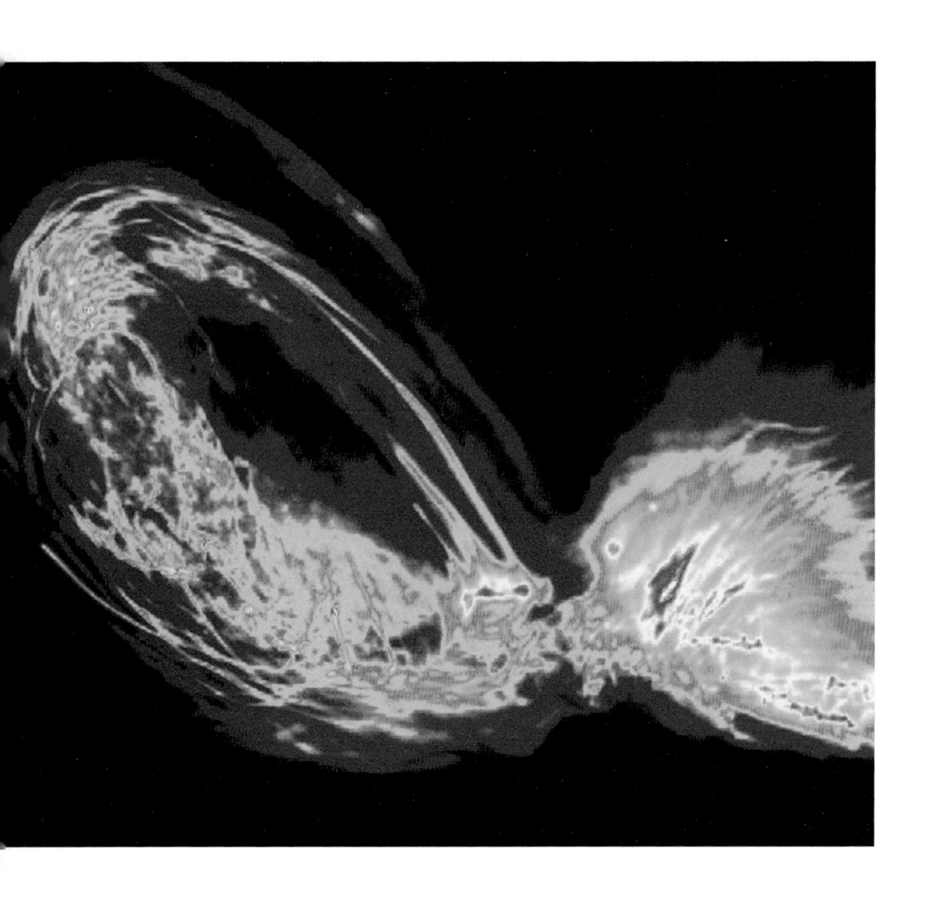

PLATE 8 False color side view image of a breaking wave (left to right), with warm colors indicating bioluminescence from dinoflagellates, used as a proxy for shear stress, a measure of turbulent energy dissipation. From Deane et al. 2016, fig. 2.

PLATE 9 Ken Charon, *Obama Rides the Big Wave*, 2008, acrylic on hemp canvas board.

PLATE 10 Judy Kirpich and Alex Berry,
The Wave of the Future, 1981, lithograph.

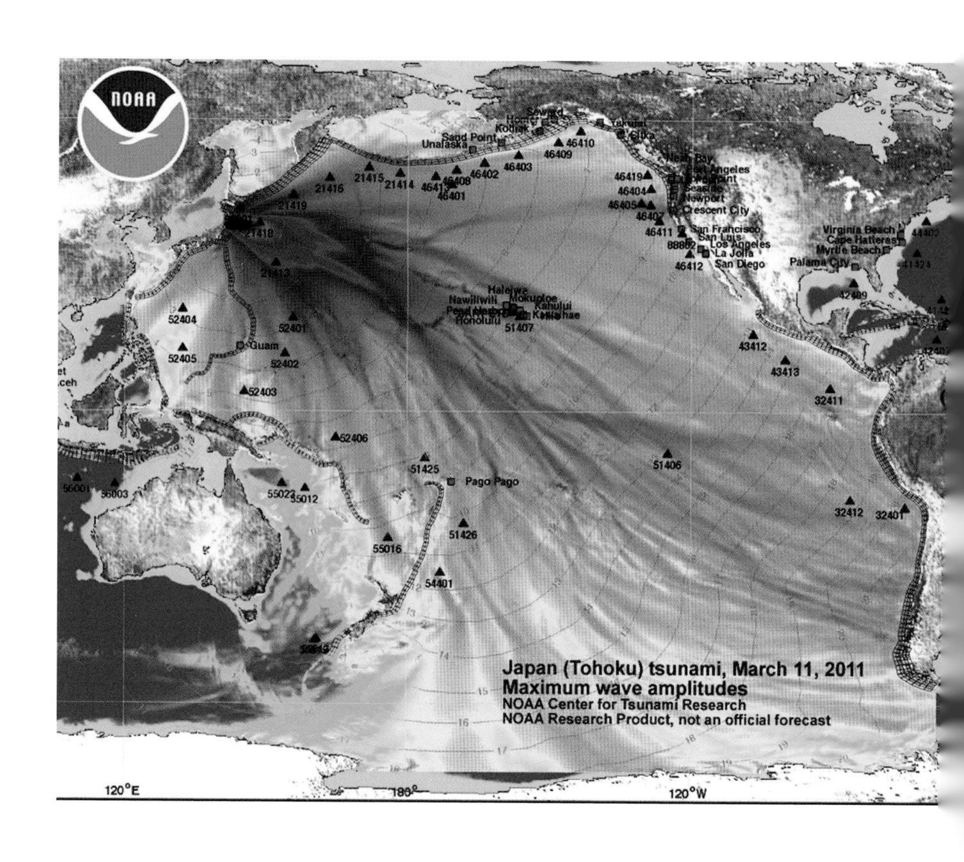

PLATE 11 "Japan (Tohoku) tsunami, March 11, 2011. Maximum wave amplitudes, NOAA Center for Tsunami Research, NOAA Research Product, not an official forecast." http://nctr.pmel.noaa.gov/honshu20110311/Energy_Polot20110311=1000_ok.jpg.

PLATE 12 Illustration from Maekawa Monzo's 大水海 (Large Water Sea), an account of the tsunami of 1896, published in 1903.

PLATE 13 Bonnie Monteleone, *Plastic Ocean: In Honor of Captain Charles Moore*. From the *What Goes Around Comes Around* collection, 2011, trash, after Katsushika Hokusai, *Under the Wave off Kanagawa*, ca. 1829, woodblock print. Copyright Bonnie Monteleone.

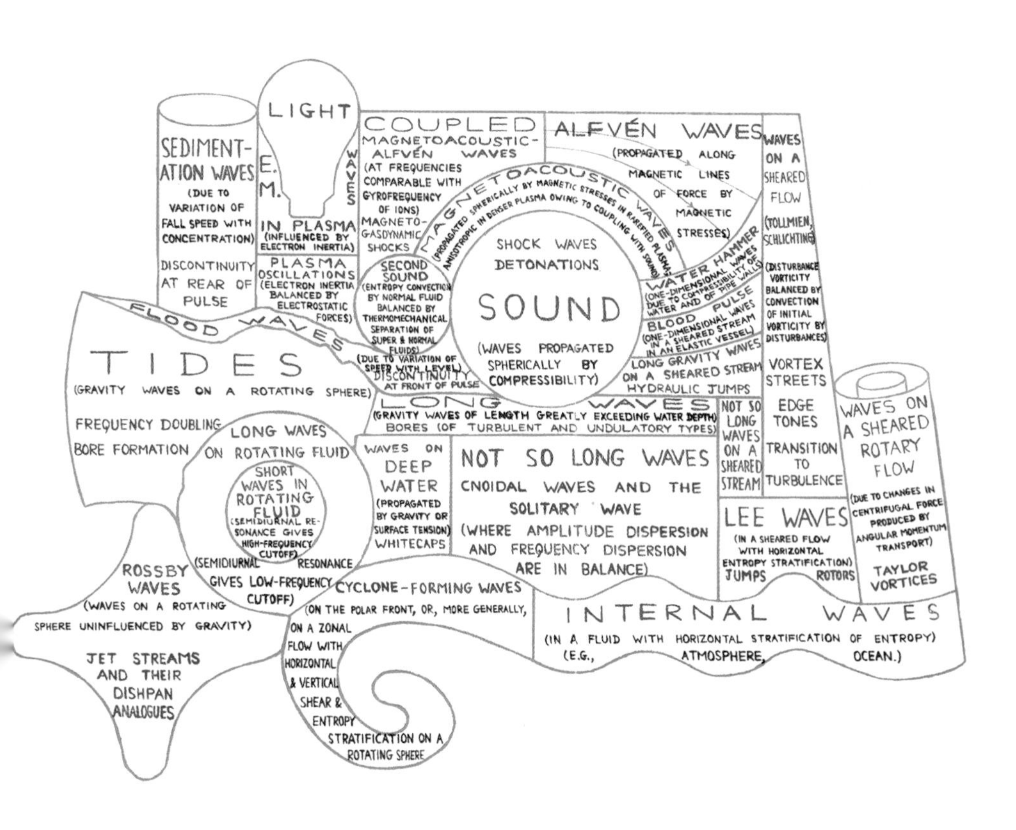

PLATE 14 Diagram. From Lighthill 1967, 269–70.

PLATE 15 Edgar Arceneaux, *The Slave Ship Zong*, 2011, acrylic, graphite on paper. Courtesy of the artist and Vielmetter Los Angeles. Photograph by Robert Wedemeyer.

Chapter Five

Wave Theory, Southern Theory

Disorienting Planetary Oceanic Futures, Indian Ocean

IN FEBRUARY 2014, Southern Hemisphere summer, I traveled to Newcastle, Australia, to attend Kiwi-Oz Waves: The First Australasian Wave Conference. The workshop brought together scientists from New Zealand and Australia (hence, Kiwi plus Oz) to center attention on oceanographic questions pertaining to this part of the globe. A wider series of conferences on meteorology and oceanography in the Southern Hemisphere had begun in 1984, to correct for some of the Northern Hemispheric bias of oceanography (e.g., a persistent marking of Northern summer as simple summer), but this was the first meeting on wave dynamics.

I had learned while attending my first wave conference, in 2013 in Banff, that Southern ocean dynamics were underrepresented in wave science. That had been common knowledge for a while. For example, in 2000, one European model had been found persistently to underestimate wave heights by 20 percent in the Southern Hemisphere (Janssen 2000). Right around the time of Kiwi-Oz, *Nature Climate Change* published about mis-estimations of Southern ocean surface temperatures in most climate models, the result of assuming too low cloud cover over Southern seas (Wang et al. 2014).

On my flight into Sydney from Boston I read through recent scholarly writings in social science that posed the question of whether dominant social theory might be provincialized as "Northern theory"—with analytics of "economy," "nation," and "modernity" understood not so much as universal frames but, rather, as terms emergent from European and American projects of expansion and progress, almost always secured through exploitation of the colonized world. The sociologist Raewyn Connell asked in *Southern Theory* (2007) whether analytic views from the Global South might more accurately capture the actual experiences and conditions of the world, both historically and now. *Global South,* a term covering countries in Asia, Latin America, Africa, and Oceania that had been earlier labeled "developing," "third world," or "peripheral," means to point not so much to a neatly bounded Southern Hemisphere as to common histories of subordination under—and resistance to—Euro-American–originated imperialism and capitalism.[1] Connell (2007, ix), following the Beninese philosopher Paulin Houtondji, suggested that social theory as Northern theory proceeded with the assumption that "data gathering and application happen in the colony, while theorizing happens in the metropole." Scholars of Global South epistemology suggest in response that "Southern theory" might offer prescient accounts of the direction world affairs might be taking and, indeed, serve as sources for social theory more relevant to the planet than ideal typical or ideological notions of social contract, utilitarian exchange, or development.[2]

I became curious, then, about whether there might be for physical oceanography something like "Southern theory" for wave theory, modes of thinking about wave dynamics that drew from processes unfolding in Southern oceans; processes that might have hemispheric significance, that might prefigure what might be in store elsewhere, or that might have global consequences, either enduringly or newly. I had been drawn to the conference in part by the words *first* and Austral*asian*. I did not necessarily expect a shot-across-the-bow postcolonial, subaltern studies takedown of Northern Hemispheric presumptions embedded in computational wave models, a call for wave modeling to be flipped upside down, like a counter-ideological south-up map (figure 5.1).[3] And anyway, New Zealand/Australia—that is to say, Euro–New Zealand/Australia—could hardly be considered a fully distinct community of scientists, especially at this workshop, since in spite of the conference's name, no one, with the

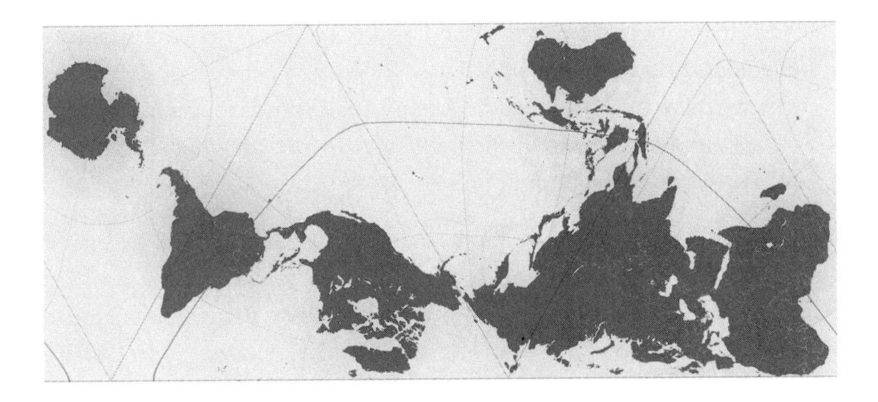

FIGURE 5.1 "South-Up" AuthaGraph Projection centered in the Pacific Ocean meant also to offer proportional areas for land and sea. The Japan-based designers say this projection, which can be adjusted to center anywhere on the Earth, can "provide a new angle of perspective to equally view the world so as to be free from existing perceptions defined by usual phrases such as 'far east,' 'go up north,' Western.'"

exception of one postdoctoral fellow from Iran, had any academic base in Asia (nor was there, at this conference, acknowledgment of the Awabakal and Worimi peoples, the traditional Indigenous custodians of the land on which Newcastle/Muloobinba sits). With scholars from France, Germany, and the United Kingdom, this was a cross-equatorial eddy from the North.

Still, there was a sharp sense at the workshop that differences between the Northern Hemisphere and the Southern Hemisphere mattered. An Australian meteorologist I had first met back in Banff outlined four major distinctions between Southern and Northern Hemisphere oceans:

> First, there is more solar radiation hitting the sea in the south than the north. One reason is that there is more ocean surface area in the Southern Hemisphere than in the Northern—20 percent more. Another, more seasonally bound reason is that the Southern Hemisphere's summer coincides with that time when Earth's orbit takes it closest to the sun.
>
> Second, there's a *lot* more uninterrupted ocean in the south, which means the proportion of the ocean traversed by *swells*—waves no longer driven by wind—is much higher. That can make for stronger waves.
>
> Third, there's less particulate plant matter in the near-sea atmosphere because there's less land.

Fourth, there is a larger area of *ice* in the Southern Hemisphere—particularly in that portion called the Southern Ocean [also known as the Antarctic Ocean and *not* to be confused with what Europeans once called the South Seas or the South Pacific or, indeed, with all oceans in the Southern Hemisphere].

He was blunt about what these Southern Hemisphere characteristics meant for oceanographic science: "Northern Hemispheric assumptions are *built right in* to the models, including of waves. That makes them challenging to use in the Southern Hemisphere." Another scientist joined our conversation, speaking about the ocean that rings Antarctica: "If you take parameters from the Northern Ocean, they don't necessarily apply to the Southern Ocean." He turned sarcastic: "Who *cares* about the Southern Ocean, except us? But people *will* start caring if it affects global climate." The message I was getting was clear: thinking from oceans in the "South," with their greater proportion of seawater and ice and with their larger span of *fetch* across which wind waves grew, might be needed to complicate the Northern assumptions built into many wave models (think back to the general model for the growth of wave spectra, originally based on North Sea data collected in the 1960s) and to account for intensified ocean storms, massive coral and mangrove depletion, and sea-ice breakup: Southern Hemisphere ocean processes with planetary effects.

One paper at the conference stood out, providing an example of how northern assumptions might distort research into other seas. The paper was delivered by Arvin Saket, a scientist working in a country peripheral, even excluded, from dominant wave science: Iran. While not a Southern Hemisphere country, Iran does fall south of the Brandt Line, that wiggly contour drawn by German Chancellor Willy Brandt in the 1980s to map divides between a wealthy North and impoverished South. Describing the challenges of characterizing waters in the Gulf of Oman, particularly in the Chabahar Bay, Saket (2014) told us that Iran was seeking to harness waves there for electricity but needed more information about their strength and direction. Unlike the Persian Gulf, the Gulf of Oman has poor buoy coverage. A simulation was required. Saket worked with data he could get, plugging them into a model of wave action called Simulating Waves near Shore (SWAN); he did not use WAVEWATCH not only because that framework is better for the open ocean, but also because trade embargoes prevent WAVEWATCH from being exported from the United States to Iran. He needed a recipe for simulated wind and borrowed from a European model. It did not work. What happened? The Melbourne-based, Russian-born wave scientist Alexander Babanin helped me figure it out, telling me, "Wave models traditionally are validated by means of observations, and that's always through some geographically linked observations. Satellites are global, but buoys are definitely regional, so if there is anything in your region and you tune your model to perform well in this region, then if you take it to another region where certain things are different, it can have biases." In the case of the Gulf of Oman, average wind speeds from the North Sea, to which European models tuned, were different from Gulf wind speeds. This was a natural-science iteration of what Dipesh Chakrabarty (2000) names the Eurocentric logic of universal knowledge, which, for social sciences, have often claimed "first in Europe, then elsewhere," with Europe, for this oceanographic case, not so much a full-blown ideological referent as a set of tiny technical details.[4]

There are other matters in play here, beyond modeling, and those have to do with the fact that oceans in the Southern Hemisphere have been less heavily instrumented and measured than northern oceans. There is a geopolitical reason for this: among the large coastal nations that might manage such systems—Argentina, Brazil, Chile, India, Indonesia, South Africa—there has not been a highly state-funded, capitalized maritime infrastructure (see Lavery 2020, 310). Even Australia is ringed by only twenty-odd buoys (the United States monitors around two hundred, which, most complain, is still too few). (Recall the Brazilian project, described in chapter 4, to train surfers to record wave heights in the manner of Datawell Waverider buoys.) Satellite data can substitute, though it can be hampered by cloud cover. Thinking in a South-wide way, there is also the question of what *kind* of data to collect—about open-ocean wave action, about the wave-agitated breakup of Antarctic ice, about storm surges in low-lying south Asian countries. These were all topics at Kiwi-Oz. The multiplicity of dynamics about which scientists would need to think to configure a covering, comprehensive "Southern wave theory" points to the obvious fact that oceans in "the South" are as multiple, as heterogeneous, as the cultural and political histories sometimes homogenized through the optic of the "Global South." It is well to be alert to what Shanti Moorthy and Ashraf Jamal (2010, 6), in their call for thinking from the Indian Ocean, name as "the dangers of Occidentalism and the inversion of binaries."

Zooming back out to the hemispheric view, however: when I turned to writing this chapter in 2019, looking at my notes from the 2014 conference, I found that claims I had heard percolating about growing wave sizes—not just under-measurements—in the Southern Hemisphere were beginning to appear in more fully researched form in scientific publications. In April 2019, *Science* published "The Ocean's Tallest Waves Are Getting Taller," reporting on a paper showing that the waves of the Southern Ocean (the ocean around Antarctica) were "getting bigger, thanks to faster winds attributed to climate change" (Barras 2019).

The article on which this story reported, published in *Nature Communications* by a team based at Oregon State University, was titled, "A Recent Increase in Global Wave Power as a Consequence of Oceanic Warming." Its abstract spelled out the results:

> Global wave power, which is the transport of the energy transferred from the wind into sea-surface motion, has increased globally (0.4% per year) and by ocean basins since 1948. We also find long-term correlations and statistical dependency with sea surface temperatures, globally and by ocean sub-basins, particularly between the tropical Atlantic temperatures and *the wave power in high south latitudes [66 degrees, 33 minutes down to the south pole], the most energetic region globally*. Results indicate the upper-ocean warming, a consequence of anthropogenic global warming, is changing the global wave climate, making waves stronger. (Reguero et al. 2019, emphasis added)

Meanwhile, in the Indian Ocean, bounded on its south by the Southern Ocean, scientists had also been finding increased wave heights. In "Wave Climate Projections along the Indian Coast," published in the *International Journal of Climatology* in 2019, oceanographers in Mumbai reported that their models found "an increase in wave heights and periods along much of the Indian coast, with the maximum wave heights increasing by more than 30% in some locations" (Chowdhury et al. 2019). That meant increasing power. (An increase in wave period means an increase in wave speed and therefore strength.) All of this work uses by now ever updating wave models that not only generate fuller data but also increasingly focus on Southern matters. What is also intriguing here is how waves become hybrid human-nonhuman agents ("anthropogenic global warming . . . is changing the global wave climate"). In *Allegories of the Anthropocene*, Elizabeth DeLoughrey (2019, 150) writes that, in discussions of climate change, "humans are positioned as a geological force, yet the ocean seems to be our proxy. In this sense, the ocean may function allegorically as the daemonic agent, a figure with more than natural power that moves between realms." Waves are devils in the details of global warming. What oceanographers call the greater "oceanicity" of the Southern Hemisphere is amplifying their force.

At the same time as I was racing to keep up with reports from the Southern Hemisphere, humanists and social scientists who had trained their attention on oceans were beginning to call for the "blue humanities" to turn South. In "Thinking from the Southern Ocean," Charne Lavery underscored many of the things I found in Australia in 2014. She wrote, drawing on the work of the historian Alessandro Antonello (2017), "Because its winds and currents can circulate endlessly without encountering a continental barrier, the Southern Ocean is characterized by perpetual movement: waves can 'lap' themselves, in theory drawing on infinitely circulating fetch." She exhorts readers to "bring together questions that pertain to the global South—from the still-decolonizing countries of the world, centering social and economic justice—while also registering the interrelations between changing global climates and the currents of the Southern Ocean" (Lavery 2020, 310, 308). In a piece cowritten with Meg Samuelson, Lavery offers that "the southern region of the globe is most readily conceived of as what is bound by the longitudinal lines of imperial and metropolitan domination or described by the curvier Brandt line as comprising the 'poorer nations.' But it might also be defined by the relatively vast maritime expanses that distinguish the Southern hemisphere" (Samuelson and Lavery 2019, 37). Such a view might, Samuelson and Lavery urge, "sketch out the oceanic South as a category that draws together the dispersed landmasses of the settler South, the decolonized and still colonized South, the 'sea of islands' comprising Indigenous Oceania, and the frozen continent of Antarctica" (38; see also Chari 2019).

Beyond gathering up an oceanic "Global South," however, such a sketch might also summon something like a *planetary* South. As Gayatri Spivak (2003), Chakrabarty (2021), and others have suggested, the figure of the "globe" coalesces out of—and still conveys—a European history of colonial and trade expansion enabled by maritime navigation, a history that has also distributed, in ideology but not fact, a notion of a universal human, the over-the-horizon figure of the global citizen as "free" to engage in untrammeled economic and cultural exchange. The "planetary," meanwhile, with its authorizing voice something like Earth systems science, weaves together geology, biology, water, and atmosphere to name patterns and combinations of human and nonhuman forces. Those forces include such oceanic phenomena as the North Atlantic Meridional Overturning Circulation, hurricanes, monsoons, and waves, as well as such astronomical phenomena as the intake of solar heat. The very etymology of the word *south* points to a Proto-Indo European root that means "the sun" (*sāwel-*), with evidence for this paging back to Sanskrit *suryah* and Latin *sol*. The *Oxford English Dictionary* (*OED*) offers that *sun* is linked to south because, in the Northern Hemisphere, "the south [is] the direction of the midday sun." As one way to think from the oceanic South, then, think about the sun—that force whose heat, imparting energy to wind and therefore to waves, has intensified effects in Southern oceans.[5] The oceanic South fuses the global and the planetary south.

Such a scaling up and over to the planetary, however—even if qualified by the political specification of "the South"—has something of a universalizing, homogenizing, flavor (it also figures South-as-sun from the North), as, indeed, may the superhuman notion of "forces" (DeLoughrey 2019, 15). Also useful, then, is to be specific, to attend to fluctuating framings of scale.

My fieldwork with wave scientists gave me a good sense of where to look. India and Bangladesh's Bay of Bengal has been an increasingly visible topic in wave science, posing challenges to models not yet fully tuned to the Indian Ocean or its coasts. In what follows, I shift the question of the oceanic or planetary South by starting from wave theories circulating and spinning off a phenomenon particular to this ocean: the monsoon. This recurring seasonal pattern of wind and water—what a humanities research group based at the United Kingdom's University of Westminster calls the "monsoon assemblage" (see Bremner 2018)—has effects that stretch from East Africa to western Australia. Figure 5.2 offers a synoptic map of the summer monsoon's winds, drawing on work conducted by the India Meteorological Department. Figure 5.3, from a 1986 issue of that department's journal, *Mausum* (मौसम), tracks the travel of wave swells in response to the monsoon.

Following wave scientists, much of what I zero in on here fixes on dynamics in the Bay of Bengal. T/here, where the Ganges, Brahmaputra-Jamuna, and Meghna Rivers arrive at the sea, where cyclone-driven waves visit continual reorderings of land and water in India and Bangladesh, and where the legacies of Dutch hydrological planning meet unfolding futures, waves can be read as manifestations of human-inhuman convolution. I follow the historian Sunil Amrith in attending to how the monsoon, once upon a time a force outside human impress, is now, with sociogenic climate change, a hybrid natural-cultural form. If, as Amrith (2018, 325) writes, "Changes in the weather also bring a sense of disorientation," disoriented and reoriented climate may be sensed in the changing of the waves.

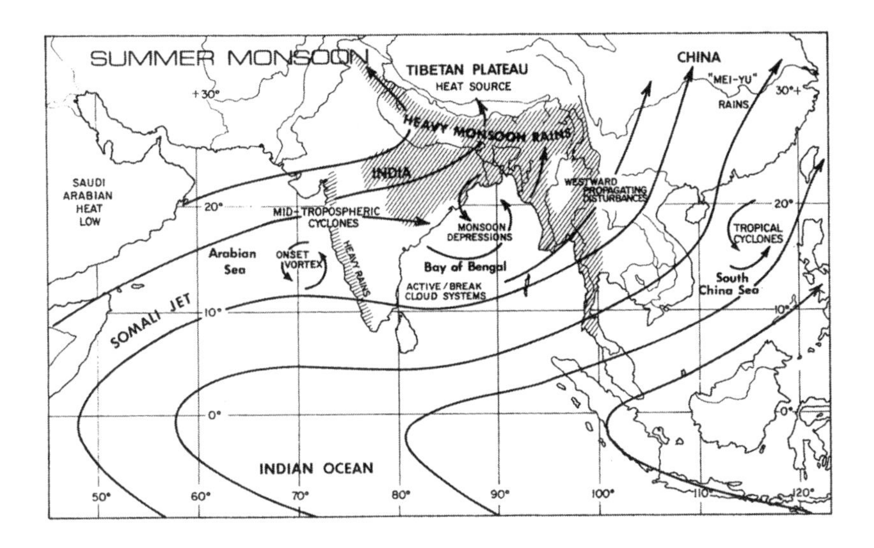

FIGURE 5.2 Map of summer monsoon winds. From Johnson and Houze 1987, fig. 10.30.

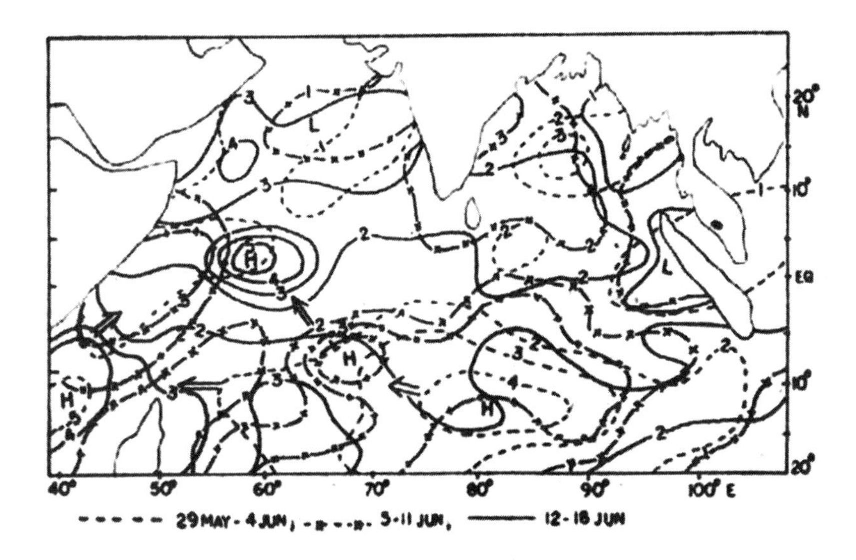

FIGURE 5.3 "Migration of Southern Hemispheric Swells." From Ray 1986, fig. 6.

Assembling Waves in the Indian Ocean

The Indian Ocean, a quarter-hemisphere-wide sweep of drifts and rotating currents, encompasses and mixes a multitude of histories, cultural and ecological. Its wind-curled waters have connected Malays, Chinese, Indians, Arabs, and Africans in maritime circulations of exchange, pilgrimage, migration, and servitude—connections that were later inhabited, rearticulated, and exploited by Portuguese, British, Dutch, French, and American merchant and imperial enterprise (Hofmeyr 2007, 6; see also Aiyar 2020; Alpers 2014; Pearson 2003, 2015). The winds of the monsoon—from the Arabic *mawsim* (season)—are noted for their seasonal reversal, blowing from the northwest between June and October and from the northeast from November to April. This pattern has shaped timings and locations of departures and trade routes over centuries, setting, for example, Turkish, Bengali, Arab, and Gujarati sailors out at nearly synchronized moments to take advantage of the monsoon. Sinnappah Arasaratnam (1994, 1–2) narrates the rotating winds and waters as a kind of infrastructure: "The monsoon winds, rainfall and oceanic currents determine in many ways the rhythms and patterns of trade, the location of ports, the flow of rivers and alluvial deposits, the depth and shallowness of the coastal waters and the shape and structure of the coastline." Technologies of navigation "such as the Chinese compass; the Dhow ships with triangular sails that allowed sailors from Persia, India, the Arabian Peninsula, and East Africa to sail upwind; Zheng He's treasure fleet; and the Islamic astrolabe used to determine latitude" (Scammacca and Zyman 2020) all fastened to this infrastructure, all created a kind of wave theory, animated (see Ghosh 2008). The historian Markus Vink (2007, 55) observes:

> The decoding of the secrets of the monsoon regime represented a quantum leap forward in "seaborne connectivity" across the Indian Ocean, leading to the emergence of an India-centred trading network involving commodities . . . and culture (initially Buddhist, later Hindu) after the third century BCE. The "Greater India" culture zone was one of the outcomes of the ensuing multifaceted, multilateral diffusionist process of "southernization," effectively integrating the societies and cultures of what the Arabs styled *al-bahr al-Hindi* ("Indian Ocean"), the Persians *darya'i akhzar* ("Green Sea"), and the Chinese the *Nan-yang* ("Southern Ocean").[6]

The spread of Islam from the seventh century onward also shaped this ocean, turning it into a "Muslim lake" (see Sherrif 2010), mixing peoples from Ottoman, Safavid, and Mughal polities and providing the medium for intersecting cosmopolitanisms, inspiring some to call the Indian Ocean "the cradle of globalization" (see Cassanelli et al. 2002; Subramanian 2010).[7]

With the nineteenth-century rise of steam power, the construction of the Suez Canal, and submarine-cabled telegraphy, a linearized technological meshwork was superimposed on/beneath the environmental-cultural infrastructure of the monsoon, with the Indian Ocean becoming a net of nodes largely conjured and leveraged by the British Empire, though also constitutive of an early twentieth-century oceanic public sphere "rooted in pan-religious movements, be these Buddhist, Muslim or Hindu," and "based in the port cities of the Indian Ocean and sustained by the intelligentsias of intersecting diasporas" (Hofmeyr 2007, 7). What counts as the modern Indian Ocean world—what Sugata Bose (2006, 6) calls "an interregional arena of political, economic, and cultural interaction"—comes into being. Though Bose refuses the idea of the Indian Ocean as a "system" (which sets it too firmly within a "world-systems" theory that makes capitalism key to all), he does urge attention to connection. His *A Hundred Horizons* uses the 2004 Indian Ocean tsunami and its emanating effects as a symbol for such interdependencies. He writes, "Just as waves in one ocean produce fluctuations in sea levels in others, the human history of the Indian Ocean is strung together at a higher level of intensity in the interregional arena while contributing to and being affected by structures, processes, and events of global significance" (3). The historian Michael Pearson (2006, 359) turns similarly to waves to describe these ricocheting relations, using Jean-Claude Penrad's figure of *ressac*, "the threefold violent movement of the waves turning back on themselves as they crash against the shore," an image he deploys to discuss shore-emanating connections for "which the to-and-fro movements of the Indian Ocean mirror coastal and inland influences that keep coming back at each other just as do waves." Mark Ravinder Frost (2010, 252), writing about early twentieth-century efforts by the Nobel Prize–winning Bengali poet Rabindranath Tagore to create an encompassing vision of the Indian Ocean world, observes: "In the writings of prominent Calcutta-based literati, Asian civilization became an 'ocean of idealism' through which, in the distant past as well as in their more immediate present, 'waves,' 'currents' and 'ripples' of Indian thought repeatedly washed up on distant 'shores' and 'beaches' to unite the region as one unified whole." In these accounts, waters, winds, and waves of the Indian Ocean become a potent store of metaphor for imagining regional commonality. In *Waves across the South*, his history of late eighteenth- and early nineteenth-century Indian and western Pacific Ocean peoples' world making in the face of the counterrevolutionary forces of Dutch, French, and British Empire, Sujit Sivasundaram (2020, 5) mobilizes the figure of waves to

> think with the push-and-pull dynamic of globalisation . . . to consider the surging advance of connection across the sea as well as turbulent disconnection and violence across waters. . . . Thinking with waves is also apposite in reminding ourselves that the physical setting of this story matters. The physicality of rain, storms, squalls, cyclones, waterspouts, fevers and earthquakes had to be combatted to make global empire work, through regimes of study, tabulation, mapping, modelling, medicine and urban fortification and planning.[8]

Monsoon winds—and more than metaphorical waves—were rendered into new technoscientific tabulation in the mid-twentieth-century, activated by international Cold War–era interests in geophysical Earth process (preceded, to be sure, by British colonial efforts to attach monsoon prediction to imperial imperatives [Cullen and Geros 2020]). At the opening of the twenty-first century, inquiries into monsoon dynamics are motivated by questions that have to do with climate change, especially as the Indian Ocean sees some of the world's steeper rises in water warming ("holding more than 70% of all the heat absorbed by the upper ocean since 2003" [Hofmeyr and Lavery 2020]), sea level (with the Arabian Sea and Bay of Bengal having risen by four inches more than other parts of the world's oceans since 1880 [Swapna et al. 2017]), and frequency of cyclones. Waves—rising, surging, rolling—can now be read as materializations of climate changes in littoral Asia.

A capsule history of the Indian Ocean as a technoscientific object for physical oceanographers could be organized around the International Indian Ocean Expedition, a 1959–65 effort modeled after the United Nations' International Geophysical Year, 1957–58, a project ostensibly meant to foster transnational scientific collaboration but also thickly enmeshed in Cold War contests to map the Earth's environmental processes, resources, and strategic affordances (figure 5.4).[9] As Amrith writes in his magisterial history of South Asia and water, *Unruly Waters* (2018, 233), "Midcentury oceanographers were drawn to the Indian Ocean for the same reason that medieval traders could cross it—the seasonal reversal of the monsoon winds. This pattern of reversing winds made the Indian Ocean unique." The Scripps oceanographer Warren Wooster (1965, 290) described "the Indian Ocean as a model of the world ocean," noting that "physical

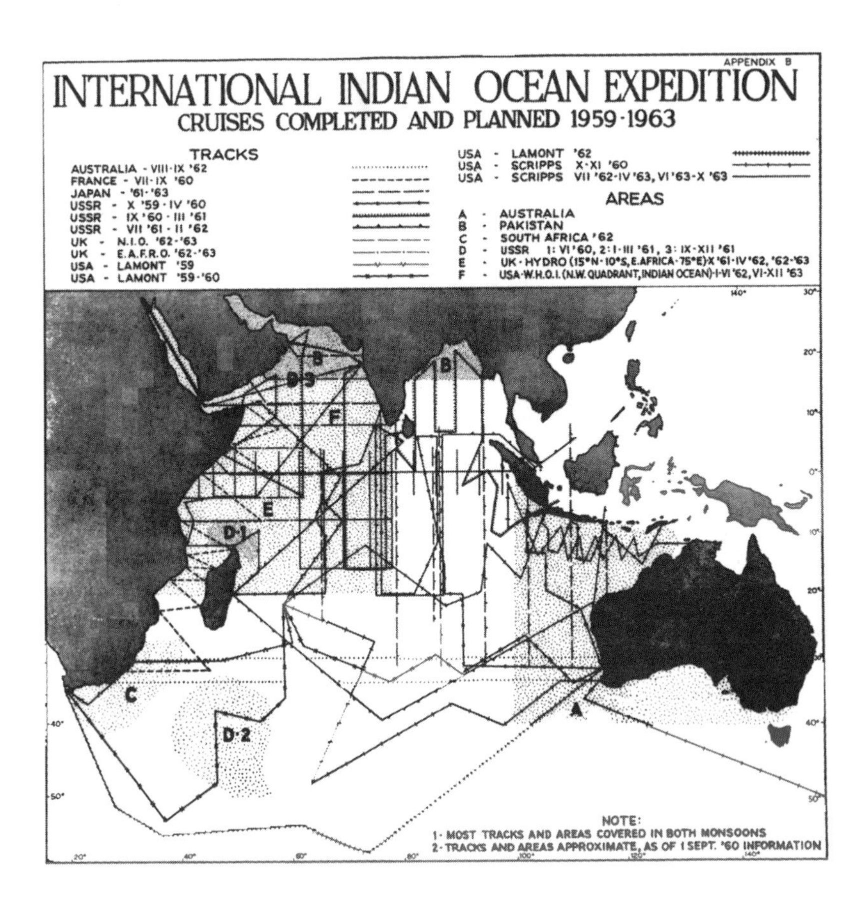

FIGURE 5.4 Completed and planned cruises, International Indian Ocean Expedition. Note the prominence of US and USSR ships, as well as note 1: "Most tracks and areas covered in both monsoons." From Snider 1961, 115.

oceanographers recognized the importance of examining a system where their wind-driven models could be tested by observing conditions under opposite regimes of surface wind stress." The expedition's goals were many and "encompassed the study of ocean currents and littoral drift; an investigation of ocean chemistry, salinity, and temperature; the exploration of marine life, and especially fisheries; the study of wind and atmospheric conditions and rainfall" (Amrith 2018, 235). Some thirteen nations were involved, and some forty ships enrolled, and participating constituencies had a range of priorities, with some invested in fisheries surveys, others looking to map potential mineral wealth at the seafloor, and many, at the largest scale, aiming to understand energy exchange between ocean and

FIGURE 5.5 Miao Ti, "One Wave Higher than the Last." From "The First Afro-Asian-Latin American Peoples' Solidarity Conference," *Peking Review*, vol. 4, January 21, 1966, 20.

atmosphere. That understanding—later amplified with the aid of satellite imagery and new computational record keeping—became the foundation for monsoon knowledge from then forward. The Indian Ocean became a *system*—something more than an assemblage of phenomena. If, as Bose (2006, 10) writes, "The spatial boundaries of the Indian Ocean have varied according to the nature of cultural, economic, and political interactions" (see also Forbes 1995), turning the ocean into a system with feedbacks made it into an object of Cold War technoscientific knowledge (see Choi 2020).[10]

That knowledge was a geopolitical precipitate. With major funding from the National Science Foundation (and scientists from the United States, hailing from Scripps and Woods Hole, in key administrative roles), along with a view, from the US vantage, of countries such as India, Indonesia, and Pakistan as "underdeveloped" (NSF 1967) (against dominant seagoing powers such as Britain, Germany, the Soviet Union, and the United States), this knowledge both drew on and helped *create* "North-South" divisions, even as those divisions were always porous, and actively struggled against. This was also an era of decolonization (figure 5.5).[11]

Work done at the International Meteorological Center at Bombay built on the work of the physicist Anna Mani, who, during the International Geophysical Year, "took charge of a network of stations to measure solar radiation across India" (Amrith 2018, 238). Farther back in Indian Ocean history, such figures as the fifteenth-century Arab navigator Ahmad ibn Mājid and the seventeenth-century Ottoman encyclopedist Kâtip Çelebi pioneered accounts of regional sea winds, accounts that later would be relied on by British imperial meteorologists seeking to predict the storm waves of cyclones (Amrith 2020, 142).

The expedition fashioned the Indian Ocean as a transnational ocean-atmosphere system whose data-driven understanding would supply improved knowledge of weather process in Asia, with relevance for everything from the prediction of cyclones and agricultural planning to the launching of satellites. As the Earth became imagined as a cybernetic system, so did its oceans. In addition to setting the outlines of an oceanic/planetary South, this reframing of the Indian Ocean proffered a *reorientation*—with all the puns that word might suggest: parts of what Europeans once considered "the Orient" became parts of "the South," and the monsoon, flipping orientations over the year, phased into being a model for planetwide systems.

As Amrith (2018, 307) observes, bringing the story up to the present, the monsoon itself is these days changing, in calibration with climate change: "Monsoon Asia means something quite different now, when the monsoon's behavior, increasingly erratic, responds to human intervention." He explains: "It is increasingly clear that monsoon rainfall is affected not only by planetary warming but also by transformations on a regional scale, including the emission of aerosols—from vehicles, crop burning, and domestic fires" (304). In 1999, the Indian Ocean Experiment led by the Scripps oceanographer V. Ramanathan reported that "the equatorial Indian Ocean is a unique natural laboratory for studying the impact of anthropogenic aerosols on climate, because pollutants from the Northern Hemisphere are directly connected to the pristine air from the Southern Hemisphere by a cross equatorial monsoonal flow into the inter-tropical convergence zone" (cited in Amrith 2020, 146; see also Middleton 2020). The effects then reach back out from the Indian Ocean, since the monsoon plays a key role in worldwide atmospheric and oceanic circulation, reaching into the hemispheric pattern known as the El Niño-Southern Oscillation

(Davis 2017). The possibility I discussed earlier—that portions of the Oceanic South may herald futures for elsewhere on the planet—is one Amrith (2018, 15) also articulates: "The history of how Indians have understood and coped with the monsoon may have wider lessons at a moment when climate can no longer be ignored, anywhere in the world—in this sense, at least, India is not behind the world but ahead of it." More than a cradle of globalization, the Indian Ocean now also becomes a cradle of climate change. As DeLoughrey (2019, 136) argues, thinking from Indian Ocean islands such as the Maldives, which are being lost to sea level rise, "Our planetary future is becoming more oceanic." The implications of changed monsoon dynamics also amplify processes set in motion in the past (think colonial construction, urban development, dams) which scale up not so much to the figure of the globe but to the *planet*. Edward Sugden (2018, 16) writes about an earlier moment when oceans phased into unsettling futures (his case is the nineteenth-century Pacific, which saw the peak of whaling, the compromised independence of Pacific island nations such as Hawai'i, the Spanish-American War), "The oceans . . . operate something like an untimely avant-garde, testing out the future before it comes into being."

Such futures arrive at less global scales, more human (if overwhelming) scales, in the form of waves. In *The Great Derangement: Climate Change and the Unthinkable,* the anthropologist and novelist Amitav Ghosh (2016, 50–51) paints a nightmare portrait of waves arriving at India's west coast in an event he imagines emerging after a category 4 or 5 cyclone: "Waves would be pouring into South Mumbai from both its sea-facing shorelines; it is not inconceivable that the two fronts of the storm surge would meet and merge. In that case, the hills and promontories of South Mumbai would once again become islands, rising out of a wildly agitated expanse of water." Such waves are more than physics-y forces; they carry, slow and fast, a charge of trash and pollution (see Anand 2020). They are waves that see the ocean field revealed as a planetary laboratory testing the variables of a choppy Anthropocene, of a *chronic ocean,* a sickened sea made of churning, confused time—thalassochronic waters of wake and break.

Zooming into the Bay of Bengal

Move to the other coast of the subcontinent, to the Bay of Bengal, where waves are again human-scale messengers of monsoons and of killing cyclones (see map 5.1, marked with places—as well as cyclone paths—discussed in this chapter). In Bangladesh, deaths from storm surges have numbered in the hundreds of thousands. As Amrith (2018, 309) reports, "Approximately 40 percent of global storm surges in the last fifty years have hit Bangladesh, including the two with the highest death tolls, in 1970 and in 1991." Back in the 1960s, the Dutch landscape form known as a polder—"a piece of low-lying land reclaimed from the sea" (*OED*) and often enclosed by embankments (see chapter 1)—was introduced to Bangladesh by the World Bank–funded Coastal Embankment Project both to control floods and to manage irrigation networks (Cornwall 2018; Dewan 2021, 42). By the late 1990s, however, it was clear that that control had also caused damage, rerouting sediment in ways that congested rivers and waterlogged agricultural lands, leaving them with reduced nutrient input or even uncultivable.[12] With much of Bangladesh a low-lying delta, sea level rise was also leading to increased flood depth and hazard from surge.

With wave effects in the Bay of Bengal the emergent results of, all at once, the monsoon; cyclones; and histories of colonial, national, and international development—and with the place carrying a strong echo of Dutch intervention, as well as a premonition of sea level futures elsewhere—these forces and forms might be vital entities through which to query and reorient wave theory.

To complete this chapter, I planned to undertake fieldwork with hydrologists and oceanographers around the Bay of Bengal and started setting up logistics for a visit in summer 2020 to wave scientists at the Bangladesh University of Engineering and Technology (BUET), in Dhaka. I corresponded with Ali Mohammad Rezaie, whom I had met in 2017 at the WAVEWATCH workshop (see chapter 4) and who began his studies at BUET, in the Institute of Water and Flood Management, before finishing a doctorate at George Mason University in 2019. In early 2020, he was back in Bangladesh working as a research coordinator at Dhaka's International Centre for Climate Change and Development, and he put me in touch with his professors and collaborators, sending me one of his papers, from late 2019, "Storm Surge and Sea Level Rise: Threat to the Coastal Areas of Bangladesh," a precis of present and impending dangers in the deltaic plain (Rezaie et al. 2019).

My plans to visit Bangladesh were not to materialize. Instead, the COVID-19 pandemic took command. Meanwhile, in Bangladesh, late May 2020 saw Cyclone Amphan strike, leading to widespread flooding, followed by monsoon rains in June that caused extreme river overflow, leaving more than a quarter of the country inundated.[13] Newspaper images flipped between satellite views of large land masses and on-the-ground photographs of people in waist-deep water, of waves hitting cars in downtown

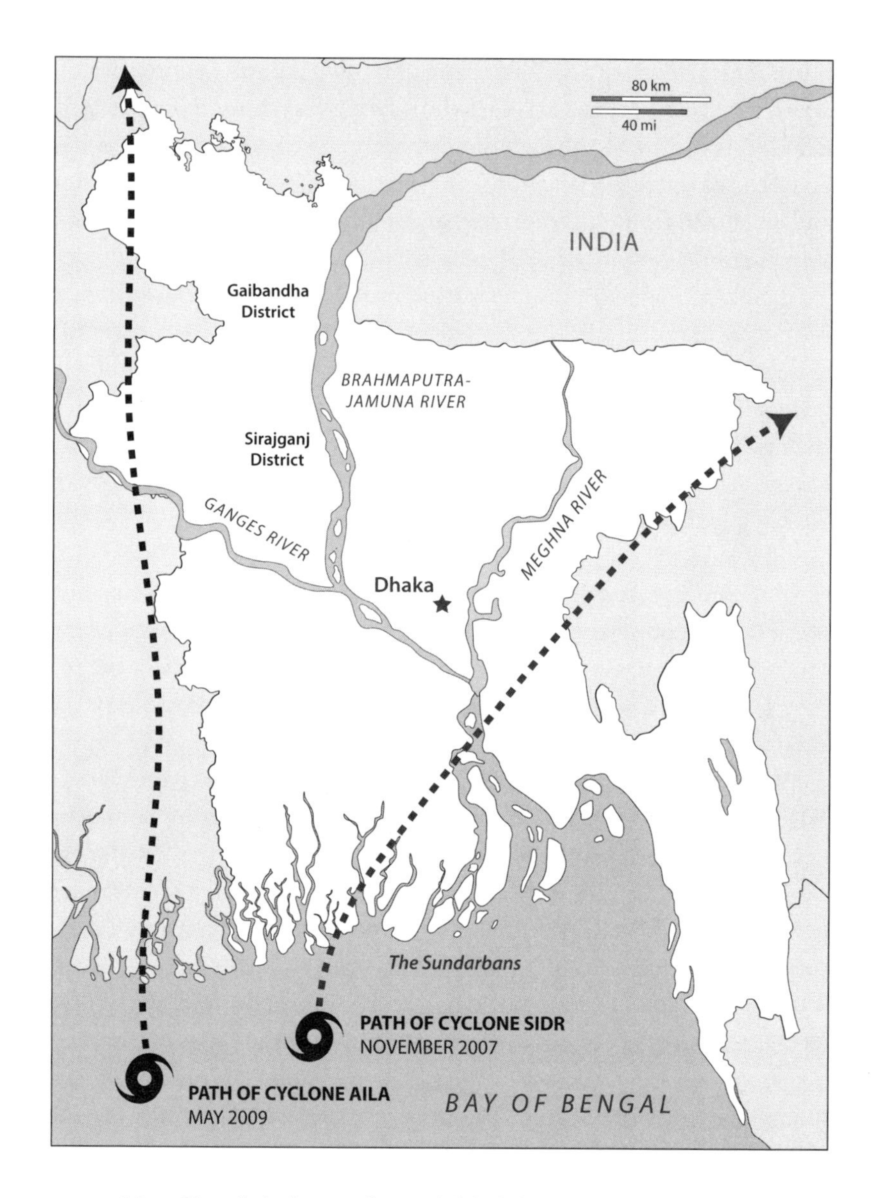

MAP 5.1 Map of Bangladesh, Bay of Bengal, labeled with sites discussed in the chapter. Created by Christine Riggio.

Dhaka. The whiplash of online representations that swung across scales, zooming out, zooming in, was clarifying and confounding all at once. Most coverage was funneled into a story of "sea level rise," even though water was moving in many different directions—down rivers, from the sea—back and forth into different states (rain, mud, silt). More than one story was being written by the waves.

Just after Amphan (which, with successful evacuations, killed about one hundred people, far fewer than earlier storms), I wrote to Rezaie, to see how he was faring, both in the wake of flooding and in the midst of the pandemic. He and colleagues from BUET were OK, and much like me, at the Massachusetts Institute of Technology, in lockdown, teaching remotely. When I reached back out to Rezaie's onetime adviser, Munsur Rahman, a professor at BUET, he filled me in on a webinar series on flood management the institution was running through October 2020. Would I like to join?

When I signed in for the first session, on Zoom, I saw scores of other participants, many from around the Indian Ocean region, with attendees calling in from Afghanistan, India, Kenya, Somalia. The inaugural session was a crash course in the hydrology of the region, organized around the Ganges, Brahmaputra-Jamuna, and Meghna (GBM) river system. The hydrologist Mashfiqus Salehin showed a satellite image of the Delta, tagged with the "multiple stresses" to which the system was subject—with river floods, cyclones, and surges named as *natural processes* and global climate change and growing population named *human processes*. Salehin's presentation zoomed in on each stress in turn, pointing to their inextricable braiding together.

Zoom. That was the name of the digital video conferencing platform so many people came to use in 2020. The word *zoom*, according to the Online Etymology dictionary, "gained popularity c. 1917 as aviators began to use it." Around then, this onomatopoetic word also became associated with the zoom lens, which enabled the "camera shot that changes smoothly from a long shot to a close-up, or vice versa, without loss of focus" (*OED*). The idea of "zooming," then, carries histories of adjustable aerial vision; it is a kind of remote presence on the move.

And Zooming came to define not just the remote fieldwork I now found myself conducting, but also the operating idiom among many hydrologists with whom I spoke, as discussions zoomed in, out, and across the oceanic, regional, coastal, community, household, and individual scales at which waves mattered. If my on-the-ground-and-beach fieldwork in the Netherlands had me moving linearly from station to station, this internet work offered an exponential path of telescoping scales. That was an artifact not only of the scalar recalibrations, from planet to locality, that Earth systems science made second nature to my interlocutors, but also of the way that pandemic ethnography had pressed me, against my predilection, into toggling between "fly-by" Googling and "parachuting-in" video fieldwork (the aviation metaphors, like *zoom*, resonating with how much anthropology as usual already depends on the technology of flight [Knox 2018.])(and see Horton 2021).

The next webinar session in the flood management series was led by the civil engineer Anisul Haque, whose presentation zoomed *way out*—into outer space, cataloging the ways Earth's moon shaped tidal flooding, through the daily variation of high tide and low tide; the fourteen-day cycle of spring and neap tides (shaped by the pull of the moon's gravity aligning with the pull of the sun [spring tide] or operating at right angles to it [neap tide]); and a 4.4 year cycle of high tidal level, consequent on the fact that the elliptical movement of the moon around the Earth itself wobbles and rotates over time. That last tide dynamic, Haque said, modulates the tide in Bangladesh by four centimeters (see also Jones 2019). The dynamics of all of these tides—and the kinds of storm surges they carry—are now contoured by sociogenic effects such as sea level rise. Here was an explicit articulation of tide effects as following from both the *globalized* and the *planetary*—even the *solar systemic*—an astronautical optic making a claim for the Global South as the planetary South.

After a few webinar sessions, I followed up with Rahman to learn more about waves. Having spent most of his career thinking about upriver processes, he had recently turned his attention to the coast. The GBM system, he said, is dominated by the ebb and flow of *tides*, so wind waves are not always at the center of models. Such waves, operating at a scale below tides, are usually input in the form of general parameters, plugged into simulations. (This is so in part because there are virtually no Bangladesh-owned buoys in the bay; what information there is comes from the Indian Meteorological Department, though because that comes a bit from the west, it is not always immediately informative.) But the presence of these parameters is important, since, where tidal rivers meet the sea, monsoon winds and depressions can create significant wave impacts.

One of Rahman's colleagues, Mohammad Asad Hussain, who works on numerical modeling of processes along coastal Bangladesh, told me more. Offering that "Bangladesh is a hydrological laboratory," a place with wide lessons for elsewhere—hence, the Indian Ocean–wide attendance at the webinar—he immediately qualified the claim, telling me that "wave measurement data is very scarce in our country." That meant that one had to work with what was available, and here conversation zoomed down to the scale of "local knowledge," by which Hussain meant accounts of what floods looked like from the position of people living in coastal villages. He told me about visiting the coast in the aftermath of Cyclone Sidr, in 2007, to interview people about how they confronted storm surges. He remembered one set of stories vividly:

> We met people who fought with the storm all night. They stayed in trees, and the waves were battering, but they hung on all night, just to wait for the water to go down. And they knew it would go down, from their forefathers. And one of the guys was an eight-year-old—and I think he will be explaining his story to his sons in the future. And that's how local knowledge will be propagated to different generations.

This was a vision of the future as a time of social recurrence—a kind of generation-to-generation complement to the solar system's repeating of the periodicity of tide dynamics. Hussain said he had come to think of intergenerational wave knowledge as critical to everyday littoral and riparian life after his years in Japan, where in 2006 he completed his doctorate at the University of Tokyo and learned about the legacy of tsunamihi, stones of mourning and warning (see chapter 4). Japan has been a key site of training for people at BUET, I learned, owing to a commitment by the Japan International Cooperation Agency (JICA) to provide development assistance and capacity building to Bangladesh, an aim linked to Japan's need to maintain robust connections to Indian Ocean ports, a path through which the country's trade with the Middle East travels.[14]

We zoomed in through another medium, as Hussain screen-shared a news video with me of footage from 2009's Cyclone Aila, showing waves hammering a small village while farmers ran along a polder carrying sheep and goats, saving them, Hussain observed, because they were vital to their livelihood.[15] Any cyclone warning system, he emphasized, must account not just for how people might get to shelter, but for what they will judge necessary to survive in waves' wakes.

Another future now appeared in our conversation—of intensifying storms, of possibly irreversible futures, and of new technological infrastructures for prediction. Flood and surge warnings can be relayed not only through on-the-ground volunteer organizations but also through radio, text, and WhatsApp. More—precisely because of the sometime paucity of data collection in the bay—such warnings may increasingly be generated through computer simulation. Rezaie, for example, had been at work on an artificial neural network that might use historical data to predict surges. We are back, here, to the becoming digital of ocean memory (chapter 4)—virtual waves, perhaps in the cloud, foretelling actual waves. In early September 2020, he told me, Google announced it would be using machine-learning artificial intelligence (AI) programs to deliver predictions of floods to people's phones.[16] Zoom out to the scale of satellites and across to globally distributed programming practice (and see Eeltink et al. 2022). Local knowledge would come to be stacked within large-scale cybernetic prognostication, an oracular book of waves written by AI.

And local knowledge, in turn, some engineers hoped, could be fed back into prediction plans. Rahman had worked in Bangladesh's Gaibandha District, a mid-country area for which he and his collaborators were developing flood risk assessment models, which they hoped they could enrich with knowledge from people living in the locality (Ahsan et al. 2019). In November 2018, Rahman had led a group of researchers in dialogue with nongovernmental organizations (NGOs) and local administrators. In a paper he cowrote, his opening address to Gaibandha District villagers was reproduced. He said, to open, "We have a very limited understanding and ideas about this area as we are not from here," continuing that "you know this area, rivers, land, and weather much better than us." A number of people reported on local patterns of erosion, and some suggested that concerted dredging might help. Here was more evidence about the failures of the imported embankment projects that Rezaie had told me about back in Maryland. International participation was not absent, however; Rahman was joined in this project by collaborators from Japan, where Rahman, like Hussain, had completed a doctorate, in 1999.

I learned more about the national government's Flood Preparedness Program (FPP) from a 2019 annual report on Bangladesh's National Resilience Programme. It chronicled some of the data gathering scientists had undertaken, under the aegis of the United Nations Development Program, to get a sense of what floods looked like from the perspective of those most exposed to them. The FPP, the document said, sought to "develop location specific inundation models harnessing existing . . . forecast products [and to] . . . take the warning products to the last mile in an understandable manner along with substantive lead time for early actions. A 'first line defense' will be created for floods like Cyclone Preparedness Programme recruiting volunteers from at risk communities to be named as FPP volunteers" (National Resilience Program 2019a, 9). The document zoomed in on the kinds of volunteers who might be tapped, zeroing in on recruiting women as point people and leaders, a mandate of that part of the program sponsored by the United Nations Entity for Gender Equality and the Empowerment of Women. A companion document named women as vulnerable "due to their lower levels of preparedness and restrictive gender norms [including on mobility]" (National Resilience Program 2019b, 3; see also Shaw 1992). Such a claim, while advanced in the name of human rights, is incomplete, reading social precariousness as a linear matter of tradition and culture rather than as also a precipitate of the ill effects of earlier international development, global capitalist market pressures, and the retreat of state infrastructures of care—dependent, sometimes, on idioms of victimization forwarded to justify economic restructuring and intervention (Dewan 2021, 18). I spoke with Assistant Professor Debanjali Saha of BUET's Institute of Water and Flood Management after reading an article she cowrote, "Gender Vulnerability Assessment due to Flood in Northern Part of Bangladesh," which emphasized that "vulnerability and its antithesis, resilience, are determined by physical, environmental, social, economic, political, cultural and institutional factors" (Leya et al. 2020, 236). Saha, who was finishing her doctorate in civil engineering at the University of Tokyo when I spoke to her, was also keen to place precarity and resilience in the context of how river-adjacent practices (e.g., rice farming) had shifted in response to changed patterns of sand flow. This, in turn, had seen women entering wage labor in sand removal, rendering them simultaneously more exposed to exploitation and more responsible for community monies. Just as "waves"—of new heights, of novel sediment content—are not sheerly natural objects, neither are "restrictive gender norms" simply given.

In *Misreading the Bengal Delta: Climate Change, Development, and Livelihoods in Coastal Bangladesh*, the anthropologist Camelia Dewan (2021) inquires into the history of early projects to embank or "polder" islands in the Bay of Bengal delta, looking at projects in wave attenuation put in place by Dutch consultants in the 1960s (see also Dewan et al. 2015). As Rezaie and Rahman prepared me to understand, these had had effects opposite to their intents. Thus, rather than preventing undesirable flooding (some monsoon floods are welcome, as they support agrarian livelihood), the four thousand kilometers of embankment that texture the Bangladesh coastline often entrapped water instead, which made storm surges, with higher water to amplify them, more damaging. As Dewan (2021, 59) writes about the

Dutch projects, "This foreign solution did not translate well into a heavily sedimented, tropical monsoon delta and has resulted in longstanding—and difficult to overcome—problems of siltation and waterlogging."

A major problem in coastal and river infrastructure projects nowadays, argues Dewan (2022, 541, quoting Cons 2018), is an overemphasis in funding proposals—and in donor expectations—on the frame of climate change: "Donors are now recasting Bangladesh with its low-lying floodplains as an 'epicentre of climate change,' making it into 'a laboratory for so-called resilient development' for the rest of the Global South to follow." Such "climate reductive translations" (using "climate change" as a "spice" [Dewan 2021, 15] in grant proposals) make it difficult to see the complex range of causes behind flood disaster. The language of "laboratory," which poses places as bounded time-spaces with universal relevance, can be ill suited to see the nonlinear, scale-breaking dynamics of history, of the field of real time.[17]

Funding pitches and World Bank pronouncements, often delivered in English rather than Bangla, toe the line, Dewan argues, of an "official transcript" that places inundation in Bangladesh squarely in the lap of sea level rise. Dewan (2022, 542) found a "hidden transcript," however—in her Bangla-language conversations with development officials—that admitted that the meta-code, the "passport for funding," of climate change, occluded the fact that embankments, "by preventing beneficial *borsha* [monsoon] floods[,] silt up key rivers and increase damaging *jalabaddho* [waterlogging] floods." Dewan learned that that knowledge did not necessarily stop Bangladeshi development workers, dependent on global donors, from affirming climate change optics. One told her, "Polders are necessary, especially in the light of climate change. We must thoroughly redesign our polders; we need to address the threat of rising sea levels. The southwest coastal zone is the most backward region of this country because of salinity, tidal surges, cyclones and sea level rise" (quoted in Dewan 2022, 544). Such discursive framings engage in "re-arranging causal events to legitimise the success of projects" (7).

My conversations with engineers from BUET revealed a kindred complexity—a sense that climate change and sea level rise were crucial to the dynamics unfolding in the GBM delta but were also not the entire story. The story was scaled all the way from the transboundary and transnational contexts of the delta—which implicated river flow in India, Nepal, and Myanmar—down to the everyday stories Rahman and Hussain had told me. In his engagement with residents of the Gaibandha District, Rahman observed that thinking locally also had its risks: "The problem is we are thinking from our local perspective. It's a three thousand-kilometer-long river and only two hundred kilometers [a]cross Bangladesh. . . . So whatever we do here, it will not change much at the upstream." Rahman was drawing attention to the necessity, as well as the inadequacy, of thinking across scales, zooming out and in. "Scale" is not necessarily always the right tool for the job, not the right grid for what I have, in this book, called *reading waves*. As Dewan (2021) argues, reading all movements of water as telling a story about sea level rise is often to engage in a *misreading*.

Anthropological research can offer different reading frames (and this segment of this book, the reader will have noted, depends on other anthropologists' research, on other ethnological books of waves). In her ethnographic work with villagers living on the ever shifting islands, or *chars*, of the Brahmaputra-Jamuna River, Naveeda Khan (2019) investigates how residents' embodied experience of river ebb, flow, and flood in the Sirajganj District (just south of Gaibandha) tethers them to large-scale riverine processes that, in turn, loop into those geospatialities that define that unevenly distributed planetary time-space called the Anthropocene. And in this looping, the scales of people, communities, rivers, deltas, and the Earth are not neatly nested within one another. Instead, as Anna Tsing and colleagues (2019) have theorized, they are patchy—or, in a more waterish idiom, churning and choppy. That always-in-process entanglement can be seen, Khan argues, in how gigantic geological processes may tell on everyday life. Just as, for example, the Bengal delta is "the product of the interrelation between different waves of sedimentation occurring on the earth's surface and tectonics or processes determining the movement of the earth's eight plates" (Khan 2019, S338), so river residents' embodied habits may come to be the partial product of the effects of these processes on, say, groundwater plumbing, the content of which (e.g., amplified arsenic) may determine health and life chances. If "the Brahmaputra-Jamuna River provides a patch between the Bengal delta and the Anthropocene," geologically considered, the waves that travel its rivers or press in as storm surge are a choppy relay among bodies, deltas, and ocean planetarity (Khan 2019, S335, 338; see also Morita and Suzuki 2019; Rahman, Ghosh, Salehin, et al. 2020). "Sediment waves" are the key to the story here. Atypical patterns of silting in rivers are the more proximate process than "sea level rise" inflecting rural livelihoods in the area.

Some of the hydrological texts to which I was referred by Bangladeshi wave scientists described those livelihoods not only as holding local knowledge, but also, sometimes, as possible "indigenous approaches" to water

FIGURE 5.6 Construction of bandal-like semipermeable structures in Sirajganj District, Bangladesh. From Rahman, Rahman, Shampa, Haque, et al. 2020.

management (Rahman, Rahman, Shampa, Haque, et al. 2020). Rezaie clarified that the term most used for *indigenous* in Bangla was *Adivasi*—old inhabitants—which, along with *tribal*, may gather together ethnic groups with very different histories, attachments to place, and relations to dominant Bengali settlement projects (and see Briggs and Sharp 2004 for a postcolonial caution about the use of the category *indigenous* in development discourse). Rezaie was skeptical that Adivasi knowledges were best apprehended as somehow constitutively "Southern." The differences seemed to him less about North and South than about power, but he did think local knowledge and traditional technologies (e.g., temporary earthen dikes) might offer alternatives to, for example, imported hydrological formations such as polders.

Rahman told me about a technology of water flow management and attenuation called the *bandal*. A lattice-like structure made of bamboo that might control riverine silt flow—and that had built into it a component of wave protection—he said, the bandal was a kind of indigenous technology. He shared with me the work of one of his students, which tracked the building of a very large bandal along a length of the Brahmaputra-Jumana River in Sirajganj District (figure 5.6). It was a construction that, inspired by traditional techniques, was aided by Doppler current profilers and echo sounders, as well as by proof-of-concept computer simulations. Waves were, again, not always the central players here, with subaqueous flow the more germane hydrological dynamic. Important were the workings and multiple politics of siltation, with sediment, a mix of land and sea, long a site of contested cartographies, sovereignties, and legalities (Bhattacharyya 2021)—colonial, postcolonial, more. Wave theory from this segment of the "South" may not always, really, be about waves—or, better, may underscore how waves are never alone, never pure.

Shift farther west in the Bay of Bengal, to the Sundarban delta of West Bengal, a massive land-and-waterscape braided through by shifting tidal streams, a waterscape divided between Bangladesh and India. It is home to some five million people distributed over twenty thousand square kilometers across some fifty islands. Over its long history, this marshy archipelago of salt-tolerant mangrove forests, this liminal land, has confronted storm surges and waves—a process given a literary rendering in Ghosh's classic Indian Ocean novel *The Hungry Tide*, set in the Sundarbans, a tale that ends with the arrival in the tide lands of massive cyclone-generated waves. A protagonist finds herself battered about in a small boat: "They ran into waves much taller than those they had already faced. When the water curled up ahead of them, they had to strain against the oars to carry the boat over the crest. . . . [T]hen suddenly she would find herself tobogganing backwards into the wave's trough" (Ghosh 2004, 370). Later comes a tidal surge—a bore, or, in Bangla, a *baan*: "Then the noise of the storm deepened and another roar made itself heard, over the rumbling din of the gale: a noise like that of a cascading waterfall. Stealing a glance through her fingers, Piya glimpsed something that looked like a wall, hurtling towards them, from downriver. . . . It was a tidal wave, sweeping in from the sea; everything in its path disappeared as it came thundering towards them" (383; see also Consolandi 2020). The Sundarbans in the past century have battled such waves with mixed success, largely owing to the double-edged affordances of artificial embankments created by colonial agents and development workers. As Megnaa Mehtta (2018, 34) explains in her ethnography of life in the Sundarbans, colonial infrastructure projects have left an unhelpful legacy that has sought to resist the ever transforming character of the islands' wetland morphology, clearing instead of fortifying those mangrove root systems that hold islands in place and whose planty bodies operate as sequestering sinks for carbon dioxide: "The design imagination of the British, based on European riparian systems, failed to adapt to the assertive rivers, monsoon and cyclonic winds of the Bay of Bengal delta." Some colonial administrators recognized this, even as they continued the work.

One wrote, in 1875, "The inundation works cruel havoc among [the] low-lying isolated villages. . . . [T]he more the forest is cleared away, the smaller the barrier placed between the cultivator and the devouring wave" (quoted in Mehtta 2018, 35). Mehtta reports that projects to create wave-blocking embankments continue apace in the present, with some people as strong advocates. She recounts a meeting with one local official who explained to her the promise of dikes, telling her, "This is nothing like you've seen before. This is based on designs from the Netherlands. This is a modern embankment, not like the usual mud embankments (*bandhs*). . . . Do you know . . . , in the Netherlands half their country would sink if they didn't build their embankments? We here in the Sundarbans need to build like them" (30–31).[18]

Here, hydrotheory from the North—enabled by "government, engineering firms, contractors, international NGOs, and climate change adaptation funding bodies"—contours and often deforms local hydrosocialities, sometimes rendering more damaging the recurring disaster of cyclones, which displace people whose lands are already subsiding owing to sea level rise and brackish saltwater flooding. As Debjani Bhattacharyya (2021) shows for the Bengal delta, floods have been far from simply natural occurrences. They have also been the legacy of colonial endeavors, material and conceptual, to make marsh into land—and into fixed property—which has amplified the vulnerability of delta areas (see also Mukherjee 2020). They have ignored, she argues, the reality of the *shoal*, the sandbar, the reef, the shallow, that zone where waves slow and diffract and that the Black studies scholar Tiffany Lethabo King (2019, 8) names (aiming to bring together the watery idiom of Black diaspora studies and the matter of land in Indigenous studies) an "alternative space always in formation, . . . and not already overwritten or captured by the conceptual constraints of the sea or the land." Imperial projects and abstractions in the Bay of Bengal too easily sideline the fact that monsoons are integral to South Asian place, missing the estuarine and terraqueous realities of "soaked cartographies" (Pombo 2018) and the importance of silt as a moving material that is enfolded in monsoon-channeled politics of development (including toxic development [Dewan 2020]), as well as of resistance.[19] Recent work to reforest mangroves—which can slow down waves appreciably (Narayan et al. 2016)—and to adapt computational wave models to mangrove littorals (I had seen some of this at the wave lab in Oregon) seeks to redress these problems, though the matter of sea level rise, which will bring higher surge and wave set-up, will require additional strategies, kinds of "water literacies" (Hastrup 2018) woven through with many other substances and stories.

Time, Tide, Waves, Theory, Books

A key passage in Ghosh's *The Hungry Tide* (2004, 224) is written in the voice of an English teacher from Kolkata who is reflecting on the sense of time in these lands:

> To me, a townsman, the tide country's jungle was an emptiness, a place where time stood still. I saw now that this was an illusion, that exactly the opposite was true. What was happening here, I realized, was that the wheel of time was spinning too fast to be seen. In other places it took decades, even centuries for a river to change course; it took an epoch for an island to appear. But here, in the tide country, transformation is the rule of life; rivers stray from week to week, and islands are made and unmade in days.

As my conversations with hydrologists at BUET helped me see, there exist many interlocking, embedded kinds of time in wave dynamics, many scales, and not all are neatly nested, neatly zoomable. The anthropologist Ashawari Chaudhuri (2019), in her work on South Asian science, tracks what she calls "braided time," a concept that describes the asynchronous winding together of different knowledge systems. This frame is useful to think not only about the braided rivers of the GBM delta, but also about the multiple time streams (hydrological, seasonal, agricultural, national) any river entwines.[20] And more: as Jessica Lehman (2021, 841–42) observes, international bids for "planetary-scale knowledge about the ocean"—attempted by projects such as the World Ocean Circulation Experiment (1990–2002)—are ever met by "scale-defying processes" such as eddies, which confound easy readings of scale at sea.

Ghosh's teacher character meditates on the multiple stories any landscape tells:

> I had a book in my hands to while away the time and it occurred to me that in a way a landscape too is not unlike a book—a compilation of pages that overlap without any two ever being the same. People open the book according to their taste and training, their memories and desires: for a geologist the compilation opens at one page, for a boatman at another and still another for a ship's pilot, a painter and so on.

> On occasion these pages are ruled with lines that are invisible to some people, while being for others, as real, as charged and as volatile as high-voltage cables. (Ghosh 2004, 224)

A seascape, too, might be a book, with waves, as I suggested earlier, its pages. One character in *The Hungry Tide* continues a meditation on water writing, looking at a tideland river: "It was as if a hand, hidden in the water's depths, were writing a message to her in the cursive script of ripples, eddies and turbulence" (352).

I first read Ghosh's *The Hungry Tide* many years ago, when, as part of an earlier project, I traveled to India for a conference on deep-sea, mid-ocean ridges held at the National Institute of Oceanography in Goa. The conference had been scheduled for January 2005 and thus came just three weeks after the devastating 2004 tsunami. Traveling through the states of Tamil Nadu and Kerala a couple of weeks after the tsunami (away from the areas worst hit by the wave, with the exception of a few days in Chennai, where I stayed near the still recovering Marina Beach), I followed news stories about the *Sagar Kanya*, the flagship research vessel of India's National Institute of Oceanography, sent out to the Andaman Islands to assess the undersea movement of the Indian tectonic plate toward the Burma plate and to assay the geological and geographic transformations visited upon the landscape of the Andaman archipelago. This rapid scientific response was more or less simultaneous with the speedy travel to the Andamans of Ghosh (2005), who wrote in the pages of *The Hindu* newspaper, about the tsunami, "It is as if the hurried history of an emergent nation had collided here with the deep time of geology."

In retrospect, and with the rising seas of climate change—not tsunami waves—in view, deep time seems itself to be getting faster. In the years after 2004, Ghosh turned his attention to this dynamic, writing in *The Great Derangement* that large-scale transformations in the organic and inorganic nonhuman world have been difficult for many writers to discern because of what he calls "the uniformitarian expectations that are rooted in the 'regularity of bourgeois life.'"[21] This regularity is installed in such forms as the novel and the calendar. If one thinks of waves either as uniformly breaking since time immemorial or as only occasionally jolting everyday time with geological time, one will miss the changing climate dynamics to which waves might point (Ghosh 2016, 35).

Ghosh's sequel to *The Hungry Tide*, *Gun Island* (2019), returns to the Sundarbans but now also follows a Bengali migrant from the islands on a journey across the Mediterranean to Venice—another site of dramatic sea level change—underscoring the undulating, tidalectic, planetarity of climate transformation. The *al-Bahr al-Hindi* (Arabic Indian Ocean), *darya'i akhzar* (Persian Green Sea), and *Nan-yang* (Chinese Southern Ocean) escapes its bounds, becoming Mediterranean, in the middle, in the muddle, of earthly change—and, in the bargain, overflows any steady containment within bins that would once-and-for-all separate Global North and Global South or West and East, disorienting, dis-Orientalizing, received readings of hemispheric oceanicity. Such recognition is critical to the work of decolonization, unfastening oceanography from earlier nation-building projects and models and opening it up to more and different kinds of waves. Such recognition can also open it up to different modes of *theory*, where theory, derived from θεωρός, ancient Greek for "to see," can be expanded to describe not just ways to see, conceptualize, and abstract but also more capacious modes of sensing and sense making, of doing *theory for*—for livable climate futures, for antiauthoritarian planning, for science study.[22]

If a seascape is not unlike a book, with waves its pages, consider this chapter's swells and breakers interleaved with sheets from what, in "Fragment in Praise of the Book," the Indian American poet Meena Alexander (2018, 42) names, in thinking of her transoceanic migrations, a "book of rice paper tossed by monsoon winds." And consider those pages shuffled, too, into what the American poet Madeline De Frees (1983) names "The Book of Sediments," a text that has "waves [that] know their burden of heavy minerals by weight."[23] Books, of course, are only one medium for thinking the sea. In the age of spectacular climate disaster film and server-overheating doomscrolling and doomsurfing, one could argue that today's public culture is spooling ever faster away from any anchoring in the "regularity of bourgeois life" (see Bould 2021). Reading waves as a kind of reading ahead is now a multimedia practice, an endeavor of constant reorientation, of never settling, of learning to scan the tilted horizon of a broken ocean in a world turned upside down.

POSTFACE
THE ENDS OF WAVES

Imagine a rippling sea. Imagine, next, the surface roiling with winds that stir up waves of new height, participating in a surge that carries the news of sea level rise.

Where I live, in eastern Massachusetts, one can now study a range of maps showing inundation markers in 2030, 2050, and 2100. A number of projects aimed at confronting ocean rise are in motion. None summon the sort of dystopian/utopian adaptations Kim Stanley Robinson imagines in his science fiction novel *New York 2140* (2017), which describes twenty-second-century streets in lower Manhattan as having become Venice-like channels: "The canals were like a perpetual physics class's wave tank demonstration—backwash interference, the curve of a wave around a right angle, the spread of a wave through a gap" (16). Instead, plans for Boston envision flexible infrastructures, floating eco-friendly constructions that provide wave-attenuation capacities. The projects are similar to the Living Breakwaters project in New York City, "which proposes a large reef of optimized oyster habitats to be constructed off the coast of Staten Island" (Marcus 2020, 255). In 2020, the American Society of Civil Engineers gave its Innovation Award in Sustainable Engineering to a project to be sited around the shores of East Boston. Called the Emerald Tutu, it proposed an

> array of fibrous floating mats loaded with biomatter that flank the waterfront areas of the city, reducing surge and waves from threatening storm events that are supercharged by climate change. Individual mats grow to be massive waterlogged units, with semi-aquatic marsh grass above and aquatic seaweeds attached below. As an interconnected system, they absorb and transfer energy through the network, acting to disrupt large incoming waves, and the submerged biomatter impedes the inland rush of atmospheric pressure-driven tides known as storm surge. (Environmental Solutions Initiative 2020)

"Emerald Tutu" riffs on Boston's "Emerald Necklace," a one thousand-plus-acre chain of green parks linked by waterways, designed in the 1870s and 1880s by the landscape architect Frederick Law Olmsted. The "Tutu" moniker is meant to be playful, a pointer to the ruffly skirt-like patterns of marshes it would install. Emphasizing the against everyday wisdom of this aqueous infrastructure, the team that designed the tutu produced publicity that featured a drag queen wearing an emerald tutu, declaring that the solution was not only "more resilient, more ecologically habitable for aquatic as well as for human life," but also "a little more queer, a little more activist, a little more anarchist." The project is, at this writing, still in the making.

There also exist more canonical construction projects, including one that would see a seawall run from coastal Quincy to Deer Island in Boston Harbor. Deer Island is noted today for its space-age wastewater treatment facility, but in the 1800s it served as a quarantine area—and burial ground—for Irish immigrants escaping the Great Famine. Earlier, in the 1670s, it was a prison to some five hundred Nipmucs, who were removed to the island from the town of Natick when settlers became frightened that these Christian Indians would join the Wampanoag sachem Metacomet (known to the English as King Philip) in his war against Puritan colonists (Lepore 1998; "Remembering Deer Island" 1994). Many Nipmucs died on the island, as well as on nearby Long Island, and their remains may be there still. In 2018, the Muhheconneuk Intertribal Committee on Deer Island, in written testimony for an October hearing by the Boston City Council's Committee on Planning, Development, and Transportation, observed that "Deer Island, Long Island and other sites remain sites of concern to Indian tribes . . . because of the burial ground sites connected with the concentration camp history" ("Legacy of Genocide Resurfaces" 2019).

The fact that the area has been cleared of this history and is now a site for waste underscores the environmental scientist Max Liboiron's (2021) argument that *pollution is colonialism*—that sites of dumping are almost always, also and earlier, sites of dispossession. Most official statements about the waters, tides, and waves of Boston Harbor have until recently included neither its Indigenous history nor the long history of the harbor's pollution. A description in a National Park Service Geologic Resources Inventory Report from 2017 offers a thin focus on "the wave dynamics and currents flowing around the Boston Harbor Islands" that shape patterns of erosion, compressing the history into a timeline that centers European settlement and claiming that "the isolation afforded by the islands . . . made them places to house people 'cast off' by society," posing the relocation of Indigenous people from Natick as a simple affordance of the island's geography rather than as a move in colonial domination (Thornberry-Ehrlich 2017, 9).

Continued work to keep the memory of Nipmuc removal alive has come in recent years with the seventeen-mile Sacred Paddle by Natick Nipmuc people down the Charles River to, as Pam Ellis, tribal historian for the Natick Nipmuc Council, has put it, "trace the journey our ancestors took on the forced removal. . . . It is a sacred journey for us, not a reenactment" (quoted in Spitz 2010).[1] That tracing, which uses the medium of the dugout canoe to revisit the wound of the wake, as well as to demonstrate ongoing presence, ongoing work of reading history, offers an argument about time—that the past is not past—and it resonates with much of what I have urged here about waves and their multiple time-lives. The waves that arrive at Deer Island are waves in history, as well as harbingers of what is to come; reading them requires constant reorientation to the significances that they carry, as matter and meaning. On the day I visited the island, in fall 2020, it was marked by blue face mask trash rolling up onto its shores.

I return to Walter Munk's thought experiment about an extraterrestrial sapient looking at Earth's waves, unable to predict their breaking based on abstraction alone. In 1958, Munk's colleague Eugene LaFond cowrote a piece with R. Prasada Rao of India's Andhra University, "The End of a Wave," which looked at *swash*—water that keeps running up a shore after a wave breaks, escaping waveform patterning. The end of a wave, here, is a kind of material and formal carrying forward, a reach of disorder into the future, a time of continuation and clutter at once, a break. In the larger social moment, now—in a break that is at once an opening, a pause, and a fracture—it is vital to attend to that which is broken, breaking, not so much simply to fix it, as if that were once-and-for-all possible, but to acknowledge its causes, its inheritances (Shange 2020). "Sea-level theory" (Gilroy 2018) cannot treat the level, now rising, as an even reference; it is not what Gilles Deleuze and Félix Guattari ([1980] 1987, 22) have defined as a *plateau*: "a continuous self-vibrating region of intensities whose development avoids any orientation toward a culmination point or external end." The sea future is that which precisely *demands* an orientation.

As I end this book, I realize I did not know what would happen to time when I began research back in the 2010s; did not know which pasts not past I would find in the chronic ocean, as American history (to take a parochial reference point) moved from the era of Barack Obama's bodysurfing to Donald Trump's seawalls. I did not know that climate and political calamity would roil the world so catastrophically, so fast, by decade's end.[2] When I started, I operated within the modest expectations of routine day-to-day life. The plan was to write an anthropological account of a community—wave scientists, with other players here and there—who had a fairly regular recipe for generating collectivity and continuity. Working at universities and other institutions, on a cyclical year-to-year clock, these researchers operated within a temporality I thought I knew, ever edging toward new questions in the field, in the lab, and in digital domains; questions that emerged at the intersection of science and society, oceans and polities.

As it happened, the world dramatically unsteadied such cadences. The most workaday science came under attack as the rise of authoritarian forces in the United States and elsewhere struck directly at appeals to evidence and, in the context of anti-immigrant policy, functioned as an attack on a professional community constituted by its transnational cast. Meanwhile, waves themselves intensified, with hurricanes and cyclones hammering world shorelines with amplifying effects. Water—what the anthropologist Diane Nelson (2019) once named a *genre*, a mode of thinking and writing—mixes with many new forms and materialities to rescript worlds and futures. If this, now, is, perhaps not yet, the time of a broken ocean, it is, to be certain, a time of churn, of breaking, the wakes of history predictable and unpredictable both.

ACKNOWLEDGMENTS

I am grateful to the scientists, engineers, and mathematicians who took time to tutor me in reading waves. In the Netherlands, I thank Anouk de Bakker, Jurjen Battjes, Evert Bouws, Anita Engelstad, Marcel van Gent, Henri van Heiden, Leo Holthuijsen, Gerbrand Komen, Ad Reniers, Gerben Ruessink, James Salmon, Marion Tissier, and Marcel Zijlema. The late Gerbrant van Vledder was an indispensable guide into Dutch wave science. Klaus and Susanne Hasselmann were kind hosts in Germany. At the Scripps Institution of Oceanography, an early conversation with the late Walter Munk set frames for my questions, as did exchanges with the late Ken Melville. Other Scripps scientists central to my thinking include Matthew Alford, Janet Becker, Falk Feddersen, Peter Franks, Drew Lucas, Sophia Merrifield, Robert Pinkel, Nick Pizzo, Kim Prather, and Amy Waterhouse. My stay on the FLoating Instrument Platform (FLIP), led by Qing Wang, puts me in debt to her and to Marc Buckley, Dave Ortiz-Suslow, Ivan Savelyev, and Ryan Yamaguchi, as well as to captain Tom Golfinos and the vessel's crew. The Scripps archivist Heather Smedburg provided indispensable aid. My research at the Hinsdale Wave Research Laboratory, in Oregon, and of Japanese wave science would not have been possible without Dan Cox, Pedro Lomónaco, Alicia Lyman-Holt,

Harry Yeh, and Solomon Yim. Wave scientists central to WAVEWATCH were Jose-Henrique Alves, Ricardo Martin Campos, Arun Chawla, Jessica Meixner, Roberto Padilla-Hernandez, and André Van der Westhuysen. I also thank Ali Abdolali, Spicer Bak, Mindo Choi, Phaedra Jessen, Danker Kolijn, Jian Kuang, Jiangyu Li, Qianquian Liu, Talea Mayo, Cigdem Ozkan, Angelica Pedraza-Diaz, Americo Ribeiro, Deanna Spindler, Teresa Vidal Juarez, Sam Wilson, and Dongming Yang. Hendrik Tolman, the originator of WAVEWATCH, was crucial to my entering the wider wave community. My discussion of wave science in Bangladesh depended on exchanges with Mohammad Asad Hussain, Munsur Rahman, Ali Rezaie, and Debanjali Saha. Other stewards of wave knowledge were Amel Alzakout, Alexander Babanin, Mark Bushell, Andrew Cox, Hernan del Valle, Yalin Fan, Nabil Gmati, Tom Gomes, Christian Grant, Mohamet Ali Hamdi, Mike Meylan, Russel Morison, Lorenzo Pezzani, Elizabeth Solomon, and Val Swail.

Colleagues in anthropology, history of science, and media studies have been vital. The project began in earnest when Dan Reichman, Eleana Kim, and Bob Foster invited me to give the 2014 Lewis Henry Morgan Lectures in the Department of Anthropology at the University of Rochester. Mike Fortun, Anand Pandian, and Nicole Starosielski were critical commentators on the nascent form of the book. I experimented with earlier ideas during a 2010 fellowship at the Institute of Advanced Study at Durham University; Ash Amin, Paul Langley, Surajit Sarkar, and Marilyn Strathern were central interlocutors. A year at the Radcliffe Institute for Advanced Study at Harvard University, coupled with a Guggenheim Fellowship in 2018–19, got the first draft of this book on register. I thank Jessica Bardsley, Lucas Bessire, Moon Duchin, Corinne Field, Francisco Goldman, Durba Mitra, Nicole Nelson, Meredith Quinn, and Evie Shockley for collegial counsel and conversation.

Many people commented closely on chapter drafts. I thank Sunil Amrith, Nikhil Anand, Andrea Ballestero, Dwai Banerjee, Wiebe Bijker, Ashley Carse, Camelia Dewan, Stephanie Dick, Clemens Driessen, Michael Fisch, Danny Fisher, Joe Genz, Friederike Gesing, Alison Glassie, Steven Gonzales Monserrate, Ian Gray, Jacob Hamblin, Naveeda Khan, Jessica Lehman, Jamie Lorimer, Adrian Mackenzie, Joe Masco, Ryo Morimoto, David Novak, Harriet Ritvo, Paul Roquet, Helen Rozwadowski, Heather Anne Swanson, Rachel Thompson, Sarah Vaughn, Brit Ross Winthereik, Ben Wurgaft, Emily Yates-Doerr, and Mei Zhan. Ashawari Chaudhuri was a crucial research assistant for chapter 5. Melody Jue sounded out many drafts of the whole text, offering expert pointers to how I might orient arguments to make them the most oceany. Melody, Tim Choy, and Hugh Raffles joined a book workshop early on that set me a clear compass.

Other colleagues put eyes and ears to key segments of the argument. I thank Francisco Alarcon, Kyrstin Mallow Andrews, Etienne Benson, Caitlin Berrigan, Franck Billé, Rosi Braidotti, Irus Braverman, Eugenie Brinkema, Jeremie Brugidou, Nils Ole Bubandt, Lisa Cartwright, Brenda Chaflin, Una Chaudhuri, Eliana Creado, Reade Davis, Marianne DeLaet, Elizabeth DeLoughrey, Kim De Wolff, Jatin Dua, Kathie Foley-Meyer, Jennifer Gabrys, Jonathan Galka, Raviv Ganchrow, Daniela Gandorfer, Ayesha Hameed, Cori Hayden, Eva Hayward, Nanna Heidenreich, Stefanie Hessler, Cymene Howe, Lilly Irani, Caroline Jones, David Jones, Steph Jordan, Michi Knecht, Christine Lee, Alix Levain, Sonia Levy, Armin Linke, Wayne Marshall, Bill Maurer, Margarida Mendes, Molly Mullin, Tim Neale, Astrida Neimanis, John Durham Peters, Kimberly Peters, Marina Peterson, Dmitry Portnoy, Tara Rodgers, Gabriel Samach, Kate Sammler, Astrid Schrader, Susan Schuppli, Hillel Schwartz, Christy Spackman, Tulasi Srivinas, Charles Stankievech, Phil Steinberg, Veronica Strang, Sal Suri, Renzo Taddei, Dave Tompkins, Christina Vagt, Thom van Dooren, Laura Watts, Jana Winderen, and Helena Wittmann. Colleagues in the Ocean Memory Project, the University of Oslo's Media Seas of the High North Atlantic group, the Danish Research Council's "BLUE: Multispecies Ethnographies of Oceans in Crisis" conversation, and the Getty Institute Pacific Standard Time Research cluster on "Oceanographic Art + Science: Navigating the Pacific" at the University of California, San Diego, have all been inspiring.

Many others contributed decisively—in conversation, in collegiality—to the lines of thinking here. I thank Stacey Alaimo, Samer Alatout, Warwick Anderson, Karen Barad, Debbora Battaglia, Naor Ben-Yehoyada, Chris Boebel, Tom Boellstorff, Andrea Bohlman, Geoffrey Bowker, Dominic Boyer, Vince Brown, Janet Browne, Xan Chacko, Aadita Chaudhury, Xenia Cherkaev, Beth Coleman, Anne Dippel, Emily Dolan, Walker Downey, Joe Dumit, Paul Edwards, Richard Fadok, Kim Fortun, Rayvon Fouché, Sarah Franklin, Michele Friedner, Joan Fujimura, Peter Galison, Ilana Gershon, Cristina Grasseni, Andy Greydon, Hugh Gusterson, Sherine Hamdy, Donna Haraway, Aída Hernandez, Jen Hsieh, Lochlann Jain, Ian Jones, Chris Kelty, Clare Kim, Grace Kim-Butler, Eben Kirksey, Rijul Kochhar, Agnieszka Kurant, Nicole Labruto, Hannah Landecker, Jia Hui Lee, Vincent Lépinay, Max Liboiron, Alexandra Lippman, Tim Loh, Victor Luftig, Adrienne Mannov, Emily Martin, Theresa MacPhail, Eden Medina, Lisa Messeri, Hélène Mialet, Annemarie Mol, Andrew Moon, Cris Moore, Paul Muldoon, Michelle

Murphy, Burcu Mutlu, Natasha Myers, Alondra Nelson, Rodrigo Ochigame, Valerie Olson, Canay Özden Schilling, Tom Özden Schilling, Trevor Paglen, Verena Paravel, Annet Paulelussen, the late Trevor Pinch, Perig Pitrou, Anne Pollock, Joanna Radin, Boyd Ruamcharoen, Alexander Rehding, Luísa Reis Castro, Liz Roberts, Sophia Roosth, Franco Rossi, Michael Rossi, Caterina Scaramelli, Natasha Schüll, Nick Seaver, Beth Semel, Nika Son, Michelle Spektor, Alma Steingart, Jonathan Sterne, Hallam Stevens, Banu Subramaniam, Ajantha Subramanian, Lucy Suchman, Benjamin Tausig, Karen-Sue Taussig, Michaela Thompson, Sharon Traweek, Rebecca Uchill, Maria Vidart, Antonia Walford, Emily Wanderer, Claire Webb, Kath Weston, Caroline White-Nockleby, Sara Wylie, Di Wu, Anya Yermakova, Juliane Yip, and Casey Zakroff. The inspiring work and words of the late Diane Nelson stay with me always.

The Massachusetts Institute of Technology (MIT)* Anthropology unit is a wonderful place to work and think. I thank Héctor Beltrán, Manduhai Buyandelger, Amah Edoh, Michael Fischer, Jean Jackson, Graham Jones, Amy Moran-Thomas, Heather Paxson, Susan Silbey, Bettina Stoetzer, and Chris Walley. The staff over the years—Carolyn Carlson, Karen Gardner, Kate Gormley, Irene Hartford, Barbara Keller, Amberly Steward—smoothed the way. Farther afield at MIT, I thank Sandy Alexandre, Tanja Bosak, Christopher Capozzola, Ian Condry, Mary Fuller, David Kaiser, Leila Kinney, Emily Richmond Pollock, Shankar Raman, Jay Scheib, Gediminas Urbonas, and Evan Ziporyn.

I presented portions of material here in many venues, including at the Center for 21st Century Studies at the University of Wisconsin, Milwaukee;

* During the time of my writing this book, MIT, like many other institutions, struggled to compose a meaningful statement of Indigenous land acknowledgment, work still in process. The institute sits on the traditional unceded territory of the Massachusett Tribe, and has, in its founding (consider the Morrill Land-Grant Act of 1862, which leveraged land acquired from Indigenous groups through treaty, cession, or seizure into college endowments) and ongoing work (e.g., in mining around North America), often participated in the dispossession of Indigenous land and resources. These are practices for which acknowledgment is insufficient, and for which repair and justice are better paths forward. Work, for example, to chip away at National Park occupation of the Boston Harbor Islands and permit islands to recover from toxic and other insult has been undertaken by a range of Indigenous organizations (e.g. the Muhheconneuk Intertribal Committee on Deer Island referred to in this book's postface) and their possible allies (e.g., the Stone Living Lab, which studies coastal resilience and ecological restoration in the Harbor Islands and includes Massachusett representation: https://stonelivinglab.org/).

Department of Anthropology at Memorial University, St. John's, Newfoundland; Environmental Humanities Network, Edinburgh, Scotland; Department of Anthropology, University of Chicago; National Research University, Higher School of Economics, Moscow; IT University of Copenhagen; Alexander Grass Humanities Institute, Johns Hopkins University, Baltimore, Maryland; Laboratoire d'Anthropologie Sociale, Collège de France, Paris; Humanities Research Institute, University of California, Irvine; State University of New York Law School, Buffalo; Sixth Brazilian Anthropology of Science and Technology Meeting, University of São Paulo; Goethe-Institut, Amsterdam; Centre for Research Architecture, Goldsmiths, University of London; Department of Anthropology, Yale University, New Haven, Connecticut; Max Planck Institute for the History of Science, Berlin; Department of Comparative Literature, Princeton University, Princeton, New Jersey; Department of Anthropology, Rice University, Houston, Texas; Science, Technology, and Society (STS), Tufts University, Medford, Massachusetts; Department of History and Sociology of Science, University of Pennsylvania, Philadelphia, Pennsylvania; Scripps Institution of Oceanography, University of California, San Diego; University of California, Santa Barbara; Akademie der Künste der Welt, Cologne; Music Department, University of Pennsylvania, Philadelphia, Pennsylvania; SciencePo, Paris; National Museum of Natural History, Brest, France; University of Manchester; Museum of Art, Architecture, and Technology, Lisbon; Institut für Ethnologie und Kulturwissenschaft, Universität Bremen; Art Laboratory, Berlin; and Dipartimento di Storia, Antropologia, Religioni, Sapienza, Università di Roma. Portions of chapters have appeared in *BOMB Magazine*; *Cabinet; Cultural Anthropology*; *Environmental Humanities*; *Media and Environment*; *Public Culture*; *Science, Technology, and Human Values*; and *WSQ: Women's Studies Quarterly*. My gratitude, too, to the team at Duke University Press: Ken Wissoker, Ihsan Taylor, Mattson Gallagher, Chad Royal, and Ryan Kendall. Paula Durbin-Westby expertly indexed.

I have, finally, to thank my family, Heather Paxson and Rufus Paxson Helmreich, as well as my parents, Mary and Gisbert Helmreich, along with the Paxson crew. Their illimitable love and wisdom has been a steady current throughout the research and writing of this book.

42.3601° N, 71.0942° W, 10 meters above sea level, 2023

NOTES

Preface

1. Writing in 2022, the Intergovernmental Panel on Climate Change notes that "the largest observed changes in coastal ecosystems are being caused by the concurrence of human activities, waves, current-induced sediment transport, and extreme storm events" (500).
2. Surf lore often has the third—sometimes the seventh or ninth—wave as the largest in a set, but there is no universal trend; much depends on locality (Deacon 1984).
3. I am inspired by analytics of diffractive reading offered by Donna Haraway (1997) and Karen Barad (2014), which tune to how texts generate multiple, superimposed meanings.

Introduction

1. This quotation and others from Munk are from my interview with him on August 25, 2015. I reproduce these words, as I do others from interviews with scientists I name, with permission. Munk died in 2019.
2. There is a thick historical literature on the shaping of oceanographic knowledge about fisheries, tides, currents, circulation, atmospheric carbon dioxide, radioactivity, and more by nation-state patronage and

international initiatives (see Benson and Rehbock 2002; Deacon 1971; Deacon et al. 2001; Hamblin 2005; Höhler 2017; Mills 1989, 2009; Oreskes 2020; Rehbock 1979; Rozwadowski 2002, 2005; Schlee 1973).

3 Ancient Greek investigations into waves are part of an Occidental tradition. Plato (429?–347 BCE), discussing social change in *The Republic*, organized key transformations into "three waves," which, classicists suggest, called on then contemporary knowledge about wave groups in the Mediterranean (Sedley 2005). Aristotle (384–22 BCE) and Plutarch (ca. 44–120 CE) pondered how wind generates waves. Still earlier, and more impressionistically, *The Iliad* (ca. 1260–1180 BCE) offered the word κύμα for wave. Jamie Morton (2001, 32) suggests that this drew on an image of the sea as procreatively female: "derived from κύω, to conceive or be pregnant, κύμα denotes something swollen."

4 Much early wave science centered attention on waves in canals rather than at sea, keying to the port logistics of imperial powers such as France, Britain, and the Netherlands (see, e.g., Green 1839; see also Darrigol 2003; Mukerji 2009).

5 Analogies of the travel of sound to the travel of waves themselves have a long history (Kilgour 1963).

6 Or not. The twentieth-century Anatolian author Cevat Sakir Kabaağaçlı scolds writers for thinking that the materiality of writing shares anything with sea substance. "If that blue is the sea's own," he dares storytellers, "then dip your pen into it and write blue" (quoted in Opperman 2019, 444). On "Self-Recording Seas," see Burnett 2011. On "wavewriting," see Philippopoulos-Mihalopoulos 2022.

7 One of Munk's colleagues, Carl Wunsch (personal communication, February 7, 2019), told me the term has caused confusion, since the word *significant* does not point to a measure such as statistical significance, as some might surmise.

8 The vision forwarded by Walcott (1978) grew out of postcolonial contestations of the sea as *aqua nullius*—the figuration offered by the Dutch jurist Hugo Grotius in *Mare Liberum* (1609)—as well as contestations of the sea grabs of powers such as the United States, which, with the Truman Proclamation of 1945–46, arrogated to itself an exclusive economic zone of two hundred miles around its coasts, tripling the country's area. This was followed by other nations that, in so doing, both enclosed territory and rendered "the high seas" outside national sovereignty and history (DeLoughrey 2017).

9 Drawing on the work of the Martinican poet Édouard Glissant, Elizabeth DeLoughrey (2017, 33) describes sea history as pointing to "a submarine temporality in which linear models of time are distorted and ruptured."

10 On the "shape" of time keyed to "loyalties of various kinds: to God, land, descent group, king, nation, employer, one's children, and so on," see Greenhouse 1989, 1632.

11 For a record of this series back to 1986, see the website of the International Workshop on Waves, Storm Surges and Coastal Hazards, http://www.waveworkshop.org.

12 Lord Kelvin postulated that winds that were strong enough could overcome the surface tension of water and raise ripples. In 1924, Sir Harold Jeffreys "proposed that the wind, moving in the same direction as a wave but somewhat faster, would tend to separate from the water surface as it flowed over a crest. In effect, the wind would leap from crest to crest, avoiding the troughs. The back side of the crest would then *shelter* the front side of the crest, so that the pressure on the front side would be lower" (Zirker 2013, 32–33). That idea turned out to be not quite right; the pressure differential was not big enough. The fluid dynamicist Owen Phillips, who studied turbulence over airplane wings, posited in the 1950s that "turbulent pressure fluctuations in the wind might generate ripples" if such fluctuations fell into resonance with water waves; *that* might generate and amplify waves. Meanwhile, the engineer John Miles, at the University of California, Los Angeles, also had his eyes on pressure, positing that "once weak waves appear, they modify the airflow and therefore the pressure distribution near the water surface, in such a way as to amplify themselves" (Zirker 2013, 35, 37). For more recent views, see Pizzo et al. 2021.

13 On fixed versus mobile reference points for ocean space, see Steinberg 2013.

14 As Paul Edwards (2010, 109) argues in *A Vast Machine*, his history of climate modeling, "Data are never an abstraction, never just 'out there.' . . . Data remain a human creation, and they are always material; they always exist in a medium. Every interface between one data process and another—collecting, recording, transmitting, receiving, correcting, storing—has a cost in time, effort, and potential error: data friction."

15 In a wave spectrum, "wave records [are] represented as a weighted sum of sine waves; the relative weightings constitute . . . the spectrum" (Irvine 2002, 379). See Kinsman 1965 for an early, authoritative rendering.

16 Spectra have something in common with computer simulations—theoretically animated models within which scientists perform virtual experimental operations (Galison 1997).

17 See Cohen 2006 on "chronotopes of the sea," which she uses to classify the ways novels stage the time-space of the ocean in tales of the open sea, the shore, and the deep.

18 The ECMWF's headquarters are in Reading, UK; Bologna, Italy; and Bonn, Germany. See also Janssen 2000.

19 Commercial users can also employ wave-prediction products by paying licensing fees, the amounts of which will depend on whether they need medium-, extended-, or long-range forecasts (ECMWF n.d.).

20 On oil and gas licensing in the 1970s enabled by maps of extreme wave heights generated by the United Kingdom's National Institute of Oceanography, see Draper 1996.

21 The Arctic is at risk, too. As the political theorist William Connolly (2017, 104) reports, as "melting ice enlarges the space of open seas during the Arctic summer, the combination of more open water and more intense winds produces larger waves than heretofore experienced."

22 Early discussions pegged the beginning of the Anthropocene to the industrial revolution, but recent deliberations—by the Anthropocene Working Group of the Subcommission on Quaternary Stratigraphy of the International Commission on Stratigraphy—date it to the beginning of the atomic age, which released bomb carbon into the atmosphere and oceans, a process that has left a definitive geological marker (see Masco 2018).

23 Donna Haraway (2015) proposes that ecopolitical messes—oceanic dead zones filled with mucilage communities, populations of jellies and slime—have muddied bright lines between evolutionary pasts and futures, between the putatively natural and the cultural. Stealing a page from the fantasist H. P. Lovecraft, Haraway suggests that eco-theory might remake, as a refigured mascot, *Cthulhu*, the tentacled monster of a repressed, abject but potent Earth. Re-spelling as *Chthulu* to call back a chthonic multiplicity against Lovecraft's racist vision, she suggests the *Chthulucene*, a heterochronic time in which the boundary between the ancient and contemporary is mucked up (see Hetherington 2019; Lorimer 2017). For a catalog of *-cenes*, see Mentz 2019a. On "Indigenizing the Anthropocene," see Todd 2015.

24 See also Zalasiewicz and Williams 2011, which notes that much knowledge about Anthropocene stratigraphy comes, in fact, from the deep sea.

25 Take the case of chemical weapons (e.g., mustard gas) dumped by Allied powers into the Baltic Sea after World War II, detritus now leaking waste upward into the sea, making military pasts into damaging presents and futures (see Neimanis et al. 2017).

26 Mentz (2015, xxii, xii) also suggests the *Naufragocene*, the age of shipwrecks, whose "contours present themselves whenever and wherever keels plough waves" as "the sudden shock and pressure of immersion fractures ships, systems, and alliances."

27 Consider something like the "choppy Anthropocene," a riff on the "patchy Anthropocene" of Tsing et al. 2019.

28 These do not leave the geological behind. Kathryn Yusoff (2018) argues that the field of geology is suffused by histories of extractivism that have called foundationally on unfree Black labor (in mining, shipping), extractivisms that, bound up with the Middle Passage, have an oceanic component.

29 Chakrabarty's work on climate builds on earlier writing (see Chakrabarty 2000) on different kinds of history—what he calls History 1 (normative, secular, Enlightenment history, told by the West) and History 2 (subaltern,

postcolonial, sometimes supernatural accounts). The literary theorist Ian Baucom (2020) proposes that climate change might require a new history, History 3—*naturalhistorical*—that, in a future-facing idiom, he calls "History 4° Celsius," keyed to an extreme but possible temperature rise over the next century. Maybe Semedo's image is one visual aid for that.

30 Wave statistics inform maritime insurance since the shipping industry embeds its financial planning within the calculus of sea-travel risk.

31 Ferdinand Lane (1947, 65) reported the scold: "Scientists, critical of untrained observers, discount such tales [of larger-than-expected waves], since gravity, they remind us, works steadily to prevent waves from vaulting above a certain height." In 1995 came the first measured instance, at a North Sea Norwegian gas pipeline-monitoring platform, the Draupner. In a field that treats waves as statistics, this one—in a sea of thirty-nine-foot waves this wave was eighty-four feet—was given an individual name: the Draupner wave. (Other waves, particularly those beloved by surfers, have names—Pipeline in Hawai'i, Mavericks in California—though those designations pick out recurring events, not once-in-a-lifetime happenings.)

32 "Waves forming out at sea off the Cape of Storms can grow almost 100ft tall from trough to crest, about the same as a ten-story building" (Pretor-Pinney 2010, 148).

33 Climate change also means warming water—which is expanding water, which can connect to intensifying wind, which can generate higher waves. Sergey Gulev and Vika Grigorieva (2004) suggest that waves in the North Atlantic have been rising fourteen centimeters each decade since the 1950s, while the Pacific has seen increases of eight to ten centimeters.

34 Reading a Kipling story in which a lighthouse keeper goes mad from hallucinating the markings of his maritime maps projected into the sea around him, Stephen Donovan (2013, 396) suggests that the late nineteenth century ushers in "a new mode of seeing in which maritime phenomena are considered primarily as artefacts of consciousness." Such seeing may also operate as distracted reading; Michel de Certeau argues that reading entails "detours and drifts across the page, imaginary or meditative flights taking off from a few words, overlapping paragraph on paragraph, page on page" (quoted in Dening 2002, 4).

35 "DUKW is a manufacturer's code based on D indicating the model year, 1942; U referring to the body style, utility (amphibious); K for all-wheel drive; and W for dual rear axles": Michael Ray, "DUKW Amphibious Vehicle," *Encyclopedia Britannica*, online ed., https://www.britannica.com/technology/DUKW.

36 Compare Sverdrup et al. (1942), in which the authors write, "In physics the general picture of surface waves is that of sequences of rhythmic rise and fall which appear to progress along the surface. . . . The actual appearance of the sea surface of the open sea, however, is mostly in the sharpest

contrast to that of rhythmic regularity . . . from the point of view of physics [these real seas] can be termed 'waves' only by stretching the definition" (quoted in Irvine 2002, 378).

37 In *We, the Navigators: The Ancient Art of Landfinding in the Pacific* (1994), David Lewis reports on one Micronesian navigator: "Kaho is said to have dipped his hand into the sea, tasted the spray and bade his son tell him the directions of certain stars. He then averred that the water was Fijian and the waves from the Lau group where they duly arrived the next day" (quoted in Mack 2011, 128). For a critical view of this tradition's representation in Western ethnography, see "Set One, Third Wave: Wave Navigation, Sea of Islands."

38 In the late nineteenth century, the physicists Albert Michelson and Edward Morley tested (and found mistaken) the then prevalent belief that light required a "luminiferous aether" as a medium in which to propagate (Holton 1969). My use of the word *media* here rather emphasizes the message-carrying qualities of signals themselves.

39 Eva Hayward's work examines how sea species know the ocean as compounds of bodies, senses, sexualities, impressions, transductions, and mediations (see, e.g., Hayward 2005).

40 See also Komel 2019, which points out the biblical quality of this moment in Saussure, who founds his linguistics in the same way God, in the book of Genesis, "moved over the face of the waters."

41 Waves are kinds of *nonhuman actors*—a term employed in recent anthropological studies of human entanglements with other creatures (including such customarily significant others as apes, dogs, and whales, but also insects, fungi, and microbes), as well as other efficacious entities (including computers, glaciers, and volcanoes). For a survey of early work in "multispecies ethnography," see Kirksey and Helmreich 2010; see also TallBear 2011. Waves, in this idiom, might be *non-animal-plant-microbe-mineral*, though, since water waves teem with organic and chemical happenings of many kinds, *more-than-animal-plant-microbe-mineral* might also fit, an amalgam that could also upend the life/nonlife distinction (Povinelli 2016), making waves—not unlike, say, electricity, light, or oil slicks—patterned processes unfolding in the zone of the animate inanimate (see Chen 2012).

42 For work in the blue humanities, see Anderson and Peters 2014; Blum 2010; Bolster 2008; Cohen 2017; Cusack 2014; DeLoughrey 2019; De Wolff et al. 2021; Gillis 2013; Gilroy 1993; Lewis and Wigen 1999; Shewry 2015.

43 Anthropologists have multiplied accounts of maritime materiality, too. Studies of fisheries, maritime governance, transoceanic migration and diaspora, ports and logistics, and piracy see the ocean not merely as a stage, but also as a place whose form and physicality matters for life at sea (see Ben-Yehoyada 2017; Chalfin 2015; Dua 2013; Ho 2006; Kahn 2019; Lien 2015;

Markkula 2011; Pauwelussen and Verschoor 2017; Subramanian 2009; ten Bos 2009).

44 On "epistemic things" in science—things not quite known, that generate questions—cf. Rheinberger 1997. Hans-Jörg Rheinberger contrasts these with "technical objects," things known well enough to be operationalized in the search for other things.

45 Helen Rozwadowski (2010a, 162) summarizes how the sea has been rhetorically posted outside human history: "Most glimpses out to sea reveal endless waves reaching to the horizon rather than any lasting evidence of human presence." Steve Mentz (2015, 48) quotes Joseph Conrad on waves as signs of the ocean's oldness: "If you would know the age of the earth, look upon the sea in a storm. The greyness of the whole immense surface, the wind furrows upon the faces of the waves, the great masses of foam, tossed about and waving, like matted white locks, give to the sea in a gale the appearance of hoary age."

46 On the "temporalization of nature," see Porter 1980.

Chapter One. From the Waterwolf to the Sand Motor

1 See Bosscher and Maljaars 2017; Nederlands Instituut voor Beeld en Geluid, *De "nieuwe waterweg" in Noordoost polder*, film, Polygoon-Profilti, 1956; Steenhuis et al. 2015. See also Abe Hoekstra's *History of Waterloopbos* website, http://waterloopbos.net. The site has been used as a stage for environmental art (see van der Molen 2012; cf. Keiner 2004).

2 He continued, "We would have wave data from the physical model in the woods, and later, we had a colleague mathematician who would put that into programs in FORTRAN. But we had no computer! So, he would travel to Groningen, bringing data with him and then typing that onto punch cards."

3 I worry I was meant to see the photo through a white European gaze, one that poses racial others as subject peoples while disavowing it is doing so (see Wekker 2016). The primary source I find on the Bangkok project—Frijlink 1963—does not name Thai participants. On the transnational travel of Thai engineers, see Morita 2013.

4 In late 2019, van Vledder died unexpectedly, from a respiratory infection. A memory of his life by Giordano Lipari is posted on the LinkedIn social media site under the title, "In Memoriam: Gerbrant van Vledder (1957–2019): A Soft-Spoken Gentleman in Permanent Discovery Mode," December 5, 2019, https://www.linkedin.com/pulse/memoriam-gerbrant-van-vledder-1957-2019-soft-spoken-gentleman-lipari/.

5 For a computational contribution, see van Vledder 2006. For a philosophical meditation, see van Vledder 2017.

6 On the domestication of "air" in colonial South Africa, cf. Flikke 2018.

7 For a satiric reading of this claim, see van Boxsel 2004, 36, which quotes the English poet Andrew Marvell in the 1650s wisecracking, "Holland, that scarce deserves the name of land/ As but the off-scouring of the British sand, . . . / This ingested vomit of the sea/ Fell to the Dutch by just propriety."

8 Shore-close seascapes show how the "new sea power of the Netherlands took form not [only] on the high seas but in coastal waters, estuaries, channels" (Siegert 2014, 10).

9 The translation is by Clemens Driessen.

10 The hydraulic engineer Eco Bijker (1996) has suggested that practices of cooperation and compromise were always suffused by dynamics of inequality, even if folk narrative has them as embodying frictionless democracy.

11 Bruno Latour (1987, 230–31) saw something similar during a visit to Bijker's father: "When Professor Bijker and his colleagues enter the Delft Hydraulics Laboratory in Holland they are preoccupied by the shape that a new dam to be built in Rotterdam harbour—the biggest port in the world—should take. . . . The engineers build a dam, measure the inflow of salt and fresh water for a few years for different weather and tide conditions; then they destroy the dam and build another one . . . [but] the years, the rivers . . . the wharfs, and the tides have been scaled down in a huge garage that Professor Bijker, like a modern Gulliver, can cross in a few strides."

12 Intriguingly, Dutch "rewilding" projects aimed, for example, at introducing Heck cattle in Flevoland so they might "de-domesticate" (Lorimer and Driessen 2013) have occasionally (and controversially) been accompanied by desires to reintroduce wolves as predators.

13 It shows up, for example, in literature that links Dutch water engineering across centuries (see de Groot 1987; Kelly 2012; Rooijendijk 2009; TeBrake 2002).

14 See also Kraijo 2016. On evacuation simulations in the Netherlands, see Traufetter 2008.

15 Children were guided by the artist Rob Cerneus. For photographs of the event, see the website at http://schoutenenterprises.com/MarkerNieuws/2016/9121.htm. See also Rijser 2016. The burning practice may have drawn on a Dutch tradition of burning Christmas trees on Epiphany, which fell close to the 1916 Marken flood.

16 For an earlier, German, book of waves, see Thorade 1931.

17 Christopher Connery (1995, 56) calls the ocean "capital's favored myth-element," imagined as "free" to exploit.

18 Battjes also developed a model for energy dissipation in random waves over gently sloping bottoms, offering a mathematical term now incorpo-

rated, the world over, in numerical models for the generation, propagation, and dissipation of coastal wind waves (Battjes and Janssen 1978).

19 This work led to savings in construction cost. Battjes, with his student Martijn de Jong, was also central to discovering the process of seiche generation in the Port of Rotterdam, creating a model for operational prediction, which aided ships in traveling smoothly into the harbor (de Jong and Battjes 2004).

20 Quoted in the museum copy for an exhibition on Ivan Aivazovsky at the Tretyakov Gallery, Moscow, July 29–November 20, 2016.

21 The Waverider became widespread after a 1972 meeting at the National Institute of Oceanography in the United Kingdom that set enduring, world buoy standards (Joosten 2013, 76).

22 This is distinct from Datawell's purchasing, research and development, and management office, in Haarlem.

23 Integrating twice: acceleration → velocity → position.

24 This enables not just the *becoming environmental of computation*, but also a baseline for the *becoming computable of environment* (see Gabrys 2016).

25 Because, in the initial design, "plastics intended for the design of the accelerometer and the platform were all heavier than water" (Joosten 2013, 50), an early Datawell inventor experimented with adding sugar to the water to increase its density (inspired, the lore goes, by the inventor having spent a childhood keeping bees). Company literature records that, in early days, "dissolving enormous amounts of sugar in hot water . . . gave Datawell to the surprised passer-by the appearance of an illegal distillery" (Datawell BV 2001, 2).

26 The central polystyrene sphere in which the Datawell sensor sits is nestled loosely within another polystyrene sphere. That permits the inner sphere to stay "still" and the outer to move with waves, preventing "the transfer of the buoy's pitch and roll motion to the [inner] sphere" (Joosten 2013, 56).

27 In its earliest manifestation this double-sphere was able to measure height, but not direction. The *Directional* Waverider, introduced in 1988, added another accelerometer, in the perpendicular, coupled with a compass to convert to a North-West coordinate system, adding *magnetoception*, a sensory mode on beyond human (present, for example, in homing pigeons). This introduced another puzzle: controlling for magnetic properties of the Datawell buoy's metal casing (Joosten 2013, 148).

28 On materials that make up undersea cables, cf. Starosielski 2015.

29 On "signal traffic," see Parks and Starosielski 2015.

30 De Bakker is now a researcher in coastal morphodynamics at Deltares.

31 Schiermonnikoog is not far from Terschelling, another Wadden island, where the New York–based artist Sarah Cameron Sunde showed video works in June 2016 based on enactments of her *36.5: A Durational Performance with the Sea*, a piece in which Sunde stands in seawater at sites

around the world for periods of thirteen hours, letting the tide rise and fall around her (https://www.365waterproject.org). Sunde's immersion (wearing street-clothes over a wetsuit) in a slowly rising and falling sea at locales in the United States, the Netherlands, and Bangladesh, is shadowed by a sense that one day the sea may rise up for good.

32 For pointers to discussions that led to the Zandmotor, see Adviescommissie voor de Zuid-Hollandse Kust 2006. See also van Dijk 2012. On the early changing morphology of the Sand Motor, see de Schipper et al. 2016.

33 On genres of "soft" sand coastal infrastructure in New Zealand/Aotearoa, see also Gesing 2021.

34 For more about how waves shape beaches, see Warren Brown, dir., *The Beach: A River of Sand*, Encyclopedia Britannica Films, 1965. For specifics of the Dutch case, see Groeneweg et al. 2006.

35 Rewilding and ecological engineering fit within recent European Union advocacies for "nature-based solutions" (see Calliari et al. 2019).

36 Lotte Bontje and Jill Slinger (2017) do this work in a cultural, folkloric register, examining how Dutch residents weave the Sand Motor into personal narratives.

37 On environmental infrastructure, see Ballestero 2019; Carse 2012; Jensen 2015; Scaramelli 2019. See also Pritchard 2011.

Set One, First Wave. The Genders of Waves

1 On "vibrant matter," see Bennett 2010. On "new materialisms," see Coole and Frost 2010.

2 I am inspired here by Anthropocene feminism (Alaimo 2016; Gibson-Graham 2011) and queer critical race accounts of the animacies of metals and toxins (Chen 2012).

3 See also Ortberg 2016.

4 Consider also the wave of Octavio Paz's 1949 short story "Mi vida con la ola" (My Life with the Wave), a wave the protagonist takes home, a tempestuous seductress that might be tamed, and finally dissolved, in a mermasculine heterosexual conquest.

Set One, Second Wave. Venice Hologram

1 For more about Judith Munk, see Sterman and Gullette 2005.

Set One, Third Wave. Wave Navigation, Sea of Islands

1 Mention of vanishing brings to mind projects to generate infrastructures to *hide* islands from waves, which are at the center of investigations into *cloaking*, a technique naval engineers have explored to make ships that can

hide their wakes and with which coastal engineers have been experimenting to generate surface calm around buoys (Cho 2012; Newman 2014). See also Alam 2012, which suggests a mode of canceling water waves beneath buoys by sculpting portions of the ocean floor so they realize shapes that attenuate underwater wave action.

2 See also Simon Penny's "Orthogonal Project" (http://simonpenny.net/orthogonal), which "combin[es] design and pedagogy with indigenous knowledge systems and with the embodied intelligences of sailing and of skilled making and tool use."

3 Beware, however, claims that one place is the future of another. Carol Farbotko (2010, 54) observes that treating Pacific Islands as "a mere sign of the destiny of the planet as a whole" may serve to contain and displace the worries of the privileged.

Chapter Two. Flipping the Ship

1 The "Is It a Duck or Is It a Rabbit?" illusion first appears in a German humor magazine, the *Fliegende Blätter*, in 1892. It was adopted by the psychologist Joseph Jastrow in 1900 to demonstrate the underdetermination of interpretation by visual stimulus and was then made famous by Wittgenstein in *Philosophical Investigations* to describe the difference between "seeing that" and "seeing as." It is picked up by in Kuhn 1962 to illustrate the notion of a paradigm shift.

2 On FLIP, see "History of FLIP," Scripps Institution of Oceanography, n.d., https://scripps.ucsd.edu/ships/flip/history. On *Alvin*, see Helmreich 2007.

3 On how *contexts of motivation* shape questions asked even before what the philosopher Hans Reichenbach once called scientists' *contexts of discovery* (which may be serendipitous) and eventual *contexts of justification* (which name how results are shored up logically), see Oreskes 2003.

4 The *Oxford English Dictionary* (OED) points to the hip-hop song "Mecca and the Soul Brother" (1991), by Pete Rock and CL Smooth, as the origin of the phrase. My definitions of *flip* are adapted from both Merriam-Webster and the OED.

5 The contrast between "matters of fact" and "matters of concern" is from Latour 2004.

6 For an account of Chinese women in science in the United States, attending to how the range of women who come to the country is inflected by histories of regional difference in China, see Gu 2016.

7 Oreskes (2003, 730), writing about Scripps and the Navy, notes: "As scientists trained students, the interests of the next generation remained weighted towards issues originally driven by Cold War concerns, even after military funding and decreased . . . Military concerns were naturalized, and the extrinsically motivated became the intrinsically interesting."

8 Non-Navy funds—from the NSF, the National Oceanic and Atmospheric Administration, and the US Department of Energy—also become dominant after the late 1970s.

9 On the history of Japanese Pacific Empire before World War II, see Tsutsui 2013.

10 On British initiatives, see Cartwright 2010; Tucker 2010. On American ones, Inman 2003; Ross 2014; Shor 1978; Sverdrup and Munk 1943.

11 On the Steere Surf Code, see Bates 1949. On the ca. 1805 Beaufort Wind Force scale, which offered descriptors for local, not global conditions (e.g., "wind felt on exposed skin"), see Huler 2005.

12 "Height of Breakers and Depth at Breaking, Preliminary Report on Results Obtained at La Jolla and Comparison with South Beach State and B.E.B. Tank Results," SIO Wave Project, report no. 8, March 11, 1944, Walter Heinrich Munk Papers, 1944–2002, accession no. 87–35, Scientific Papers, Manuscripts and Talks, Scripps Institution of Oceanography Archives, (hereafter, Munk Papers), box 23, Special Collections and Archives, UCSD, La Jolla, CA.

13 The pressure sensors were called "brass boxes"—though, as they were neither made of brass nor were shaped like boxes, a later report by Munk and colleagues offered that "brass box is a hell of a name for a bronze cylinder": "Brass Box Records," SIO Wave Project, report no. 157B, Munk Papers.

14 "Effect of Bottom Slope on Breaker Characteristics as Observed along the Scripps Institution Pier," SIO Wave Project, report no. 24, October 23, 1944, Munk Papers.

15 SIO Subject Files, 1890–1982, "Waves in Shallow Water," 1944 Woods Hole Oceanographic Institution, US Navy Bureau of Ships, box 50, folder 6 (hereafter, "Waves in Shallow Water"), Special Collections and Archives, UCSD, La Jolla, CA.

16 "An Agreement between the Woods Hole Oceanographic Institution and the Regents of the University of California," January 10, 1944, and Harald Sverdrup to John C. Hammond, letter of introduction, February 16, 1944, both in "Waves in Shallow Water."

17 John C. Hammond to Harald Sverdrup, memorandum, April 16, 1944, "Waves in Shallow Water."

18 "Proposed Uniform Procedure for Observing Waves and Interpreting Instrument Records," Munk Papers. And see, later, Bigelow and Edmondson 1947.

19 "Overtly using the islands as laboratories and spaces of radiological experiment, British, American, and French militaries configured those spaces deemed by Euro-American travelers as isolated and utopian into a constitutive locus of a dystopian nuclear modernity"; "nuclear annihilation is not a threat looming in the future, but an experience of the past" (DeLoughrey

2019, 67). "The myth of the island isolate perpetuated by the US Atomic Energy Commission (AEC) and adopted by ecologists and anthropologists alike, helped to justify the detonation of hundreds of thermonuclear weapons in the Marshall Islands (Micronesia) and French Polynesia" (171). This would "figure the Pacific island as a 'natural' laboratory devoid of human history, subject to the 'god's eye' view of the ubiquitous cameras of the AEC" (172).

20 Rainger notes that Scripps oceanographers' willingness to collaborate with Miyake's students was part of an effort to internationalize ocean science, but also may have worked to mute or co-opt Japanese radiation research.

21 Though with imperialism and war such drivers of pollution and fuel use—which extend beyond capitalist countries (it's not *just* the Capitalocene)—it might also be named the *Militari-ocene*.

22 On Munk and Bascom's privileged "nuclear sensorium," see Shiga 2019.

23 See also Dennis O'Rourke, dir., *Half Life: A Parable for the Nuclear Age*, documentary film, Camerawork Pty, Cairns, Australia, 1985.

24 "Endless Holiday," by John Knauss, with lyrics by Ellen Revelle and Helen Raitt, 1952, Constance Mullin Papers 1962, 1986–2003, accession no. 2003–02, SIO Centennial Show, 2003, Scripps Institution of Oceanography Archives, box 1, folder 13. Rozwadowski (2010a, 170) observes that, "In naval parlance, a holiday is an area in which work has been left undone, but oceanographer-planners relished the double meaning."

25 It also risks obscuring the work of women scientists and technicians at Scripps such as Margaret Robinson, whose shoreside work on the relation between subsurface sea temperature provided accountings of the movements of water mass that were vital to underwater acoustics. As Oreskes (2000, 385) notes, "men went 'out' into the field to collect geophysical data; women stayed 'home' to process it."

26 On demographics and labor relations in oceanography, see Steinhardt 2018. For more about #BlackInMarineScience, follow #BIMSRollCall on Twitter.

27 Ortiz-Suslow is now assistant research professor in meteorology at the Naval Postgraduate School.

28 The spectral model appeared earlier, in 1938, in "The Spectrum of Turbulence," in which G. I. Taylor starts with an analogy to light and prisms. The Corps of Engineers sought in the 1950s to record spectra on magnetic tape (Beach Erosion Board 1955).

29 Robert Dierbeck, dir., *Waves across the Pacific*, film, McGraw-Hill Text-Films, 1967.

30 Summary of *Waves Across the Pacific* from American Archive of Public Broadcasting, http://americanarchive.org/catalog/cpb-aacip_75-9995xh8c. See also Dierbeck, *Waves across the Water*.

31 On an explicit Scripps policy enacted in 1949 (and overturned in 1960) that forbade women from taking passage on overnight cruises, see Day 1999.

32 Rozwadowski (2010a, 170) underscores the significance of play for these men: "While at sea and stationed on Pacific islands, Scripps oceanographers integrated into their scientific work diving for pleasure, enjoying sunsets, and collecting art and objects from natives."

33 See the biography of Helen Hill Raitt (Day 1997), wife of the geophysicist Russell Raitt, who cofounded UCSD's International Center and who in 1952 flew to meet her husband during a Pacific voyage, after which she developed a career of advocating for Tongan-language library infrastructure. See also letters from Judith Horton Munk (Walter Munk's wife) to her parents, Winter Davis Horton, and Edith Kendall Horton, written in 1963 from Tutuila, in Judith Horton Munk Papers, 1962–1968, accession no. 93–18, Scripps Institution of Oceanography Archives, box1, folder 3.

34 On anthropology and wonder, see Srinivas 2018. On wonder in sea science, see Adamowsky 2015.

35 The idea of the "turn" has been used to describe shifts of attention in the humanities and social sciences. The "linguistic turn" in philosophy (pivoting to questions of naming rather than being), the "global turn" in art history (moving from national to transnational contexts), and the "practice turn" in anthropology (looking to in-the-world activity rather than idealized cultural schemes) are examples. As Tom Boellstorff (2016, 390) points out, however, one "entailment of the turn metaphor is that turning takes place around an axis, a still center held constant."

36 See Oceanography: The Making of a Science, People, Institutions and Discovery Collection, 2000 (SMC 0087), box 1, folder 3: Douglas Inman—Interview by Ronald Rainger, February 16, 2000.

37 There were some in-place formal speculations, notably from Munk who with colleague Christopher Garrett forwarded a possible spectrum for internal waves, one they believed might be universal (Garrett and Munk 1972, 1979). Munk's first publication, in 1939, was on internal waves.

38 Oceanography: The Making of a Science, People, Institutions and Discovery Collection, 2000 (SMC 0087), box 1, folder 3: Douglas Inman—Interview by Ronald Rainger, February 16, 2000.

39 Oceanography: The Making of a Science, People, Institutions and Discovery Collection, 2000 (SMC 0087), box 1, folder 4: Walter Munk—Interview by Naomi Oreskes and Ron Rainger, February 16, 2000.

40 While internal waves are no longer ghostly state secrets, knowledge about them can permit scientists to peek into volumes that cross ocean governance regimes. Nation-states have rights over the waters of their exclusive economic zones [EEZs], as well as over the seabed located on their adjacent continental shelves; the International Seabed Authority has jurisdiction over the seabed outside EEZs; and the High Seas, the water column outside

EEZs, are subject to no such authority. US- and Taiwan-based researchers, for example, funded by the Taiwan National Science Council, in collaboration with the United States Office of Naval Research, often gather data at sites in the South China Sea controlled and claimed by Taiwan—and are then able to reconstruct internal wave processes unfolding within Chinese-claimed waters. On "volumetric sovereignty," see Billé 2017.

41 When I met the biological oceanographer Drew Lucas, I took note of a tattoo on his leg, one that summarized this coming together of the physical and biological: an equation from vector calculus inked against a flowing rendering of kelp.

42 Franks tested—and confirmed—his vision of internal waves as mechanisms to form dense plankton aggregation in a project called "Spontaneous Patch Formation in Robotic Plankton," deploying a swarm of autonomous underwater explorers (AUEs) to stand as proxies for phytoplankton (Jaffe et al. 2017).

43 In California, off the coast of Palos Verdes Peninsula, an undersea location the *Los Angeles Times* once designated as "the nation's largest ocean dumping ground of DDT," internal waves have been found to stir up sedimented DDT as well as PCB contamination, the legacy of DDT manufacture and dumping from 1947 to 1983 by the Montrose Chemical Corporation (Ferrel 1992; see also Ferré et al. 2010). DDT and PCBs have worked their way through microbes, fish, seabirds, and marine mammals—and in ways facilitated by internal waves off the Palos Verdes Shelf, which have suspended such toxic elementalities higher in the water than they might otherwise be.

44 Questions about open-ocean wave breaking have been in the asking for a long time. The head of Scripps's Air-Sea Interaction Lab, the late physical oceanographer Ken Melville, recalled to me that the Cambridge University physicist George Stokes in 1847 had posited that the shape of stable, propagating waves could be described as "a trochoid, the curve traced in space by a point on the rim of a rolling wheel" (Zirker 2013, 52). One property of such waves is that the orbitals of water beneath the crests do not stay stationary, but get carried forward a bit, opening into elongating spirals (imagine stretching a spring), an effect known as "Stokes drift." In 1967, mathematicians Brooke Benjamin and Jim Feir, trying artificially to generate Stokes waves in a wave tank, discovered that these waves were *unstable*, that, as Melville put it, "if you had a theoretically uniform wave train, of all the waves exactly the same, that after enough time, the wave field would become modulated—just like an AM/FM radio uses modulated waves to send a signal."

Melville took me to Scripps's indoor wave tank, and asked a student to generate Stokes waves down the tank. The waves marched forward regularly. But as the train of waves began to wobble, there started to appear "parasitic capillary waves," little ripples growing on the forward faces of pointy waves. These were the result of those carrier waves produced by the Benjamin-Feir instability developing a curvature at their crests large enough for surface tension to become important. Such parasitic waves are dissipative of the wave energy associated with the longer waves on which they ride, dissipation that may lead to open ocean wave breaking.

The question of how to think about air flow over waves—inspired by thinking of ocean waves, like airplane wings, as interfaces between the aero- and hydrodynamic—was one that came to occupy Melville. He wanted to know whether breaking waves "could lead to the separation of the airflow," and did experiments in a wave tank with wind blowing over it, even adding a smoke generator "to visualize the flow; we saw indeed that there was separation in the air when you had breaking" (Banner and Melville 1976; see also Melville 1983). "The breaking of surface waves," he told me, "is probably the primary way of generating currents in the ocean, so oceanographers talk about wind-generated currents, but really the waves come between the wind and the water column. The momentum from the wind goes into the waves, the waves break, and then goes out through the currents" (see Melville and Rapp 1985).

45 This work seeks to add a field case to lab work reported in Buckley and Veron 2019.

46 On attuning to atmosphere, see Choy 2012; McCormack 2018; Stewart 2011.

47 For an early summary of work in science studies on this topic, see Smit 1995.

48 While most oceanographic work was not classified starting in the 1990s, this was not true in the 1940s and '50s. And most Scripps oceanographers from the 1930s to 1960s were aware of, and usually approving of, military uses of their work (Oreskes 2020).

49 On high-energy physicists' calibration of their scientific, biographical career times to the "beamtime" they are able to secure on particle detectors, cf. Traweek 1988.

50 After COVID-19 became a matter of concern in the United States, Prather turned her expertise on aerosols to study how to reduce person-to-person transmission of SARS-CoV-2.

51 Just after I finished writing this chapter, UCSD founded the Institute for Indigenous Futures, part of the charge of which was the revitalization of Kumeyaay seafaring technologies (see Rodriguez et al. 2021). The foreclosed future imagined in apocalyptic science fiction about the ruins of FLIP is only one possible prospect for where oceanography around Scripps might head.

Set Two, First Wave. Being the Wave

1 Duane 2019 tracks the rise of "surf Nazis," mostly white California kids who, seeking confrontationally to protect their "local" spots in the 1980s, spray painted swastikas on their boards. This protectionism was entangled with histories of white supremacy and beach segregation in California.

2 Keith Malloy, dir., *Come Hell or High Water: A Body Surfing Film Documenting the Plight of the Torpedo People* (Santa Monica, CA: Woodshed Films, 2011).

3 Nemani 2015 contests dismissive judgments of bodyboarding, cataloging the ways they calibrate to gendered and racialized hierarchies that persist in posting white male board- and bodysurfers at the top.

Set Two, Second Wave. Radio Ocean

An audio companion to this essay—a sample-based composition by Wayne Marshall and Stefan Helmreich entitled "Wave Count," is available online at https://www.thewire.co.uk/audio/tracks/wire-mix-wave-count-a-montage-by-wayne-marshall-and-stefan-helmreich.

1 Seismometers employed to detect earthquakes picked up crashing waves heralding the arrival of Hurricane Irma (Wilts 2017).

2 Compare John Luther Adams's Pulitzer Prize–winning *Become Ocean* (2013), a single-movement orchestral composition made of rising swells of strings, woodwinds, and brass, played over a bed of rippling piano, meant to put listeners in mind of sea level rise and of ice melting at the poles.

3 Compare "Sound-Wave" (2012), a composition by Alexis Kirke in which Kirke wired a conductor's baton to wave-generating paddles in a research tank at Plymouth University in England.

4 Also in an experimental register, listen to Luigi Nono's 1976 ". . . Sofferte onde serene . . ." (. . . Suffering Serene Waves . . .) for piano and tape, crafted, in part, from the recorded sound of bells reflected off of the fog and waves of the canals of Venice.

5 Tristan Murail, "*Le partage des eaux*: Note," Works, n.d., http://www.tristanmurail.com/en/oeuvre-fiche.php?cotage=27533.

6 For a collection of eighty-four pieces of hydrophonic sound art, variously tuned to different underwater frequencies, listen to *freq_wave*, curated by the Thyssen-Bornemisza Art Contemporary Academy: https://ocean-archive.org/collection/72.

7 Michelle Dougherty and Daniel Hinerfeld, dirs., *Sonic Sea*, Natural Resources Defense Council, New York, 2016.

Chapter Three. Waves to Order and Disorder

1 The logging industry arrived in the 1880s, along with white American settler colonialism, which displaced Oregon Indians to reservations such as Grand Ronde, Siletz, Warm Springs, and Klamath. The Oregonian businessman O. H. Hinsdale, for whom the OSU lab is named, hailed from a family that once owned the Umpqua River Steam Navigation Company, involved in the timber business (with the name "Umpqua" lifted from Native inhabitants). The Hinsdale webpage features a poem about him, written by

a friend, Peggy Hoecker, part of which reads: "A generous philanthropist, O. H. Hinsdale had a dream, Of recreating ocean waves by the use of a machine" (https://wave.oregonstate.edu/history).

2 On "derangements of scale," see Clark 2012.

3 For a report on the experiment, see Rueben et al. 2011.

4 On the substitution of computer simulations for physical experiments, see Winsberg 2010.

5 A lot of the work is automated; in fact, the lab itself has only four permanent staff—the director, a research associate (managing lab software, hardware, and infrastructure), an education outreach coordinator, and a maintenance person.

6 Interestingly, this poster is a 1981 ad for a software company, and the pixel images in its middle were rendered by hand, in pencil.

7 One of the artificial features of these spectra is that, while waves in the ocean acquire their energy largely from *wind*, in this lab *there is no wind*; spectral parameters in a model fold wind effects into an artificially generated wave. (Other wave labs do use wind, introducing additional variables.)

8 The printing idiom may be fitting, since Corvallis, Oregon, is the home of the segment of the Hewlett-Packard company that invented the ink jet printer.

9 On "real time," see Riles 2004; Weston 2002.

10 "An algorithm was developed to detect the landward-most statistically significant bright point and then a polynomial curve was fitted to the raw edge, producing an initial estimate of the wave edge" (Rueben et al. 2011, 235).

11 See Ludwin et al. 2005, which documents the work of Indigenous scholars, anthropologists, and geologists to gather historical examples of Indigenous and First Nations narratives about Pacific waves and floods during this period. See also Finkbeiner 2015.

12 On political rhetoric of invasive species, see Subramaniam 2001.

13 For more ethnographic accounts of the 3.11 event, see Gill et al. 2013.

14 Good (2016, 146) notes, "After 3.11, the National General Association for Stone Shops in Japan began to erect 500 coastal stone monuments similar to past *tsunamihi* but modernized to include English translation and QR (Quick Response) codes linking to images and video of the disaster."

15 Ryo Morimoto, personal communication, August 6, 2019.

16 Lori Tobias in the *Oregonian*, quoted in Carr and Fisher 2016, 133.

17 Available through links in Read 2015.

18 Echoing wartime propaganda maps from the West that have Japan represented as an octopus reaching into Pacific territory. See Meier 2017, which reproduces a stark Dutch version from 1944.

19 On how the tsunami put linear models of time into crisis, see Morimoto 2012. As he writes in a later argument, "The wave of semiosis repeatedly goes from remembering to forgetting and vice versa" (Morimoto 2015, 560).

20 Nao considers herself a wave, reflecting on her position as a student who has not yet graduated (*ronin*): "The way you write ronin is 浪人 with the character for wave and the character for person, which is pretty much how I feel, like a little wave person, floating around on the stormy sea of life" (Ozeki 2013, 42).

21 When I was writing *Alien Ocean*, I attended a deep-sea conference in India weeks after the 2004 tsunami, and the questions on the minds of scientists were, indeed, about how fast they could understand what had happened and how they could calibrate different kinds of time—geological, oceanic, bureaucratic. Some held the tsunami and victims at a distance—available by speaking in terms of long, geological time frames—and, if they did not themselves know anyone present at the event, placed victims (craft fishers, people in coastal poverty) in what Johannes Fabian (1983, 104) once called "the Time and the Other," a "time" that metropolitans use to deny the coevalness of usually far-away and marginalized communities. Other scientists found themselves transformed by the disaster, eager to find some way to calibrate geological, oceanic, and social time.

22 See the online Tsunami Digital Library, at http://tsunami-dl.jp.

23 And, in fact, they do not always necessarily suggest waves. As Parry (2017, 74, 144) writes, "Rather than comprising a single wave, the tsunami had consisted of repeated pulses of water, washing in and washing out again, weaving over, under, and across one another. . . . The one thing it did not resemble in the least was a conventional ocean wave. . . . [T]he tsunami was a thing of a different order, darker, stranger, massively more powerful and violent, without kindness or cruelty, beauty or ugliness, wholly alien."

24 For an account of living through the 2004 tsunami, which killed her parents, husband, and two children, see Deraniyagala 2013. See also Goldman 2011.

25 As Greg Siegel (2014, 18) writes, "Modern reason secularizes the accident, disenchants it, casts out its primeval ghosts and goblins, while modern technology effectively humanizes it, brands it as anthropogenic failure rather than a 'natural disaster.'" Precisely this partitioning—natural versus anthropogenic—was at stake in framings of 3.11.

26 See also Fisch 2022.

27 Michael Fisch, personal communication, September 12, 2019.

Set Three, First Wave. Massive Movie Waves

1 On life, animation, and film, cf. Kelty and Landecker 2004.

2 The original Japanese version was *Japan Sinks* (日本沈没 [1973]), by Shiro Moritani.

3 Franklin, in "Inside *Interstellar*," bonus feature included with, *Interstellar*, directed by Christopher Nolan (2014, Paramount Pictures; 2015), DVD.

4 Nicole Starosielski (2013) suggests that underwater settings in 1950s films portray the deep as monstrous and in the 1960s move to narratives about the sea as a place to colonize. In disaster movies, crashing waves confuse the realms of above and below.

5 Compare *Hereafter* and *The Impossible* to Indian tsunami films such as Prabhu Solomon's 2014 *Kayal*, a Tamil romantic comedy that features the tsunami as a backdrop. The events around the 2011 *Tōhoku* tsunami in Japan have only started entering fiction film, and most (see Sion Sono's *Land of Hope* [希望の国] [2012]) focus on Fukushima rather than the wave (see Schilling 2015).

6 The blog *Mana Moana* describes itself as an online "home for indigenous critique of Disney's *Moana*": "Nau Mai," *Mana Moana*, n.d., https://manamoana.wordpress.com. In *Surf's Up*, a cartoon about surfing penguins, animators hoped to render waves in the film using computer simulation, but, finding that such virtual waves do not pack enough emotion, they worried that they wouldn't look "real." Eventually, animators produced a virtual wave "puppet," so that waves could be, as they put it, "characters."

7 On the algorithms behind *Moana*'s waves, see Lafrance 2017.

Set Three, Second Wave. Hokusai Now

1 Ocean scientists have been adamant that Hokusai's wave *not* be mistaken for a tsunami (Cartwright and Nakamura 2009). As Christine Guth (2015, 200) remarks, "It is a measure of the celebrity of 'Under the Wave off Kanagawa' that what kind of wave it represents is a question that has been taken up by geophysicists." Some have claimed the Great Wave may represent a *rogue wave*. The physicist and optics theorist John M. Dudley, science writer and diver Véronique Sarano, and mathematician Frédéric Dias postulate just this in "On Hokusai's *Great Wave off Kanagawa*: Localization, Linearity and a Rogue Wave in Sub-Antarctic Waters" (2013).

2 See Yuko Shimizu, "Climate Change and the City," *Yuko Shimizu* (blog), February 11, 2013, https://yukoart.com/blog/climate-change-and-the-city.

3 Guth (2015, 197) writes, "Waves are widely understood to connote the precariousness of human existence, but this one has translated the consequences of human actions into the workings of nature." On modernity as foam, see Sloterdijk [2004] 2016.

4 Turner's studies were aesthetic, not scientific: "Turner . . . was friends with scientists of his day and he was certainly interested in theory, particularly of colour. But it is hard to see him driven by the theories of science. What he set out to do, from his early days of embracing the romantic theories of

the 'sublime' in nature, was to depict sensation—the sensations experienced at sunrise and sunlight, in storm and dead calm, in rain and mist" (Hamilton 2012). The artist is a medium, and Turner went to extremes to gather experience, even having himself "tied to a mast of a boat so that he could draw waves smashing onto the deck" (Knight 2006, 66). Evocations take a sharper cast in the vivid red waves of Turner's *Slave Ship* (1840 [see "Set Four, First Wave: Middle Passages"]).

5 For an ethnography of the trash vortex and scientists and activists arrayed around it, see De Wolff 2017. See also Decker 2014.

Set Three, Third Wave. Blood, Waves

1 Unless the blood has been spilled in war or violence—mixing up the order of things—in which case the trope of the "sea of blood" often emerges. One notable appearance of a wave of blood in popular culture is in Stanley Kubrick's film adaptation of Stephen King's novel *The Shining* (1980). Bill Blakemore (1987), noting that Kubrick marketed the film in Europe, after its success in America, with the line, "The wave of terror which swept across America," argues that a key scene in the movie, in which a wave of blood floods out of a hotel elevator, symbolizes the blood spilled by European settlers in the genocide of Native Americans. He notes the Calumet, Navajo, and Apache symbols that decorate the movie's Overlook Hotel and writes, "The blood squeezes out in spite of the fact that the red doors are kept firmly shut within their surrounding Indian artwork embellished frames. We never hear the rushing blood. It is a mute nightmare. It is the blood upon which this nation, like most nations, was built, as was the Overlook Hotel."

2 "This is true even when the researchers compared people with the same medical conditions, the same age, and the same insurance coverage" (DeNoon 2007, summarizing Hernandez et al. 2007). See also Pollock 2008.

Chapter Four. World Wide Waves, *In Silico*

1 HOK Design, "A Breath of Fresh Air for NOAA's New Home," n.d., https://www.hok.com/design/type/science-technology/noaa-national-center-for-weather-and-climate-prediction.

2 Think of this as a descendant of the United States' Cold War–era Semi-Automatic Ground Environment (SAGE) computer network, which coordinated radar data to create representations of airspace (Edwards 1996).

3 On fire prediction, cf. Neale and May 2020. On sea-ice melt forecasting, cf. Vardy 2020.

4 On how mapping in the twentieth century came to couple God's-eye vantages with smaller-scale, sometimes embedded GPS views, see Rankin 2016.

5 Hans Blumenberg ([1979] 1997, 7) explores the history of keeping seas at observational distance, a view from "the lamenting but uninvolved spectator" who may (or not) accept their implication in the larger scene. Blumenberg quotes the nineteenth-century historian Jacob Burckhardt, reflecting in 1869 on the French revolutions of his day: "As soon as we rub our eyes, we clearly see that we are on a more or less fragile ship, borne along on one of the million waves that were put in motion by the revolution. We are ourselves these waves. Objective knowledge is not made easy for us" (Blumenberg [1979] 1997, 69–70).

6 For the notion of "ocean memory," I am indebted to the Ocean Memory Project, a science-art-humanities collaboration funded by the National Academies Keck Futures Initiative dedicated to multidisciplinary work on apprehending Earth's seas as a space of recollection (https://oceanmemoryproject.com/).

7 Ubuntu itself is not embargoed, however, and users of Linux have been keen to keep the operating system free of US embargoes (see Marti 2003). Adrian Mackenzie (2008b, 156) writes that Ubuntu, through enlisting programmers in Europe, North America, India, and East Asia, "introduces a multinational dimension to the internationalization of software, but the software itself remains universal in its aims and expectations because code and software themselves are presumed to be universal as text and practice. . . . [S]oftware now garners universality from that other universal, 'human beings,' free individuals." Chun (2011, 21) writes,

> Free software does not mean escaping from power, but rather engaging it differently, for free and open software profoundly privatizes the public domain: GNU copyleft—which allows one to use, modify, and redistribute source code and derived programs, but only if the original distribution terms are maintained—seeks to fight copyright by spreading licenses everywhere. More subtly, the free software movement, by linking freedom and freely available source code, amplifies the power of source code both politically and technically. It erases the vicissitudes of execution and the institutional and technical structures needed to ensure the coincidence of source code and its execution.

8 Mackenzie (2006, 169) writes that, in software, "relations are assembled, dismantled, bundled and dispersed within and across contexts."

9 In the novel *The Man without Qualities*, by Robert Musil ([1930–43] 1996, 76), the mathematician protagonist Ulrich struggles with the oddness of logical descriptions of water:

> And there now was water, a colourless liquid, blue only in dense layers, odourless and tasteless (as one had repeated in school so often that one could never forget it again), although physiologically it also included bacteria, vegetable matter, air, iron, calcium sulphate and calcium bicarbonate, and this archetype of all liquids was, physically speaking,

fundamentally not a liquid at all but, according to circumstances, a solid body, a liquid or a gas. Ultimately the whole thing dissolved into systems of formulae that were all somehow connected with each other, and in the whole wide world there were only a few dozen people who thought alike about even as simple a thing as water; all the rest talked about it in languages that were at home somewhere between today and several thousands of years ago.

10 One approach to tuning surprised me—one animated by a "genetic algorithm," an optimization technique based on an analogy to artificial selection (see Tolman and Grumbine 2013). For analysis of genetic algorithms, see Helmreich 1998.

11 Chun (2011, 19) writes, "Software . . . turns process in time into process in (text) space." On computers as infrastructures that fabricate what counts as "real time" (e.g., through the use of "Unix epoch" time stamps keyed to the number of seconds elapsed since January 1, 1970), see Cox and Lund 2021.

12 Here, perhaps, is code as "source, code as the true representation of action, indeed code as conflated with, and substituting for, action" (Chun 2011, 19).

13 For an Indigenous critique of the universality of the stack, see Lewis 2019.

14 This was before Trump, in September 2019, defended a map of the path of a hurricane that had been doctored with a Sharpie pen in order to align with his erroneous claims about its path, which contradicted National Weather Service predictions (Sobczyk 2020).

Set Four, First Wave. Middle Passages

1 On rhetoric describing these people as moving in "swarms," see Sanyal 2021.

2 Del Valle also reports that some thirty-five thousand people have died crossing the Mediterranean since 2000. See, for further statistics, the website of the Missing Migrants Project, organized by the International Organization for Migration (IOM): https://missingmigrants.iom.int/. See also the Search and Rescue information page of Doctors without Borders: http://searchandrescue.msf.org.

3 In the nine-screen art installation *Ten Thousand Waves*, which ran at Victoria Miro Gallery in London from October 7 to November 13, 2010, Isaac Julien (2010) took as a subject the drowning of twenty-three Chinese migrant cockleshell pickers in 2004 in Morecambe Bay, in northwestern England. The pickers' work leader was misinformed about the direction in which the tide would rise. Julien weaves this event together with a tale of the Chinese goddess of the sea, Mazu, from Fujian, the province from which the cockle pickers began their migration to the United Kingdom. Julien juxtaposes an aestheticized mythspace of wave goddess with the emergency media of grainy police helicopter footage, telling of the closely

coupled apparatuses of surveillance and neglect that track global maritime work in and around waves.

4 In an audio recording of the poem, which begins with the word *water* broken up, Philip "multiplies the phoneme 'wa' and enunciates it with increasing speed as if to both underscore its significance and to mimic the sound of an echo, a reverberation from which we can no longer trace an origin.... Water in the poem resists graphic and aural cohesion while demanding repetition.... [The poem] aurally and graphically represents waves by breaking lines and words" (Fehskens 2012, 408).

Set Four, Second Wave. Wave Power

1 And which is not in dialogue with Native Hawaiian notions of place or sea.

2 See the Save the Waves website, at https://savethewaves.org.

3 On coproduction, see Jasanoff 2004. For a reflection on coproduction that points to how resistance, opposition, and friction characterize science-society relations as much as collaborative coproduction, *see* Filipe et al. 2017.

Set Four, Third Wave. Wave Theory ~ Social Theory

1 This idea got an update in 1960, when Elias Canetti ([1960] 1962, 80) wrote, "The sea is multiple, it moves, and it is dense and cohesive. Its multiplicity lies in its waves.... The dense coherence of the waves is something which men in a crowd know well." See also Cody 2020.

2 Accounts of waves as forces to "surf" can reach toward a beachy sublime. As the midcentury psychedelic drug advocate Timothy Leary told *Surfer* magazine, "Everything is made of waves. At the level of electrons and neutrons... it's part of a wave theory. Historical waves—cultural waves... sequential, cyclical, moving, ever-changing forms" (quoted in Pezman 1978).

3 "Insurrectionary wave" first appears in accounts of the European revolutions of 1848 (Knight 1855, 217). In *Waves of Decolonization*, David Luis-Brown (2008) calls for a recognition of the power of waves of mid-twentieth-century decolonization and their concomitant movements of immigrants. Picking up on an essay by W. E. B. Du Bois, "The Souls of White Folk" (1920, 31), in which Du Bois wrote, "Wave on wave, each with increasing virulence, is dashing this new religion of whiteness on the shore of our time," Luis-Brown calls for a rescripting of waves of (non-white) immigration as waves of liberation from whiteness.

Chapter Five. Wave Theory, Southern Theory

1 The North-South dichotomy was early articulated by the Italian Marxist Antonio Gramsci to describe exploitation by northern Italian capitalists of southern Italian peasantry (see Dados and Connell 2012).

2 Boaventura de Sousa Santos (2018) points to concepts such as *ubuntu* and *ahimsa* as alternative optics through which to make social relations. On science from the South, see also Kervran et al. 2018.

3 In the usual projection, "Europe hovers over the lands below, from a position of scopic and epistemological privilege" (Wenzel 2014, 20). For the inverted map in figure 5.1, see "The Revolutionary AuthaGraph Projection" 2017.

4 "The common logic is that a system of categories is created by metropolitan intellectuals and read outwards to societies on the periphery, where the categories are filled in empirically" (Connell 2007, 66).

5 Another term to add is *tropical*, which emerges from European matrices to name hot equatorial climates and, when brought into descriptions of Indian Ocean worlds, sometimes mixes with the figure of the "Orient" (Arnold 2006).

6 See also Dasgupta and Pearson 1987.

7 Isabel Hofmeyr (2010) updates the "cradle of globalization" argument to suggest that today's Indian Ocean gathers up trends of global transformation—postnational (and post-American) politics, amplified oil extraction, the rise of Chinese capital, and the emergence of new African transnationalism.

8 *Waves across the South* opens with reproductions of nineteenth-century European paintings of Indian boats battling waves as they are tasked with carrying colonial cargo. Think of such boats and their hydrodynamics as a kind of wave theory—as many colonialists did, making close study of, for example, Burmese boats of war.

9 Sivasundaram (2020, 248, 245) documents an earlier technoscientific making of the Indian Ocean, managed by the British East India Company starting in 1792 at the Madras Observatory. The observatory's "central nautical aim, given its location at the heart of the Indian Ocean, was as a calibration point for ships across this vast ocean." It generated "calculations of longitude, tidal determinations and coastal marking points"—and did so by both drawing on and erasing local contributions. Part of what proceeded from the observatory's work was a representational smoothing of the irregularity of the Earth's equatorial bulge, a result that made "the Earth more like a spherical globe."

10 See also the listing "International Indian Ocean Expedition: Collected Reprints, VI," United Nations Educational, Scientific, and Cultural Organization, UNESDOC Digital Library, https://unesdoc.unesco.org/ark:/48223/pf0000148948.

11 See also Shanti S. Varma, dir., *Indian Ocean Expedition*, Films Division, Bombay, 1963. The Bandung conference, convened in Indonesia in 1955, aimed at fostering Afro-Asian resistance to neocolonialism (see Hofmeyr 2010).

12 “This unsuitability [was] the result of modeling the embankments after Dutch dikes (polders) in the Rhine delta, which sees only 1 percent of the sediment in the Ganges-Brahmaputra delta” (Dewan 2021, 59).

13 On how COVID-19 complicated the delivery of post-Amphan aid to coastal villages, see Mehtta 2020.

14 On JICA, see Leheny and Warren 2010. The appeal of Japan as a place for Bangladeshis to study increased after 9/11, when the United States became restrictive in issuing visas to people from Muslim-majority countries.

15 See *Tropical Storms: Bangladesh's Cyclone Aila* (Pumpkin TV, Bristol, UK, 2010), film.

16 On “hydronets,” or “neural network models designed to exploit both basin specific rainfall-runoff signals, and upstream network dynamics, which can lead to improved predictions at longer horizons,” see Moshe et al. 2020. Zach Moshe and his colleagues have tested their models on data sets from the Ganga and Brahmaputra.

17 More, as Jason Cons (2018, 267, 270) argues, discourses that stage the Bangladesh delta region as a “laboratory” for “resilience” (and for adapting “in place”) are often formatted by a Northern—and sometimes an Indian—“development-security nexus” that worries about Bangladeshi outmigration, about people in this part of the delta becoming climate refugees, framing “particular spaces at once as sites of experimental management for future crisis and as representational zones that enact spectacles of containment.”

18 For a Netherlands-based research project to compare Dutch with Bangladeshi deltas, see the Hydro Social Deltas website, at https://hydro-social-deltas.un-ihe.org.

19 On toxins in the Bay of Bengal, see the work of Ravi Agarwal, founding director of the nongovernmental organization Toxics Link (https://toxicslink.org) of New Delhi.

20 On “braided sciences,” see also Mukharji 2016. On river time, see Gearey 2018. On a Sundarban river's “whirlpools, braids, striations, and many sorts of ripples,” see also Ghosh 2019, 104; Jue 2017b.

21 The “regularity of bourgeois life” phrase is from the literary theorist Franco Moretti (2013, 81).

22 The work is badly needed; as I finished copyediting this chapter, in summer 2022, Pakistan was suffering from a supercharged monsoon disaster—a consequence of legacies of development- and nationalist-driven dam construction, of global warming, and more—leaving some ten percent of the country flooded. On the question of what science is *for*, see Fortun 2023.

23 Other books that bear mention might be “The Book of Sand,” imagined by the Argentine writer Jorge Luis Borges (1976), a text that never shows the

same page twice, and "the book of silt," conjured by the Scottish poet John Burnside (2002), his meditation on finding shore-side signs of worn-away life. In "Of Sea Changes and Other Futurisms," Ayesha Hameed (2022) asks, referring to the Indian Ocean, "What would a book be if it was invented at sea in the context of indentured and enslaved labour, following trade winds, navigating mangroves, and the sound of the co-mingling of languages and ecosystems?"

Postface

1 See also Sarah Kanouse and Nicholas Brown, dirs., *Ecologies of Acknowledgment,* video, 2019, https://www.nicholasanthonybrown.net/projects#/ecologies-of-acknowledgement. Listen also to Kristen Wyman, "Remembering Deer Island." Thanks to Elizabeth Solomon and Faries Grey, members of the Massachusett Tribe at Ponkapoag who in 2019 started leading an Indigenous Boston Harbor Tour, and whose 2022 tour was vital to my account here.

2 In a famous meditation on time, Walter Benjamin ([1940] 1968, 257–58) imagines the angel of history: "His face is turned toward the past. Where we perceive a chain of events, he sees one single catastrophe which keeps piling wreckage upon wreckage and hurls it in front of his feet." The wind of history, blowing from the past "propels him into the future to which his back is turned, while the pile of debris before him grows skyward." Contrast this with the figure of a wave coming toward us, the face of a wave that, erasing and forgetting, *also* gathers up that which was and will shape what will be.

REFERENCES

Abbott, B. P. et al. (LIGO Scientific Collaboration and Virgo Collaboration). 2016. "Observation of Gravitational Waves from a Binary Black Hole Merger." *Physical Review Letters* 116, no. 6. https://journals.aps.org/prl/pdf/10.1103/PhysRevLett.116.061102.

Abrams, Sarah. 2019. "The Power of Fear." *Radcliffe Magazine*, Winter, 30–35.

Adamowsky, Natascha. 2015. *The Mysterious Science of the Sea, 1775–1943*. London: Routledge.

Adams, Allan, and Sasha Chapman. 2016. "What Gravitational Waves Sound Like." *The Atlantic*, February 11. https://www.theatlantic.com/science/archive/2016/02/what-gravitational-waves-sound-like/462357.

Adler, Antony. 2019. *Neptune's Laboratory: Fantasy, Fear, and Science at Sea*. Cambridge, MA: Harvard University Press.

Adviescommissie voor de Zuid-Hollandse Kust. 2006. *Kustboekje: Groeien naar Kwaliteit*. The Hague: Hoogheemraadschap van Delfland.

Agostinho, Daniela. 2015. "Flooded with Memories: Risk Cultures, the Big Flood of 1953 and the Visual Resonance of World War Two." In *Hazardous Future: Disaster, Representation and the Assessment of Risk*, edited by Isabel Capeloa Gil and Christoph Wulf, 265–86. Berlin: Walter de Gruyter.

Aguilera, Mario. 2007. "Research Highlight: Saving Venice." *Scripps Institution of Oceanography News*, March 2. https://scripps.ucsd.edu/news/research-highlight-saving-venice.

Ah-King, Malin, and Eva Hayward. 2014. “Toxic Sexes: Perverting Pollution and Queering Hormone Disruption.” *O-Zone* 1: 1–12.

Ahmad, Aalya. 2015. “Feminism beyond the Waves.” *Briarpatch*, July–August. https://briarpatchmagazine.com/articles/view/feminism-beyond-the-waves.

Ahmed, Sara. 2006. *Queer Phenomenology: Orientations, Objects, Others*. Durham, NC: Duke University Press.

Ahsan, Reazul, Hajime Nakagawa, Kenji Kawaike, Masakazu Hashimoto, et al. 2019. “Informing and Involving the Flood Exposed Community in Fulcharri Upazila at Ghaibandha District Bangladesh on Flood Risks and Mitigation.” *Disaster Prevention Research Institute Annual*, no. 62B. https://www.dpri.kyoto-u.ac.jp/nenpo/no62/ronbunB/a62b0p30.pdf.

Aiyar, Sana. 2020. “Colonialism’s Afterlives and Invocations of Oceanic History.” In *Indian Ocean Current: Six Artistic Narratives*, edited by Prasannan Parthasarathi, 75–83. Boston: McMullen Museum of Art, Boston College.

Alaimo, Stacy. 2016. *Exposed: Environmental Politics and Pleasures in Posthuman Times*. Minneapolis: University of Minnesota Press.

Alaimo, Stacy. 2017. “The Anthropocene at Sea: Temporality, Paradox, Compression.” In *The Routledge Companion to the Environmental Humanities*, edited by Ursula K. Heise, Jon Christensen, and Michelle Niemann, 153–61. London: Routledge.

Alam, Mohammad-Reza. 2012. “Broadband Cloaking in Stratified Seas.” *Physical Review Letters* 108: 084502.

Alarcon, Francisco, 2019. “Water Games: An Archaeology of the Representation of Water in Video Games.” https://watercgi.com/.

Alarcon, Francisco, and Stefan Helmreich. 2020. “Ocean Amplification: Rising Waves and North-South Asymmetries.” *Strelka*, June 30. https://strelkamag.com/en/article/ocean-amplification-rising-waves-and-north-south-asymmetries. Reposted at https://mail.archiecho.com/ocean-amplification-rising-waves-and-north-south-asymmetries after Russian invasion of Ukraine.

Aleem, A. A. 1967. “Concepts of Currents, Tides and Winds among Medieval Arab Geographers in the Indian Ocean.” *Deep-Sea Research* 14, no. 4: 459–63.

Alexander, Meena. 2018. “Fragment in Praise of the Book.” In *Atmospheric Embroidery*, by Meena Alexander, 42. Evanston: Northwestern University Press.

Alford, Matthew H., Jennifer A. MacKinnon, Zhongxiang Zhao, Robert Pinkel, et al. 2007. “Internal Waves across the Pacific.” *Geophysical Research Letters* 34, no. 24. https://doi.org/10.1029/2007GL031566.

Alford, Matthew H., Thomas Peacock, Jennifer A. MacKinnon, Jonathan D. Nash, et al. 2015. “The Formation and Fate of Internal Waves in the South China Sea.” *Nature* 521: 65–69.

Al Hosani, Naeema. 2005. “Arab Wayfinding on Land and at Sea: An Historical Comparison of Traditional Navigation Techniques.” Master of arts thesis in geography, University of Kansas, Lawrence.

Alpers, Edward A. 2014. *The Indian Ocean in World History*. Oxford: Oxford University Press.

Alves, José Henrique. 1992. "O encanto e a fúria das ondas bizarras." *O Globo*, August 30, 38.

Amrith, Sunil. 2018. *Unruly Waters: How Rains, Rivers, Coasts, and Seas Have Shaped Asia's History*. New York: Basic Books.

Amrith, Sunil. 2020. "Climate Change and the Future of the Indian Ocean." In *Indian Ocean Current: Six Artistic Narratives*, edited by Prasannan Parthasarathi, 141–50. Boston: McMullen Museum of Art, Boston College.

Anand, Nikhil. 2020. "Before the Next Disaster: What Mumbai Needs to Learn from Cyclone Nisarga." *Indian Express*, June 4. https://indianexpress.com/article/opinion/mumbai-cyclone-nisarga-lessons-6441869.

Anderson, Jon, and Kimberly Peters, eds. 2014. *Water Worlds: Human Geographies of the Ocean*. Farnham, UK: Ashgate.

Anderson, Warwick, Miranda Johnson, and Barbara Brookes, eds. 2018. *Pacific Futures: Past and Present*. Honolulu: University of Hawai'i Press.

Andrews, Tamra. 2000. *A Dictionary of Nature Myths: Legends of the Earth, Sea, and Sky*. Oxford: Oxford University Press.

Anidjar, Gil. 2011. "The Blood of Freedom." In *Experiences of Freedom in Postcolonial Literatures and Cultures*, edited by Annalisa Oboe and Shaul Bassi, 122–31. London: Routledge.

Antonello, Alessandro. 2017. "The Southern Ocean." In *Oceanic Histories*, edited by David Armitage, Alison Bashford, and Sujit Sivasundaram, 296–318. Cambridge: Cambridge University Press.

Arasaratnam, Sinnappah. 1994. *Maritime India in the Seventeenth Century*. New York: Oxford University Press.

Arnold, David. 2006. *The Tropics and the Traveling Gaze: India, Landscape, and Science, 1800–1856*. Seattle: University of Washington Press.

Asscher, Maarten. 2009. *H2Olland: Op zoek naar de bronnen van Nederland*. Amsterdam: Amsterdam University Press.

Athens, Elizabeth, Brandon K. Ruud, and Martha Tedeschi. 2017. *Coming Away: Winslow Homer and England*. New Haven, CT: Yale University Press.

Atwater, Brian F., Satoko Musumi-Rokkaku, Kenji Satake, Yoshinobu Tsuji, et al. 2005. *The Orphan Tsunami of 1700: Japanese Clues to a Parent Earthquake in North America*. Seattle: University of Washington Press.

Babanin, Alexander. 2011. *Breaking and Dissipation of Ocean Surface Waves*. Cambridge: Cambridge University Press.

Bachelard, Gaston. (1942) 1983. *Water and Dreams: An Essay on the Imagination of Matter*. Translated from the French by Edith Farrell. Dallas: Dallas Institute of Humanities and Culture.

Bachelard, Gaston. (1934) 1984. *The New Scientific Spirit*. Translated from the French by Arthur Goldhammer. Boston: Beacon.

Baehr, Leslie G. 2014. "The Waves within the Waves." *Oceanus*, December 18. https://www.whoi.edu/oceanus/feature/the-waves-within-the-waves.

Bahng, Aimee. 2019. "Pacific Pessimisms: Decolonizing Transpacific Optimism." Paper presented at the Association for Asian American Studies Annual Conference, April 25–27, Madison, WI.

Bailey, Cathryn. 1997. "Making Waves and Drawing Lines: The Politics of Defining the Vicissitudes of Feminism." *Hypatia* 12, no. 3: 17–28.

Baker-Yeboah, Sheekela, D. Randolph Watts, and Deirdre A. Byrne. 2009. "Measurements of Sea Surface Height Variability in the Eastern South Atlantic from Pressure Sensor-Equipped Inverted Echo Sounders: Baroclinic and Barotropic Components." *Journal of Atmospheric and Oceanic Technology* 26, no. 12: 2593–609.

Bakhtin, Mikhail M. 1982. *The Dialogic Imagination: Four Essays*. Translated from the Russian by Caryl Emerson and Michael Holquist. Austin: University of Texas Press.

Bakker, Maarten, Satoko Kishimoto, and Christa Nooy. 2017. *Social Justice at Bay: The Dutch Role in Jakarta's Coastal Defence and Land Reclamation*. Report, Transnational Institute, April 21. https://www.tni.org/en/publication/social-justice-at-bay.

Ballestero, Andrea. 2019. *A Future History of Water*. Durham, NC: Duke University Press.

Banner, M. L., and W. K. Melville. 1976. "Separation of Air-Flow over Water-Waves." *Journal of Fluid Mechanics* 77: 825–42.

Barad, Karen. 2008. *Meeting the Universe Halfway: Quantum Physics and the Entanglement of Matter and Meaning*. Durham, NC: Duke University Press.

Barad, Karen. 2014. "Diffracting Diffraction: Cutting Together-Apart." *Parallax* 20, no. 3: 168–87.

Barad, Karen. 2019. "After the End of the World: Entangled Nuclear Colonialisms, Matters of Force, and the Material Force of Justice." *Theory & Event* 22, no. 3: 524–50.

Barber, N. F., F. Ursell, J. Darbyshire, and M. J. Tucker. 1946. "A Frequency Analyser Used in the Study of Ocean Waves." *Nature*, no. 4010: 329–32.

Barclay, David R., Fernando Simonet, and Michael J. Buckingham. 2009. "Deep Sound: A Free-falling Sensor Platform for Depth-profiling Ambient Noise in the Deep Ocean." *Marine Technology Society Journal* 43, no. 5: 144–50.

Barker, Holly M. 2004. *Bravo for the Marshallese: Regaining Control in a Post-nuclear, Post-colonial World*. Belmont, CA: Wadsworth Learning.

Barras, Colin. 2019. "The Ocean's Tallest Waves Are Getting Taller." *Science*, April 25, www.sciencemag.org/news/2019/04/ocean-s-tallest-waves-are-getting-taller.

Barthes, Roland. (1970) 1974. *S/Z*. Translated from the French by Richard Howard. New York: Hill and Wang.

Bartusiak, Marcia. 2000. *Einstein's Unfinished Symphony: Listening to the Sounds of Space-Time*. New York: Berkley.

Bascom, Willard. 1964. *Waves and Beaches: The Dynamics of the Ocean Surface*. Garden City, NY: Anchor.

Bascom, Willard. 1988. *The Crest of the Wave: Adventures in Oceanography*. New York: Harper and Row.

Baskins, Wade, trans. and comp. 2010. *The Wisdom of Leonardo da Vinci*. New York: Philosophical Library.

Bates, Charles. 1949. "Utilization of Wave Forecasting in the Invasions of Normandy, Burma, and Japan." *Annals of the New York Academy of Sciences* 51, no. 3: 545–72.

Battaglia, Debbora. 2012. "Coming in at an Unusual Angle: Exo-surprise and the Fieldworking Cosmonaut." *Anthropological Quarterly* 85, no. 4: 1089–106.

Battjes, Jurjen. 1972. "Long-term Wave Height Distribution at Seven Stations around the British Isles." *Deutsche Hydrographische Zeitschrift* 25, no. 4: 179–89.

Battjes, Jurjen. 1974. "Surf Similarity." *Proceedings of 14th Conference on Coastal Engineering, Copenhagen, Denmark*, edited by Morrough P. O'Brien, 466–80. New York: American Society of Civil Engineers.

Battjes, Jurjen. 1982. "Effects of Short-Crestedness on Wave Loads on Long Structures." *Applied Ocean Research* 4, no. 3: 165–72.

Battjes, Jurjen, and J. P. F. M. Janssen. 1978. "Energy Loss and Set-up due to Breaking of Random Waves." *Proceedings of 16th Conference on Coastal Engineering, Hamburg, Germany*, 569–87. New York: American Society of Civil Engineers.

Baucom, Ian. 2020. *History 4° Celsius: Search for a Method in the Age of the Anthropocene*. Durham, NC: Duke University Press.

Beach Erosion Board, Corps of Engineers. 1955. "A Magnetic Tape Wave Recorder and Energy Spectrum Analyzer for the Analysis of Ocean Wave Records." Technical Memorandum No. 58.

Bear, Laura. 2014. *Doubt, Conflict, Mediation: The Anthropology of Modern Time*. Chichester, West Sussex: John Wiley and Sons.

Beauregard. Guy. 2015. "On Not Knowing: *A Tale for the Time Being* and the Politics of Imagining Lives After March 11." *Canadian Literature* 227: 96–112.

Beck, Ulrich, Anthony Giddens, and Scott Lash. 1994. *Reflexive Modernization: Politics, Tradition and Aesthetics in the Modern Social Order*. Stanford, CA: Stanford University Press.

Becker, J. M., M. A. Merrifield, and M. Ford. 2014. "Water Level Effects on Breaking Wave Setup for Pacific Island Fringing Reefs." *Journal of Geophysical Research: Oceans* 119, no. 2: 914–32.

Beer, Gillian. 1996. "Wave Theory and the Rise of Literary Modernism." In *Open Fields: Science in Cultural Encounter*, by Gillian Beer, 295–318. Oxford: Oxford University Press.

Beer, Gillian. 2014. "On Virginia Woolf's *The Waves*." *Daedalus* 143, no. 1: 54–63.

Bell, K. L., H. L. Van Trees, B. R. Osborn, and R. E. Zarnich. 2011. "A Range-coherent Model for Discriminating Targets and Wave Clutter in High Range Resolution Surface Radar." *2011 IEEE Radar Conference: In the Eye of the Storm* (May 23–27): 740–45.

Beltrán, Héctor. Forthcoming. *Code Work: Hacking across the US/México Techno-Borderlands*. Princeton, NJ: Princeton University Press.

Benford, Gregory. 1980. *Timescape*. New York: Simon and Schuster.

Benjamin, Walter. (1940) 1968. "Theses on the Philosophy of History." In *Illuminations*, edited and with an introduction by Hannah Arendt. Translated from the German by Harry Zohn, 253–64. New York: Harcourt, Brace and World.

Bennett, Jane. 2010. *Vibrant Matter: A Political Ecology of Things*. Durham, NC: Duke University Press.

Benson, Keith R., and Philip F. Rehbock, eds. 2002. *Oceanographic History: The Pacific and Beyond*. Seattle: University of Washington Press.

Ben-Yehoyada, Naor. 2017. *The Mediterranean Incarnate: Region Formation between Sicily and Tunisia since World War II*. Chicago: University of Chicago Press.

Bergson, Henri. (1907) 1911. *Creative Evolution*. Translated from the French by Arthur Mitchell. New York: Henry Holt.

Bergson, Henri. (1934) 1946. *The Creative Mind: An Introduction to Metaphysics*. Translated from the French by Mabelle L. Andison. New York: Philosophical Library.

Best, Stephen, and Sharon Marcus. 2009. "Surface Reading: An Introduction." *Representations* 108, no. 1: 1–21.

Bezgodov, Alexey, and Dmitrii Esin. 2014. "Complex Network Modeling for Maritime Search and Rescue Operations." *Procedia Computer Science* 29: 2325–35.

Bhattacharyya, Debjani. 2021. "A River Is Not a Pendulum: Sediments of Science in the World of Tides." *Isis* 112, no. 1: 141–49.

Bigelow, Henry B., and W. T. Edmondson. 1947. *Wind Waves at Sea, Breakers and Surf*. Washington, DC: US Government Printing Office.

Bijker, Eco W. 1996. "History and Heritage in Coastal Engineering in the Netherlands." In *History and Heritage of Coastal Engineering*, edited by Nicholas C. Kraus, 390–412. New York: American Society of Civil Engineers.

Bijker, Wiebe. 2002. "The Oosterschelde Storm Surge Barrier: A Test Case for Dutch Water Technology, Management, and Politics." *Technology and Culture* 43, no. 3: 569–84.

Billé, Franck. 2017. "Introduction: Speaking Volumes." Theorizing the Contemporary, *Fieldsights*, October 24. https://culanth.org/fieldsights/introduction-speaking-volumes.

Bitner-Gregersen, Elzbieta M., and Anne Karin Magnusson. 2013. "Intrinsic Variability on Wave Parameters and Effect on Wave Statistics." Paper presented at Forecasting Dangerous Sea States, the Thirteenth International Workshop on Wave Hindcasting and Forecasting, Banff, AB, October 27–November 1.

Bitner-Gregersen, Elzbieta M., and Alesandro Toffoli. 2013. "Probability of Occurrence of Rogue Sea States and Consequences for Design." Paper presented at Forecasting Dangerous Sea States, the Thirteenth International Workshop on Wave Hindcasting and Forecasting, Banff, AB, October 27–November 1.

Blakemore, Bill. 1987. "Kubrick's 'Shining' Secret: Film's Hidden Horror Is the Murder of the Indian." *Washington Post*, July 12.

Bloomfield, Leonard. 1933. *Language*. New York: Henry Holt.

Blum, Hester. 2010. "The Prospect of Oceanic Studies." *Proceedings of the Modern Language Association* 125, no. 3: 670–77.

Blumenberg, Hans. (1979) 1997. *Shipwreck with Spectator: Paradigm of a Metaphor for Existence*. Translated from the German by Steven Rendall. Cambridge, MA: MIT Press.

Bock, Y., S. Wdowinski, A. Ferreti, F. Novali, and A. Fumagalli. 2012. "Recent Subsidence of the Venice Lagoon from Continuous GPS and Interferometric Syn-

thetic Aperture Radar." *Geochemistry, Geophysics, Geosystems* 13, no. 3. https://doi.org/10.1029/2011GC003976.
Boellstorff, Tom. 2016. "For Whom the Ontology Turns: Theorizing the Digital Real." *Current Anthropology* 57, no. 4: 387–98.
Bolster, W. Jeffrey. 2008. "Putting the Ocean in Atlantic History: Maritime Communities and Marine Ecology in the Northwest Atlantic, 1500–1800." *American Historical Review* 113: 19–47.
Bolter, Jay, and Richard Grusin. 1999. *Remediation: Understanding New Media*. Cambridge, MA: MIT Press.
Bontje, Lotte E., and Jill H. Slinger. 2017. "A Narrative Method for Learning from Innovative Coastal Projects: Biographies of the Sand Engine." *Ocean and Coastal Management* 142: 186–97.
Booth, Douglas. 1999. "Surfing: The Cultural and Technological Determinants of a Dance." *Culture, Sport, Society* 2, no. 1: 36–55.
Borges, Jorge Luis. 1976. "The Book of Sand." Translated from the Spanish by Norman Thomas di Giovanni. *New Yorker*, October 25, 38–39.
Borovoy, Amy. 2016. "Robert Bellah's Search for Community and Ethical Modernity in Japan Studies." *Journal of Asian Studies* 75, no. 2: 467–94.
Bose, Sugata. 2006. *A Hundred Horizons: The Indian Ocean in the Age of Global Empire*. Cambridge, MA: Harvard University Press.
Bosscher, Frans, and Margot Maljaars. 2017. *Het Waterloopbos*. Wageningen, Netherlands: Blauwdruck.
Bould, Mark. 2021. *The Anthropocene Unconscious: Climate Catastrophe Culture*. London: Verso.
Bowen, David. n.d. "Telepresent Water." David Bowen artist website: https://www.dwbowen.com/telepresentwater.
Bowker, Geoffrey. 2005. *Memory Practices in the Sciences*. Cambridge, MA: MIT Press.
boyd, danah, and Kate Crawford. 2012. "Critical Questions for Big Data." *Information, Communication and Society* 15, no. 5: 662–79.
Braidotti, Rosi. 2013. *The Posthuman*. Cambridge: Polity.
Brathwaite, Kamau. 1999. *Conversations with Nathaniel Mackey*. Staten Island, NY: We.
Bratton, Benjamin. 2015. *The Stack: On Software and Sovereignty*. Cambridge, MA: MIT Press.
Braverman, Irus. 2018. *Coral Whisperers: Scientists on the Brink*. Berkeley: University of California Press.
Bremner, Lindsay, ed. 2018. *Monsoon [+ other] Waters*. London: Monsoon Assemblages.
Briggs, John, and Joanne Sharp. 2004. "Indigenous Knowledges and Development: A Postcolonial Caution." *Third World Quarterly* 25, no. 4: 661–76.
Briggs, M. J., H. Yeh, and D. T. Cox. 2010. "Physical Modeling of Tsunami Waves." In *Handbook of Coastal and Ocean Engineering*, edited by Y. C. Kim, 1073–1105. Singapore: World Scientific.
Brinkema, Eugenie. 2014. *The Forms of the Affects*. Durham, NC: Duke University Press.

Bronson, Earl D., and Larry Glosten. 1985. "FLIP: FLoating Instrument Platform." Scripps Institution of Oceanography Reference 85–21. Office of Naval Research and Naval Sea Systems Command, October.

Brøvig-Hanssen, Ragnhild, and Anne Danielsen. 2016. *Digital Signatures: The Impact of Digitization on Popular Music Sound*. Cambridge, MA: MIT Press.

Brown, Able. 2011. "BS'N All the Time." *Surfer's Journal*, online journal entry, October 13, www.surfersjournal.com/journal_entry/bs%E2%80%99n-all-time, available via Internet Archive Wayback Machine; last snapshotted December 16, 2011.

Brown, William. 2013. *Supercinema: Film-Philosophy for the Digital Age*. New York: Berghahn.

Brugidou, Jeremie, and Clouette Fabien. 2018. "'AnthropOcean': Oceanic Perspectives and Cephalopodic Imaginaries Moving beyond Land-centric Ecologies." *Social Science Information* 57, no. 3: 359–85.

Bryant, Rebecca. 2016. "On Critical Times: Return, Repetition, and the Uncanny Present." *History and Anthropology* 27, no. 1: 19–31.

Bryant, Rebecca, and Daniel M. Knight. 2019. *The Anthropology of the Future*. Cambridge: Cambridge University Press.

Bryld, Mette. 1996. "Dialogues with Dolphins and Other Extraterrestrials: Displacements in Gendered Space." In *Between Monsters, Goddesses, and Cyborgs: Feminist Confrontations with Science, Medicine, and Cyberspace*, edited by Nina Lykke and Rosi Braidotti, 47–71. London: Zed.

Buckley, Marc P., and Fabrice Veron. 2019. "The Turbulent Airflow over Wind Generated Surface Waves." *European Journal of Mechanics-B/Fluids* 73 (January): 132–43.

Bulkens, Maartje, Hamzah Muzaini, and Claudio Minca. 2016. "Dutch New Nature: (Re)landscaping the Millingerwaard." *Journal of Environmental Planning and Management* 59, no. 5: 808–25.

Burnett, D. Graham. 2011. "Self-Recording Seas." In *Oceanomania: Souvenirs of Mysterious Seas from the Expedition to the Aquarium: A Mark Dion Project*, 134–45. Monaco: Nouveau Musée National.

Burnside, John. 2002. "History." In *The Light Trap*, by John Burnside, 40–42. London: Jonathan Cape.

Butler, Judith. 1993. *Bodies That Matter: On the Discursive Limits of "Sex."* New York: Routledge.

Bystrom, Kerry, and Isabel Hofmeyr. 2017. "Oceanic Routes: (Post-It) Notes on Hydro-Colonialism." *Comparative Literature* 69, no. 1: 1–6.

Calliari, Elisa, Andrea Staccione, and Jaroslav Mysiak. 2019. "An Assessment Framework for Climate-proof Nature-based Solutions." *Science of the Total Environment* 656: 691–700.

Calvino, Italo. (1983) 1985. *Mr. Palomar*. Translated from the Italian by William Weaver. Orlando, FL: Harcourt.

Calvino, Italo. (1967) 1969. "Blood, Sea." In *t zero*, by Italo Calvino, 39–51. Translated from the Italian by William Weaver. Orlando, FL: Harcourt.

Canetti, Elias. (1960) 1962. *Crowds and Power*. Translated from the German by Carol Stewart. London: Gollancz.

Carr, E. Summerson, and Brooke Fisher. 2016. "Interscaling Awe, De-escalating Disaster." In *Scale: Discourse and Dimensions of Social Life*, edited by E. Summerson Carr and Michael Lempert, 133–56. Berkeley: University of California Press.

Carse, Ashley. 2012. "Nature as Infrastructure: Making and Managing the Panama Canal Watershed." *Social Studies of Science* 42, no. 4: 539–63.

Carson Rachel. 1951. *The Sea around Us*. Oxford: Oxford University Press.

Cartwright, David. 2010. "Waves, Surges and Tides." In *Of Seas and Ships and Scientists: The Remarkable Story of the UK's National Institute of Oceanography*, edited by Anthony Laughton, John Gould, 'Tom' Tucker, and Howard Roe, 171–80. Cambridge: Lutterworth.

Cartwright, Julyan H. E., and Hisami Nakamura. 2009. "What Kind of a Wave Is Hokusai's *Great Wave off Kanagawa*?" *Notes and Records of the Royal Society* 63, no. 2: 119–35.

Casper, Drew. 2011. *Hollywood Film 1963–1976: Years of Revolution and Reaction*. Oxford: Wiley-Blackwell.

Cassidy, Rebecca, and Molly Mullin, eds. 2007. *Where the Wild Things Are Now: Domestication Reconsidered*. Oxford: Berg.

Cassanelli, Lee, Brian Spooner, and Robert Nichols. 2002. "Cover Letter." Indian Ocean: Cradle of Globalization Summer Institute, University of Pennsylvania. http://ccat.sas.upenn.edu/indianocean/index.html.

Cavaleri, Luigi. 2000. "The Oceanographic Tower *Acqua Alta*—Activity and Prediction of Sea States at Venice." *Coastal Engineering* 39, no. 1: 29–70.

Chakrabarty, Dipesh. 2000. *Provincializing Europe: Postcolonial Thought and Historical Difference*. Princeton, NJ: Princeton University Press.

Chakrabarty, Dipesh. 2021. *The Climate of History in a Planetary Age*. Chicago: University of Chicago Press.

Chalfin, Brenda. 2015. "Governing Offshore Oil: Mapping Maritime Political Space in Ghana and the Western Gulf of Guinea." *South Atlantic Quarterly* 114, no. 1: 101–18.

Chao, Sophie, Karin Bolender, and Eben Kirksey, eds. 2022. *The Promise of Multispecies Justice*. Durham, NC: Duke University Press.

Chari, Sharad. 2019. "Subaltern Sea? Indian Ocean Errantry against Subalternization." In *Subaltern Geographies*, edited by Tariq Jazeel and Stephen Legg, 191–209. Athens: University of Georgia Press.

Chaudhuri, Ashawari. 2019. "The Kernel of Doubt: Agricultural Biotechnology, Braided Temporalities, and Agrarian Environments in India." PhD diss., Massachusetts Institute of Technology, Cambridge, MA.

Chen, Mel Y. 2012. *Animacies: Biopolitics, Racial Mattering, and Queer Affect*. Durham, NC: Duke University Press.

Cheng, Irene. 2006. "The Beavers and the Bees: Intelligent Design and the Marvelous Architecture of Animals." *Cabinet* 23. https://www.cabinetmagazine.org/issues/23/cheng.php.

Chin, Andrea. 2015. "Daan Roosegaarde Floods Amsterdam's Museumplein with Waterlicht." *Designboom*, May 13. https://www.designboom.com/art/daan-roosegaarde-waterlicht-05-13-2015.

Cho, Adrian. 2012. "Proposed Cloaking Device for Water Waves Could Protect Ships at Sea." *Science*, March 2. https://www.sciencemag.org/news/2012/03/proposed-cloaking-device-water-waves-could-protect-ships-sea.

Choi, Vivian. 2020. "Exploring the 'Unknown': Indian Ocean Materiality as Method." In *Reimagining Indian Ocean Worlds*, edited by Smriti Srinivas, Bettina Ng'weno, and Neelima Jeychandran, 227–42. London: Routledge.

Chowdhury, Piyali, Manasa Ranjan Behera, and Dominic E. Reeve. 2019. "Wave Climate Projections along the Indian Coast." *International Journal of Climatology* 39: 4531–42.

Choy, Timothy. 2011. *Ecologies of Comparison: An Ethnography of Endangerment in Hong Kong*. Durham, NC: Duke University Press.

Choy, Timothy. 2012. "Air's Substantiations." In *Lively Capital: Biotechnologies, Ethics, and Governance in Global Markets*, edited by Kaushik Sunder Rajan, 121–53. Durham, NC: Duke University Press.

Chun, Wendy Hui Kyong. 2011. *Programmed Visions: Software and Memory*. Cambridge, MA: MIT Press.

Clark, Timothy. 2012. "Scale." In *Telemorphosis: Theory in the Era of Climate Change*, vol. 1, edited by Tom Cohen, 148–66. Ann Arbor, MI: Open Humanities.

Cochran, William G., Frederick Mosteller, and John W. Tukey. 1953. "Statistical Problems of the Kinsey Report." *Journal of the American Statistical Association* 48, no. 264: 673–716.

Cody, Francis. 2020. "Wave Theory: Cash, Crowds, and Caste in Indian Elections." *American Ethnologist* 47, no. 4: 402–16.

Coen, Deborah R. 2018. *Climate in Motion: Science, Empire, and the Problem of Scale*. Chicago: University of Chicago Press.

Cohen, Ashley L. 2017. "The Global Indies: Historicizing Oceanic Metageographies." *Comparative Literature* 69, no. 1: 7–15.

Cohen, Jeffrey Jerome. 2015. *Stone: An Ecology of the Inhuman*. Minneapolis: University of Minnesota Press.

Cohen, Margaret. 2006. "Chronotopes of the Sea." In *The Novel v. 2*, edited by Franco Moretti, 647–66. Princeton, NJ: Princeton University Press.

Comer, Krista. 2010. *Surfer Girls in the New World Order*. Durham, NC: Duke University Press.

Connell, Raewyn. 2007. *Southern Theory: The Global Dynamics of Knowledge in Social Science*. Sydney: Allen and Unwin.

Connery, Christopher L. 1995. "Pacific Rim Discourse: The U.S. Global Imaginary in the Late Cold War Years." In *Asia/Pacific as Space of Cultural Production*, edited by Rob Wilson and Arif Dirlik, 30–56. Durham, NC: Duke University Press.

Connolly, William. 2017. *Facing the Planetary: Entangled Humanism and the Politics of Swarming*. Durham, NC: Duke University Press.

Cons, Jason. 2018. "Staging Climate Security: Resilience and Heterodystopia in the Bangladesh Borderlands." *Cultural Anthropology* 33, no. 2: 266–94.

Consolandi, Pietro. 2020. "Our Wooden Shelter: Mangrove Forests along the Coasts of India." Ocean-Archive.org. https://ocean-archive.org/collection/119.

Coole, Diana, and Samantha Frost, eds. 2010. *New Materialisms: Ontology, Agency, and Politics*. Durham, NC: Duke University Press.

Corbin, Alain. (1988) 1995. *The Lure of the Sea: The Discovery of the Seaside in the Western World 1750–1840*. Translated from the French by Jocelyn Phelps. Harmondsworth, UK: Penguin.

Cornish, Vaughn. 1934. *Ocean Waves and Kindred Geophysical Phenomena*. Cambridge: Cambridge University Press.

Cornwall, Warren. 2018. "As Sea Levels Rise, Bangladeshi Islanders Must Decide between Keeping the Water Out—or Letting It In." *Science*, March 1. https://www.sciencemag.org/news/2018/03/sea-levels-rise-bangladeshi-islanders-must-decide-between-keeping-water-out-or-letting.

Cox, Geoff, and Jacob Lund. 2021. "Time.now." In *Uncertain Archives: Critical Keywords for Big Data*, edited by Nanna Bonde Thylstrup, Daniela Agostinho, Annie Ring, Catherine D'Ignazio, and Kristin Veel, 523–31. Cambridge, MA: MIT Press.

Craik, Alex D. D. 2004. "The Origins of Water Wave Theory." *Annual Review of Fluid Mechanics* 36: 1–28.

Crary, Jonathan. 2001. *Suspensions of Perception: Attention, Spectacle, and Modern Culture*. Cambridge, MA: MIT Press.

Creado, Eliana Santos Junqueira, and Stefan Helmreich. 2018. "A Wave of Mud: The Travel of Toxic Water, from Bento Rodrigues to the Brazilian Atlantic." *Revista do Instituto de Estudos Brasileiros* 69: 33–51.

Crucefix, Martyn, and Amel Alzakout. 2019. *Cargo of Limbs*. London: Hercules.

Crutzen, Paul J., and Eugene F. Stoermer. 2000. "The 'Anthropocene.'" *Global Change Newsletter* 41: 17–18.

Cullen, Beth, and Christina Leigh Geros. 2020. "Constructing the Monsoon: Colonial Meteorological Cartography, 1844–1944." *History of Meteorology* 9: 1–26.

Cusack, Tricia. 2014. *Framing the Ocean, 1700 to the Present: Envisaging the Sea as Social Space*. London: Routledge.

Dabydeen, David. 1995. *Turner: New and Selected Poems*. London: Jonathan Cape.

Dados, Nour, and Raewyn Connell. 2012. "The Global South." *Contexts* 11, no. 1: 12–13.

Darrieussecq, Marie. 2008. *Précisions sur les vagues*. Paris: P.O.L.

Darrigol, Olivier. 2003. "The Spirited Horse, the Engineer, and the Mathematician: Water Waves in Nineteenth-Century Hydrodynamics." *Archive for History of the Exact Sciences* 58: 21–95.

Dasgupta, Ashin, and Michael Pearson, eds. 1987. *India and the Indian Ocean*. Calcutta: Oxford University Press.

Daston, Lorraine. 1992. "Objectivity and the Escape from Perspective." *Social Studies of Science* 22, no. 4: 597–618.

Daston, Lorraine, and Peter Galison. 2007. *Objectivity*. New York: Zone Books.

Datawell BV. 2001. "History of Datawell." Brochure, Datawell BV website, n.d. Accessed June 16, 2022. http://www.datawell.nl/Portals/0/Documents/Brochures/datawell_brochure_history_b-13-01.pdf.

Davenant, Charles. (1698) 1771. "Discourses on the Public Revenue and on the Trade of England." In *The Political and Commercial Works of That Celebrated Writer Charles D'avenant, Collected and Revised by Sir Charles Whitworth*, vol. 1, 127–459. London: R. Horsfield.

Davies, Sarah R. 2021. "Atmospheres of Science: Experiencing Scientific Mobility." *Social Studies of Science* 5, no. 2: 214–32.

Davis, Adam. 2014. "Never Quite the Right Size: Scaling the Digital in CG Cinema." *Animation* 9, no. 2: 124–37.

Davis, Heather. 2022. *Plastic Matter*. Durham, NC: Duke University Press.

Davis, Heather, and Etienne Turpin, eds. 2015. *Art in the Anthropocene: Encounters among Aesthetics, Politics, Environments, and Epistemologies*. London: Open Humanities.

Davis, Mike. 2017. *Late Victorian Holocausts: El Niño Famines and the Making of the Third World*. New York: Verso.

Davis, Robert, and Garry R. Marvin. 2004. *Venice, the Tourist Maze: A Cultural Critique of the World's Most Touristed City*. Berkeley: University of California Press.

Dawson, Kevin. 2018. *Undercurrents of Power: Aquatic Culture in the African Diaspora*. Philadelphia: University of Pennsylvania Press.

Day, Deborah. 1997. "Helen Hill Raitt Biography." La Jolla, CA: University of California, San Diego, Libraries. https://library.ucsd.edu/scilib/biogr/Raitt_Helen_Biogr.pdf.

Day, Deborah. 1999. "Overview of the History of Women at Scripps Institution of Oceanography." *UC San Diego: Library—Scripps Digital Collection*. https://escholarship.org/uc/item/85t1s746.

Deacon, Margaret. 1971. *Scientists and the Sea, 1650–1900*. Aldershot, UK: Ashgate.

Deacon, Margaret. 1984. "The Biggest Wave." *Folklore* 95, no. 2: 254–55.

Deacon, Margaret, Tony Rice, and Colin Summerhayes, eds. 2001. *Understanding the Oceans*. London: UCL Press.

Dean, Robert G., and Robert A. Dalrymple. 1991. *Water Wave Mechanics for Engineers and Scientists*. Singapore: World Scientific.

Deane, Grant B., Dale Stokes, and Adrian H. Callaghan. 2016. "Turbulence in Breaking Waves." *Physics Today* 69, no. 10: 86–87.

de Bakker, A. T. M., J. A. Brinkkemper, F. van der Steen, M. F. S. Tissier, and B. G. Ruessink. 2016. "Cross-shore Sand Transport by Infragravity Waves as a Function of Beach Steepness." *Journal of Geophysical Research: Earth Surface* 121, no. 10: 1786–99.

Decker, Julie, ed. 2014. *Gyre: The Plastic Ocean*. London: Booth-Clibborn.

De Frees, Madeline. 1983. "The Book of Sediments." *Paris Review* 89: 136–37.

de Groot, Wouter T. 1987. "Van Waterwolf tot partner: Het culturele aspect van het niuewe waterbeleid." *Milieu* 2, no. 3: 84–86.

de Jong, Martijn P. C., and Jurjen Battjes. 2004. "Analysis and Prediction of Seiches in Rotterdam Harbor Basins." *Proceedings of 29th Conference on Coastal Engineering, Lisbon, Portugal*, edited by Jane McKee Smith, 1238–50. Singapore: World Scientific.

de la Cadena, Marisol. 2015. *Earth Beings: Ecologies of Practice across Andean Worlds.* Durham, NC: Duke University Press.
Delap, Lucy. 2011. "*The Woman's Dreadnought*: Maritime Symbolism in Edwardian Gender Politics." In *The* Dreadnought *and the Edwardian Age,* edited by Robert Blythe, Andrew Lambert, and Jan Rüger, 95–108. Farnham, UK: Ashgate.
Deleuze, Gilles (1985) 1995. "Mediators." In *Negotiations: 1972–1990*. Translated from the French by Martin Joughin, 121–34. New York: Columbia University Press.
Deleuze, Gilles. (1983) 1986. *Cinema 1: The Movement-Image*. Translated from the French by Hugh Tomlinson and Barbara Habberjam. Minneapolis: University of Minnesota Press.
Deleuze, Gilles. (1985) 1989. *Cinema 2: The Time-Image*. Translated from the French by Hugh Tomlinson and Robert Galeta. Minneapolis: University of Minnesota Press.
Deleuze, Gilles, and Félix Guattari. (1980) 1987. *A Thousand Plateaus: Capitalism and Schizophrenia*. Translated from the French by Brian Massumi. Minneapolis: University of Minnesota Press.
DeLoughrey, Elizabeth. 2017. "The Oceanic Turn: Submarine Futures of the Anthropocene." *Comparative Literature* 69, no. 1: 32–44.
DeLoughrey, Elizabeth. 2019. *Allegories of the Anthropocene*. Durham, NC: Duke University Press.
Dening, Greg. 2002. "Performing on the Beaches of the Mind." *History and Theory* 41, no. 1: 1–24.
DeNoon, Daniel J. 2007. "ICD Gap for Women, African-Americans," *WebMD*, posted October 2. https://www.webmd.com/heart-disease/news/20071002/icd-gap-for-women-and-african-americans.
Deraniyagala, Sonali. 2013. *Wave*. New York: Knopf.
Desrosières, Alain. 2002. *The Politics of Large Numbers: A History of Statistical Reasoning*. Translated from the French by Camille Naish. Cambridge, MA: Harvard University Press.
de Vriend, Huib J., and Mark van Koningsveld. 2012. *Building with Nature: Thinking, Acting and Interacting Differently*. Dordrecht: Ecoshape.
Dewan, Camelia. 2020. "Living with Toxic Development." *Anthropology Today* 36, no. 6: 9–12.
Dewan, Camelia. 2021. *Misreading the Bengal Delta: Climate Change, Development, and Livelihoods in Coastal Bangladesh*. Seattle: University of Washington Press.
Dewan, Camelia. 2022. "'Climate Change as a Spice': Brokering Environmental Knowledge in Bangladesh's Development Industry." *Ethnos* 87, no. 3: 538–59.
Dewan, Camelia, Aditi Mukherji, and Marie-Charlotte Buisson. 2015. "Evolution of Water Management in Coastal Bangladesh: From Temporary Earthen Embankments to Depoliticized Community-Managed Polders." *Water International* 40, no. 3: 401–16.
De Wolff, Kim. 2017. "Plastic Naturecultures: Multispecies Ethnography and the Dangers of Separating Living from Nonliving Bodies." *Body & Society* 23, no. 3: 23–47.

De Wolff, Kim, Rina C. Faletti, and Ignacio López-Calvo, eds. 2021. *Hydrohumanities: Water Discourse and Environmental Futures*. Berkeley: University of California Press.

Dick, Stephanie. 2020. "Coded Conduct: Making MACSYMA Users and the Automation of Mathematics." *British Journal for the History of Science Themes* 5: 205–24.

Di Maio, Alessandra 2013. "The Mediterranean, or Where Africa Does (Not) Meet Italy: Andrea Segre's *A Sud di Lampedusa* (2006)." In *The Cinemas of Italian Migration: European and Transatlantic Narratives*, edited by Sabine Schrader and Daniel Winkler, 41–52. Newcastle upon Tyne: Cambridge Scholars.

DiNitto, Rachel. 2014. "Narrating the Cultural Trauma of 3/11: The Debris of Post-Fukushima Literature and Film." *Japan Forum* 26, no. 3: 340–60.

Disco, Cornelis. 2002. "Remaking 'Nature': The Ecological Turn in Dutch Water Management." *Science, Technology, & Human Values* 27, no. 2: 206–35.

Donovan, Stephen. 2013. "Shockwaves: The Interrupted Sea-Journeys of Rudyard Kipling and Morgan Robertson." *Forum for Modern Language Studies* 49, no. 4: 393–405.

Draper, Laurence. 1996. "The History of Wave Research at Wormley, a Personal View." *Ocean Challenge* 6, no. 2: 24–27.

Drenthen, Martin. 2009. "Developing Nature along Dutch Rivers: Place or Non-place." In *New Visions of Nature: Complexity and Authenticity,* edited by Martin Drenthen, Jozef Keulartz, and James Proctor, 205–28. Dordrecht: Springer Netherlands.

Dua, Jatin. 2013. "A Sea of Trade and a Sea of Fish: Piracy and Protection in the Western Indian Ocean." *Journal of East African Studies* 7, no. 2: 353–70.

Duane, Daniel. 2019. "The Long, Strange Tale of California's Surf Nazis." *New York Times*, September 28. https://www.nytimes.com/2019/09/28/opinion/sunday/surf-racism.html.

Du Bois, W. E. B. 1903. *The Souls of Black Folk*. Chicago: A. C. McClurg.

Du Bois, W. E. B. 1920. "The Souls of White Folk." In *Darkwater: Voices from Within the Veil*, by W. E. B. Du Bois, 29–52. New York: Harcourt, Brace and Howe.

Dudley, John M., Véronique Sarano, and Frédéric Dias. 2013. "On Hokusai's *Great Wave off Kanagawa*: Localization, Linearity and a Rogue Wave in Sub-Antarctic Waters." *Notes and Records of the Royal Society* 67, no. 2: 159–64.

Dupuy, Jean-Pierre. 2005 (2015). *A Short Treatise on the Metaphysics of Tsunamis*. Translated from the French by Malcolm B. DeBevoise. East Lansing: Michigan State University Press.

Durkheim, Émile. (1897) 1951. *Suicide: A Study in Sociology*. Translated from the French by John A. Spaulding and George Simpson. New York: Free Press.

Eckstein, Lars, and Anja Schwarz. 2019. "The Making of Tupaia's Map: A Story of the Extent and Mastery of Polynesian Navigation, Competing Systems of Wayfinding on James Cook's *Endeavour*, and the Invention of an Ingenious Cartographic System." *Journal of Pacific History* 54, no. 1: 1–95.

ECMWF (European Center for Medium-Range Weather Forecasts). n.d. "Ocean Wave Model High Resolution 10-Day Forecast (Set II—HRES-WAM), https://www.ecmwf.int/en/forecasts/datasets/set-ii.

Edwards, Paul. 1996. *The Closed World: Computers and the Politics of Discourse in the Cold War America*. Cambridge, MA: MIT Press.
Edwards, Paul. 2010. *A Vast Machine: Computer Models, Climate Data, and the Politics of Global Warming*. Cambridge, MA: MIT Press.
Eeltink, D., H. Branger, C. Luneau, and Y. He et al. 2022. "Nonlinear Wave Evolution with Data-driven Breaking." *Nature Communications* 13. https://doi.org/10.1038/s41467-022-30025-z.
Ehrenreich, Barbara. 1987. "Foreword." In *Male Fantasies, Volume 1: Women, Floods, Bodies, History*, by Klaus Theweleit, ix–xvii. Minneapolis: University of Minnesota Press.
Elliott, Ralph Nelson. (1938) 1994. *The Wave Principle*, reprinted in *R.N. Elliott's Masterworks: The Definitive Collection*, edited by Robert R. Prechter Jr., 85–151. Gainesville, GA: New Classics Library.
Emmelhainz, Irmgard. 2015. "Conditions of Visuality under the Anthropocene and Images of the Anthropocene to Come." *e-flux*, no. 63. https://www.e-flux.com/journal/63/60882/conditions-of-visuality-under-the-anthropocene-and-images-of-the-anthropocene-to-come/.
Engelstad, A., B. G. Ruessink, D. Wesselman, P. Hoekstra, et al. 2017. "Observations of Waves and Currents during Barrier Island Inundation." *Journal of Geophysical Research: Oceans* 122, no. 4: 3152–69.
Environmental Solutions Initiative. 2020. "'The Emerald Tutu' Wins NSF Grant for Design to Protect Boston's Coastline." *MIT News*, September 2. https://news.mit.edu/2020/emerald-tutu-design-wins-nsf-grant-protect-boston-coastline-0903.
Eshun, Kodwo. 1998. *More Brilliant Than the Sun: Adventures in Sonic Fiction*. London: Quartet.
Evers, Clifton. 2009. "'The Point': Surfing, Geography and a Sensual Life of Men and Masculinity on the Gold Coast, Australia." *Social & Cultural Geography* 10, no. 8: 893–908.
Fabian, Johannes. 1983. *Time and the Other: How Anthropology Makes Its Object*. New York: Columbia University Press.
Farbotko, Carol. 2010. "Wishful Sinking: Disappearing Islands, Climate Refugees and Cosmopolitan Experimentation." *Asia Pacific Viewpoint* 51, no. 1: 47–60.
Farmer, Ricky J. 1992. "Surfing: Motivations, Values, and Culture." *Journal of Sport Behaviour* 15, no. 3: 241–57.
Fay, Jennifer. 2018. *Inhospitable World: Cinema in the Time of the Anthropocene*. Oxford: Oxford University Press.
Feeley-Harnik, Gillian. 2021. "Lewis Henry Morgan: American Beavers and Their Works." *Ethnos* 86, no. 1: 21–43.
Fehskens, Erin M. 2012. "Accounts Unpaid, Accounts Untold: M. NourbeSe Philip's *Zong!* and the Catalog." *Callaloo* 35, no. 2: 407–24.
Felkin, Robert. 1889. *On the Geographical Distribution of Some Tropical Diseases*. Edinburgh: Young J. Pentland.
Fenton, John. 1985. "A Fifth-Order Stokes Theory for Steady Waves." *Journal of Waterway, Port, Coastal, and Ocean Engineering* 111, no. 2: 216–34.

Ferrel, David. 1992. "Off Palos Verdes, a DDT Dumping Ground Lingers." *Los Angeles Times*, September 9. http://articles.latimes.com/1992-09-09/news/mn-117_1_palos-verdes.

Ferré, Bénédicte, Christopher R. Sherwood, and Patricia L. Wiberg. 2010. "Sediment Transport on the Palos Verdes Shelf, California." *Continental Shelf Research* 30: 761–80.

Filipe, Angela, Alicia Renedo, and Cicely Marston. 2017. "The Co-production of What? Knowledge, Values, and Social Relations in Health Care." *PLOS Biology* 15, no. 5: e2001403.

Finkbeiner, Ann. 2015. "The Great Quake and the Great Drowning." *Hakai Magazine*, September 14. http://www.hakaimagazine.com/article-long/great-quake-and-great-drowning.

Finney, Ben. 2003. *Sailing in the Wake of the Ancestors: Reviving Polynesian Voyaging*. Honolulu: Bishop Museum Press.

Finney, Ben, and James D. Houston. 1996. *Surfing: A History of the Ancient Hawaiian Sport*. Rohnert Park, CA: Pomegranate Artbooks.

Fisch, Michael. 2021. "Staging Encounters with the End in Pre-Apocalyptic-Post-3.11 Japan." *Ethnos* (May). https://doi.org/10.1080/00141844.2020.1867608.

Fisch, Michael. 2022. "Japan's Extreme Infrastructure: Fortress-ification, Resilience, and Extreme Nature." *Social Science Japan Journal* 25, no. 2: 331–52.

Fisher, Fred H., and Fred Spiess. 1963. "Flip—FLoating Instrument Platform." *Journal of the Acoustical Society of America* 35, no. 10: 1633–44.

Fisher, R. A. 1937. "The Wave of Advance of Advantageous Genes." *Annals of Eugenics* 7: 355–69.

Fletcher, C. A., and T. Spencer, eds. 2005. *Flooding and Environmental Challenges for Venice and Its Lagoon: State of Knowledge*. Cambridge: Cambridge University Press.

Flikke, Rune. 2018. "Domestication of Air, Scent, and Disease." In *Domestication Gone Wild: Politics and Practices of Multispecies Relations*, edited by Heather Anne Swanson, Marianne Elisabeth Lien, and Gro B. Ween, 176–95. Durham, NC: Duke University Press.

Forbes, Vivian L. 1995. *The Maritime Boundaries of the Indian Ocean Region*. Chicago: University of Chicago Press.

Ford, Nick, and David Brown. 2006. *Surfing and Social Theory: Experience, Embodiment, and Narrative of the Dream Glide*. New York: Routledge.

Forensic Architecture. 2020. "Shipwreck at the Threshold of Europe, Lesvos, Aegean Sea," February 19. https://forensic-architecture.org/investigation/shipwreck-at-the-threshold-of-europe.

Fortun, Mike. 2023. *Genomics with Care: Minding the Double Binds of Science*. Durham, NC: Duke University Press.

Fortun, Mike, and Herbert J. Bernstein. 1998. *Muddling Through: Pursuing Science and Truths in the Twenty-first Century*. Washington, DC: Counterpoint.

Fortun, Mike, and Kim Fortun. 2005. "Scientific Imaginaries and Ethical Plateaus in Contemporary US Toxicology." *American Anthropologist* 107, no. 1: 43–54.

Frake, Charles. 1995. "A Reinterpretation of the Micronesian 'Star Compass.'" *Journal of the Polynesian Society* 104, no. 2: 147–58.
Frich, Morten, and Sille Veilmark. 2017. "Manden i bølgerne." *Information*. https://www.information.dk/mofo/manden-boelgerne.
Frijlink, H. C. 1963. "Activities of Dutch Engineers Abroad." In *Selected Aspects of Hydraulic Engineering*, edited by A. A. Van Douwen, 71–95. Delft, Netherlands: Technological University of Delft.
Fritz, Angela. 2019. "The National Weather Service Is 'Open,' but Your Forecast Is Worse Because of the Shutdown." *Washington Post*, January 7. https://www.washingtonpost.com/weather/2019/01/07/national-weather-service-is-open-your-forecast-is-worse-because-shutdown/.
Fritz, Hermann M., David A. Phillips, Akio Okayasu, Takenori Shimozono, et al. 2012. "The 2011 Japan Tsunami Current Velocity Measurements from Survivor Videos at Kesennuma Bay Using LiDAR." *Geophysical Research Letters* 39, no. 7. https://doi.org/10.1029/2011GL050686
Frost, Mark. 2010. "'The Guilty Ship': Ruskin, Turner and Dabydeen." *Journal of Commonwealth Literature* 45, no. 3: 371–88.
Frost, Mark Ravinder. 2010. "'That Great Ocean of Idealism': Calcutta, the Tagore Circle, and the Idea of Asia, 1900–1920." In *Indian Ocean Studies: Cultural, Social, and Political Perspectives*, edited by Shanti Moorthy and Ashraf Jamal, 251–79. London: Routledge.
Gabrys, Jennifer. 2016. *Program Earth: Environmental Sensing Technology and the Making of a Computational Planet*. Minneapolis: University of Minnesota Press.
Gabrys, Jennifer, and Helen Pritchard. 2018. "Sensing Practices." In *Posthuman Glossary*, edited by Rosi Braidotti and Maria Hlavajova, 394–95. London: Bloomsbury.
Galbraith, Kate. 2015. "Walter Munk, the 'Einstein of the Oceans.'" *New York Times*, August 24. https://www.nytimes.com/2015/08/25/science/walter-munk-einstein-of-the-oceans-at-97.html.
Galison, Peter. 1997. *Image and Logic: A Material Culture of Microphysics*. Chicago: University of Chicago Press.
Gandorfer, Daniela. 2016. "Deleuze & Guattari's *A Thousand Plateaus*, Law, and Synesthesia." *Nietzsche 13/13* (blog), November 5. http://blogs.law.columbia.edu/nietzsche1313/daniela-gandorfer-deleuze-guatarris-a-thousand-plateaus-law-and-synesthesia.
Gandorfer, Daniela. 2019. "Reading Law: A Beginner's Guide to the Manual of How to Read the Book of Law Yet to Come." In *Law and New Media: West of Everything*, edited by Christian Delage, Peter Goodrich, and Marco Wan, 222–45. Edinburgh: Edinburgh University Press.
Gardner, Robert. 1972. *The Art of Body Surfing*. New York: Chilton.
Garrett, Christopher, and Walter Munk. 1972. "Space-Time Scales of Internal Waves." *Geophysical Fluid Dynamics* 2: 225–64.
Garrett, Christopher, and Walter Munk. 1979. "Internal Waves in the Ocean." *Annual Review of Fluid Mechanics* 11: 339–69.

Garrison, Ednie Kaeh. 2005. "Are We on a Wavelength Yet? On Feminist Oceanography, Radios, and Third Wave Feminism." In *Different Wavelengths: Studies of the Contemporary Women's Movement*, edited by Jo Reger, 237–56. New York: Routledge.

Garrison, Tom. 2005. *Oceanography: An Invitation to Marine Science*, 5th ed. Belmont, CA: Thomson Brooks/Cole.

Gearey, Mary. 2018. "Chronos and Kairos: Time, Water, Memory and the Wayfaring Riverbank." In *Monsoon [+ other] Waters*, edited by Lindsay Bremner, 155–66. London: Monsoon Assemblages.

Gell, Alfred. 1996. "Vogel's Net: Traps as Artworks and Artworks as Traps." *Journal of Material Culture* 1, no. 1: 15–38.

Genz, Joseph. 2016. "Resolving Ambivalence in Marshallese Navigation: Relearning, Reinterpreting, and Reviving the 'Stick Chart' Wave Models." *Structure and Dynamics* 9, no. 1: 8–40.

Genz, Joseph. 2017. "Without Precedent: Shifting Protocols in the Use of Rongelapese Navigational Knowledge." *Journal of the Polynesian Society* 126, no. 2: 209–32.

Genz, Joseph, Jerome Aucan, Mark Merrifield, and Ben Finney et al. 2009. "Wave Navigation in the Marshall Islands: Comparing Indigenous and Western Scientific Knowledge of the Ocean." *Oceanography* 22, no. 2: 234–45.

Gerstner, Franz. 1804. *Theorie der Wellen*. Prague: Gottlieb Haase.

Gesing, Friederike. 2021. "Towards a More-than-Human Political Ecology of Coastal Protection: Coast Care Practices in Aotearoa New Zealand." *Environment and Planning E: Nature and Space* 4, no. 2: 208–29.

Ghosh, Amitav. 2004. *The Hungry Tide*. Delhi: Ravi Dayal.

Ghosh, Amitav. 2005. "Overlapping Faults." *The Hindu*, January 11, 11.

Ghosh, Amitav. 2008. "Of Fanás and Forecastles: The Indian Ocean and Some Lost Languages of the Age of Sail." *Economic and Political Weekly*, June 21, 56–62.

Ghosh, Amitav. 2016. *The Great Derangement: Climate Change and the Unthinkable*. Chicago: University of Chicago Press.

Ghosh, Amitav. 2019. *Gun Island*. New York: Farrar, Straus and Giroux.

Gibson-Graham, J. K. 2011. "A Feminist Project of Belonging for the Anthropocene." *Gender, Place & Culture* 18, no. 1: 1–21.

Gill, Tom, Brigitte Steger, and David H. Slater, eds. 2013. *Japan Copes with Calamity: Ethnographies of the Earthquake, Tsunami, and Nuclear Disasters of March 2011*. Bern: Peter Lang.

Gillis, John. 2013. "The Blue Humanities." *Humanities* 34, no. 3. https://www.neh.gov/humanities/2013/mayjune/feature/the-blue-humanities.

Gilroy, Paul. 1993. *The Black Atlantic: Modernity and Double Consciousness*. Cambridge, MA: Harvard University Press.

Gilroy, Paul. 2018. "'Where Every Breeze Speaks of Courage and Liberty': Offshore Humanism and Marine Xenology, or, Racism and the Problem of Critique at Sea Level." *Antipode* 50, no. 1: 3–22.

Glassie, Alison, 2020. "Ruth Ozeki's Floating World: *A Tale for the Time Being*'s Spiritual Oceanography." *Novel: A Forum on Fiction* 53, no. 3: 452–71.

Goldman, Francisco. 2011. "The Wave." *New Yorker*, February 7, 54–60, 62–65.
Gonzalez Monserrate, Steven. 2022. "The Cloud Is Material: On the Environmental Impacts of Computation and Data Storage." *MIT Case Studies in Social and Ethical Responsibilities of Computing* (Winter [January]). https://doi.org/10.21428/2c646de5.031d4553.
Goode, Lauren. 2018. "Kelly Slater's Artificial Surf Pool Is Really Making Waves." *Wired*, September 5. https://www.wired.com/story/kelly-slaters-artificial-surf-pool-is-really-making-waves/
Good, Megan. 2016. "Shaping Japan's Disaster Heritage." In *Reconsidering Cultural Heritage in East Asia*, edited by Akira Matsuda and Luisa Elena Mengoni, 139–61. London: Ubiquity.
Goodwin, Charles. 1995. "Seeing in Depth." *Social Studies of Science* 25, no. 2: 237–74.
Gorodé, Déwé. 2004. "Wave-Song." In *Sharing as Custom Provides*, edited and translated from the French and Païci by Raylene Ramsay and Deborah Walker, 42–43. Canberra: Pandanus.
Graeber, David. 2013. *The Democracy Project: A History, a Crisis, a Movement*. New York: Spiegel and Grau.
Graff, Agnieszka, 2003. "Lost between the Waves? The Paradoxes of Feminist Chronology and Activism in Contemporary Poland." *Journal of International Women's Studies* 4, no. 2: 100–16.
Grasseni, Cristina. 2004. "Skilled Vision: An Apprenticeship in Breeding Aesthetics." *Social Anthropology* 12, no. 1: 41–55.
Gray, Ian. 2020. "Damage Functions." *e-flux Architecture*, September 2. https://www.e-flux.com/architecture/accumulation/337972/damage-functions.
Green, George. 1839. "Note on the Motion of Waves in Canals." *Transactions of the Cambridge Philosophical Society* 7: 87–96.
Greenhouse, Carol. 1989. "Just in Time: Temporality and the Cultural Legitimation of Law." *Yale Law Journal* 98, no. 8: 1631–51.
Gregory, Alice. 2013. "Mavericks: Life and Death Surfing." *n+1*, no. 17 (Fall). https://nplusonemag.com/issue-17/essays/mavericks-surf.
Groen, Pier 1949. *Zeegolven*. Netherlands: Staatsdrukkerij.
Groeneweg, Jacco, Mathijs van Ledden, and Marcel Zijlema. 2006. "Wave Transformation in Front of the Dutch Coast." *Proceedings of 30th Conference on Coastal Engineering, San Diego, California*, edited by Jane McKee Smith, 552–64. Singapore: World Scientific.
Grosz, Elizabeth. 1994. *Volatile Bodies: Toward a Corporeal Feminism*. Bloomington: Indiana University Press.
Grotius, Hugo. 1609. *Mare Liberum*. Leiden: Elzevier.
Grusin, Richard, ed. 2015. *The Nonhuman Turn*. Minneapolis: University of Minnesota Press.
Gu, Diane Yu. 2016. *Chinese Dreams? American Dreams? The Lives of Chinese Women Scientists and Engineers in the United States*. Rotterdam: Sense.
Gulev, Sergey, and Vika Grigorieva. 2004. "Last Century Changes in Ocean Wind Wave Height from Global Visual Wave Data." *Geophysical Research Letters* 31: L24302. https://doi.org/10.1029/2004GL021040.

Gullander-Drolet, Claire. 2018. "Translational Form in Ruth Ozeki's *A Tale for the Time Being*." *Journal of Transnational American Studies* 9, no. 1: 293–314.

Guth, Christine M. E. 2015. *Hokusai's Great Wave: Biography of a Global Icon*. Honolulu: University of Hawai'i Press.

Hacking, Ian. 1982. "Biopower and the Avalanche of Printed Numbers." *Humanities in Society* 5: 279–95.

Hackett, Robin. 2004. *Sapphic Primitivism: Productions of Race, Class, and Sexuality in Key Works of Modern Fiction*. New Brunswick, NJ: Rutgers University Press.

Hadjioannou, Markos. 2008. "Into Great Stillness, Again and Again: Gilles Deleuze's Time and the Constructions of Digital Cinema." *Rhizomes: Cultural Studies in Emerging Knowledge* 16. http://www.rhizomes.net/issue16/hadji.

Hage, Ghassan. 2017. *Is Racism an Environmental Threat*? Cambridge: Polity.

Hamblin, Jacob Darwin. 2005. *Oceanographers and the Cold War: Disciples of Marine Science*. Seattle: University of Washington Press.

Hameed, Ayesha. 2014. "Black Atlantis: Three Songs." In *Forensis: The Architecture of Public Truth*, edited by Forensic Architecture, 712–19. Berlin: Sternberg.

Hameed, Ayesha. 2022. "Of Sea Changes and Other Futurisms." Ioan Davies Memorial Lecture, delivered via Zoom at York University, March 21. https://www.torontomu.ca/graduate/programs/comcult/news-events/2022/03/2022-ioan-davies-memorial-lecture/.

Hamilton, Adrian. 2012. "Turner: On the Crest of a Wave." *Independent*, February 6. http://www.independent.co.uk/arts-entertainment/art/features/turner-on-the-crest-of-a-wave-6534511.html.

Hamilton, Kevin, and Ned O'Gorman. 2021. "Seeing Experimental Imperialism in the Nuclear Pacific." *Media+Environment* 3, no. 1. https://doi.org/10.1525/001c.18496.

Hamilton-Paterson, James. 2007. *Seven-Tenths: The Sea and Its Thresholds*. London: Faber and Faber.

Hance, Susan. 2014. "Tackling Ocean Trash: Turning Plastic Pollution into Art for Awareness." *Wilma: Wilmington's Successful Woman*, April. https://www.wilmamag.com/tackling-ocean-trash/.

Hansen, Mark B. N. 2004. *New Philosophy for New Media*. Cambridge, MA: MIT Press.

Haram, Linsey, James T. Carlton, Gregory M. Ruiz, and Nikolai A. Maximenko. 2020. "A Plasticene Lexicon." *Marine Pollution Bulletin* 150: 110714.

Haraway, Donna J. 1991a. "'Gender' for a Marxist Dictionary." In *Simians, Cyborgs, and Women: The Reinvention of Nature*, by Donna J. Haraway, 127–48. New York: Routledge.

Haraway, Donna J. 1991b. "Situated Knowledges: The Science Question in Feminism and the Privilege of Partial Perspective." In *Simians, Cyborgs, and Women: The Reinvention of Nature*, by Donna J. Haraway, 183–202. New York: Routledge.

Haraway, Donna J. 1992. "The Promises of Monsters: A Regenerative Politics for Inappropriate/d Others." In *Cultural Studies*, edited by Lawrence Grossberg, Cary Nelson, and Paula Treichler, 295–336. New York: Routledge.

Haraway, Donna J. 1997. *Modest_Witness@Second_Millennium.FemaleMan©_Meets_OncoMouse™: Feminism and Technoscience*. New York: Routledge.
Haraway, Donna J. 2015. "Anthropocene, Capitalocene, Plantationocene, Chthulucene: Making Kin." *Environmental Humanities* 6: 159–65.
Hardenberg, Wilko Graf von. 2020. "Measuring Zero at Sea: On the Delocalization and Abstraction of the Geodetic Framework." *Journal of Historical Geography* 68: 11–20.
Hardy, Penelope K., and Helen M. Rozwadowski. 2020. "Maury for Modern Times: Navigating a Racist Legacy in Ocean Science." *Oceanography* 33, no. 3: 10–15.
Hasselmann, D. E., Dunckel, M., and Ewing, J. A. 1980. "Directional Wave Spectra Observed during JONSWAP 1973." *Journal of Physical Oceanography* 10: 1264–80.
Hasselmann, Klaus. 1962. "On the Nonlinear Energy Transfer in a Gravity-Wave Spectrum, Part 1: General Theory." *Journal of Fluid Mechanics* 12: 481–500.
Hasselmann, K., T. P. Barnett, E. Bouws, H. Carlson, et al. 1973. *Measurements of Wind-Wave Growth and Swell Decay during the Joint North Sea Wave Project (JONSWAP)*. Hamburg: Deutsches Hydrographisches Institut.
Hastrup, Kirsten. 2018. "Water Literacy: Challenges of Living with Troubled Waters." In *Monsoon [+ other] Waters*, edited by Lindsay Bremner, 167–84. London: Monsoon Assemblages.
Hau'ofa, Epeli. 1994. "Our Sea of Islands." *Contemporary Pacific* 6, no. 1: 148–61.
Hawkes, R. 2014. "Interstellar: Behind the VFX." *The Telegraph*, December 5. http://www.telegraph.co.uk/culture/film/film-news/11274246/Interstellar-behind-the-VFX.html.
Hayashi, Michio. 2015. "Reframing the Tragedy: Lessons from Post-3/11 Japan. In *In the Wake: Japanese Photographers Respond to 3/11*, curated by Anne Nishimura Morse and Anne Havinga, 166–79. Boston: Museum of Fine Arts Publications.
Hayward, Eva. 2005. "Enfolded Vision: Refracting the Love Life of the Octopus." *Octopus* 1: 29–44.
Hayward, Eva. 2014. "Transxenoestrogenesis." *TSQ: Transgender Studies Quarterly* 1, no. 1: 255–58.
Hearn, Chester G. 2002. *Tracks in the Sea: Matthew Fontaine Maury and the Mapping of the Oceans*. Camden, ME: International Marine and McGraw Hill.
Heller, Charles, and Lorenzo Pezzani. 2012. "The Left-to-Die Boat: The Deadly Drift of a Migrants' Boat in the Central Mediterranean." Forensic Architecture, Goldsmiths, University of London. http://www.forensic-architecture.org/case/left-die-boat.
Heller, Charles, and Lorenzo Pezzani. 2014. "Liquid Traces: Investigating the Deaths of Migrants at the EU's Maritime Frontier." In *Forensis: The Architecture of Public Truth*, edited by Forensic Architecture, 657–84. Berlin: Sternberg.
Helmreich, Stefan. 1998. "Recombination, Rationality, Reductionism, and Romantic Reactions: Culture, Computers, and the Genetic Algorithm." *Social Studies of Science* 28, no. 1: 39–71.
Helmreich, Stefan. 2007. "An Anthropologist Underwater: Immersive Soundscapes, Submarine Cyborgs, and Transductive Ethnography." *American Ethnologist* 34, no. 4: 621–41.

Helmreich, Stefan. 2009. *Alien Ocean: Anthropological Voyages in Microbial Seas*. Berkeley: University of California Press.
Hemer, Mark A., Yalin Fan, Nobuhito Mori, Alvaro Semedo, and Xiaolan L. Wang. 2013. "Projected Changes in Wave Climate from a Multi-model Ensemble." *Nature Climate Change* 3: 471–76.
Henderson, Bonnie. 2014. *The Next Tsunami: Living on a Restless Coast*. Corvallis: Oregon State University Press.
Henderson, Caspar. 2014. "Why Light Inspires Ritual." *Nautilus* 11. https://nautil.us/why-light-inspires-ritual-234840/.
Henderson, Margaret. 2001. "A Shifting Line Up: Men, Women, and *Tracks* Surfing Magazine." *Continuum* 15, no. 3: 319–32.
Henriques, Julian. 2003. "Sonic Dominance and the Reggae Sound System Session." In *The Auditory Culture Reader*, edited by Michael Bull and Les Back, 451–80. Oxford: Berg.
Henry, Archdeacon of Huntingdon. (1133–55) 1996. *Historia Anglorum: The History of the English People*. Edited and translated from the Anglo-Saxon and Latin by Diana Greenway. Oxford: Clarendon.
Hereniko, Vilsoni. 1999. "Representations of Pacific Islanders in Film and Video." *Documentary Box* 14. https://www.yidff.jp/docbox/14/box14-3-e.html.
Hernandez, Adrian F., Gregg C. Fonarow, and Li Liang et al. 2007. "Sex and Racial Differences in the Use of Implantable Cardioverter-Defibrillators among Patients Hospitalized with Heart Failure." *Journal of the American Medical Association* 298, no. 13: 1525–32.
Herzog, Lena. 2014. *Strandbeest: The Dream Machines of Theo Jansen*. Cologne, Germany: Taschen.
Hessler, Stefanie, ed. 2018. *Tidalectics: Imagining an Oceanic Worldview through Art and Science*. Cambridge, MA: MIT Press.
Hetherington, Kregg, ed. 2019. *Infrastructure, Environment, and Life in the Anthropocene*. Durham, NC: Duke University Press.
Hewitt, Nancy, ed. 2010. *No Permanent Waves: Recasting Histories of U.S. Feminism*. New Brunswick, NJ: Rutgers University Press.
Hiltzik, Michael. 2017. "Trump's EPA Has Started to Scrub Climate Change Data from Its Website." *Los Angeles Times*, May 1. http://www.latimes.com/business/hiltzik/la-fi-hiltzik-epa-climate-20170501-story.html.
Hlebica, Joe, and Steve Cook. 1993. *Legacy of Exploration: Scripps Institution of Oceanography since 1903*. San Diego, CA: Scripps Institution of Oceanography.
Ho, Enseng. 2006. *The Graves of Tarim: Genealogy and Mobility across the Indian Ocean*. Berkeley: University of California Press.
Hodder, Ian. 1990. *The Domestication of Europe*. Oxford: Basil Blackwell.
Hoeksema, Robert. 2006. *Designed for Dry Feet: Flood Protection and Land Reclamation in the Netherlands*. Reston, VA: American Society of Civil Engineers.
Hofmeyr, Isabel. 2007. "The Black Atlantic Meets the Indian Ocean: Forging New Paradigms of Transnationalism for the Global South—Literary and Cultural Perspectives." *Social Dynamics* 33, no. 2: 3–32.
Hofmeyr, Isabel. 2010. "Universalizing the Indian Ocean." *PMLA* 125, no. 3: 721–29.

Hofmeyr, Isabel, and Charne Lavery. 2020. "Exploring the Indian Ocean as a Rich Archive of History." *The Conversation*, June 7. https://theconversation.com/exploring-the-indian-ocean-as-a-rich-archive-of-history-above-and-below-the-water-line-133817.

Höhler, Sabine. 2017. "Local Disruption or Global Condition? El Niño as Weather and as Climate Phenomenon." *Geo: Geography and Environment* 4, no. 1: e00034.

Holthuijsen, Leo H. 2007. *Waves in Oceanic and Coastal Waters*. Cambridge: Cambridge University Press.

Holton, Gerald. 1969. "Einstein, Michelson, and the 'Crucial' Experiment." *Isis* 60, no. 2: 132–97.

Horton, Sarah Bronwen. 2021. " On Pandemic Privilege: Reflections on a 'Home-Bound Pandemic Ethnography.'" *Journal for the Anthropology of North America* 24, no. 2: 98–107

Hoyt, P. D. 2000. "'Rogue States' and International Relations Theory." *Journal of Conflict Studies* 20, no. 1: 69–79.

Hughes, J. Donald. 2009. "Ripples in Clio's Pond: Holland against the Sea." *Capitalism Nature Socialism* 20, no. 3: 96–103.

Huler, Scott. 2005. *Defining the Wind: The Beaufort Scale and How a 19th-Century Admiral Turned Science into Poetry*. New York: Three Rivers.

Hutchins, Edwin. 1995. *Cognition in the Wild*. Cambridge, MA: MIT Press.

ibn Mājid, Ahmad (1490) 1971. *Arab Navigation in the Indian Ocean before the Coming of the Portuguese*, being a translation from the Arabic of *Kitab al-Fawa'id fi usul al-bahr wa'l-qawa'id* by Gerald Randall Tibbets. London: Royal Asiatic Society of Great Britain and Ireland.

Idhe, Don. 1991. *Instrumental Realism: The Interface between the Philosophy of Science and the Philosophy of Technology*. Bloomington: Indiana University Press.

Igo, Sarah. 2007. *The Averaged American: Surveys, Citizens, and the Making of a Mass Public*. Cambridge, MA: Harvard University Press.

Ingersoll, Karin Amimoto. 2016. *Waves of Knowing: A Seascape Epistemology*. Durham, NC: Duke University Press.

Inman, Douglas L. 2003. "Scripps in the 1940s: The Sverdrup Era." *Oceanography* 16, no. 3: 20–28.

Intergovernmental Panel on Climate Change. 2022: *Climate Change 2022: Impacts, Adaptation, and Vulnerability*. Contribution of Working Group II to the Sixth Assessment Report of the Intergovernmental Panel on Climate Change, edited by H.-O. Pörtner, D. C. Roberts, M. Tignor, E. S. Poloczanska, K. Mintenbeck, A. Alegría, M. Craig, et al. Cambridge: Cambridge University Press.

Irigaray, Luce. (1977) 1985. *This Sex Which Is Not One*. Translated from the French by Catherine Porter with Carolyn Burke. Ithaca, NY: Cornell University Press.

Irvine, David. 2002. "The Role of Spectra in Ocean Wave Physics." In *Oceanographic History: The Pacific and Beyond*, edited by Keith R. Benson and Philip F. Rehbock, 378–86. Seattle: University of Washington Press.

Ishiwata, Eric. 2002. "Local Motions: Surfing and the Politics of Wave Sliding." *Cultural Values* 6, no. 3: 257–72.

Ivakhiv, Adrian. 2013. "An Ecophilosophy of the Moving Image: Cinema as Anthrobiogeomorphic Machine." In *Ecocinema Theory and Practice*, edited by Stephen Rust, Salma Monani, and Sean Cubitt, 87–105. New York: Routledge.

Jaffe, Jules S., Peter J. S. Franks, Paul L. D. Roberts, Diba Mirza, et al. 2017. "A Swarm of Autonomous Miniature Underwater Robot Drifters for Exploring Submesoscale Ocean Dynamics." *Nature Communications* 8, 14189. https://doi.org/10.1038/ncomms14189.

Jänicke, Stefan, Greta Franzini, Muhammad Faisal Cheema, and Gerik Scheuermann. 2015. "On Close and Distant Reading in Digital Humanities: A Survey and Future Challenges." Paper presented at the Eurographics Conference on Visualization, Cagliari, Italy, May 25–29. http://www.informatik.uni-leipzig.de/~stjaenicke/Survey.pdf.

Janssen, Peter. 2000. "ECMWF Wave Modeling and Satellite Altimeter Wave Data." In *Satellites, Oceanography, and Society*, edited by David Halpern, Elsevier Oceanography Series 63, 35–56. Amsterdam: Elsevier.

Jasanoff, Sheila, ed. 2004. *States of Knowledge: The Co-production of Science and Social Order*. London: Routledge.

Jauss, Hans Robert. 1970. "Literary History as a Challenge to Literary Theory." *New Literary History* 2, no. 1: 7–37.

Jensen, Casper Bruun. 2015. "Experimenting with Political Materials: Environmental Infrastructures and Ontological Transformations." *Distinktion* 16, no. 1: 17–30.

Jensen, Casper Bruun. 2020. "A Flood of Models: Mekong Ecologies of Comparison." *Social Studies of Science* 20, no. 1: 76–93.

Jensen, Robert E., Val Swail, Tyler J. Hesser, and Boram Lee. 2013. "Are Wave Measurements Actually Ground Truth?" Paper presented at Forecasting Dangerous Sea States, the Thirteenth International Workshop on Wave Hindcasting and Forecasting, Banff, AB, October 27–November 1.

Jetñil-Kijiner, Kathy. 2017. "Two Degrees." In *Iep Jaltok: Poems from a Marshallese Daughter*, by Kathy Jetñil-Kijiner, 76–79. Tucson: University of Arizona Press.

Joas, Christian, and Christoph Lehner. 2009. "The Classical Roots of Wave Mechanics: Schrödinger's Transformations of the Optical-Mechanical Analogy." *Studies in History and Philosophy of Modern Physics* 40, no. 4: 338–51.

Jobson, Ryan Cecil. 2021. "Dead Labor: On Racial Capital and Fossil Capital." In *Histories of Racial Capitalism*, edited by Destin Jenkins and Justin Leroy, 215–30. New York: Columbia University Press.

Johns, Adrian. 1998. *The Nature of the Book: Print and Knowledge in the Making*. Chicago: University of Chicago Press.

Johnson, Richard H., and Robert A. Houze Jr. 1987. "Precipitating Cloud Systems of the Asian Monsoon." In *Monsoon Meteorology*, edited by C.-P. Chang and T. N. Krishnamurty, 298–353. New York: Oxford University Press.

Jonas, Hans. 1966. *The Phenomenon of Life: Toward a Philosophical Biology*. New York: Harper and Row.

Jones, David, and Stefan Helmreich. 2020. "The Shape of Epidemics." *Boston Review*, June 26. http://bostonreview.net/science-nature/david-s-jones-stefan-helmreich-shape-epidemics.

Jones, Madison, Aaron Beveridge, Julian R. Garrison, Abbey Greene, Hannah MacDonald. 2022. "Tracking Memes in the Wild: Visual Rhetoric and Image Circulation in Environmental Communication." *Frontiers in Communication* 7. https://www.frontiersin.org/articles/10.3389/fcomm.2022.883278/full.

Jones, Mike. 2007. "Vanishing Point: Spatial Composition and the Virtual Camera." *Animation* 2, no. 3: 225–43.

Jones, Owain. 2019. "Monsoon plus Tide." Paper presented at the third Monsoon Assemblages Symposium, University of Westminster, London, March 21–22. Video available at https://www.oceanichumanities.com/post/conference-paper-permeability-ocean-concrete.

Jones, Ron. 1981. "The Third Wave." In *No Substitute for Madness: A Teacher, His Kids, and Lessons of Real Life*, by Ron Jones, 1–24. Covelo, CA: Island.

Joosten, H. P. 2013. *Datawell, 1961–2011: Riding the Waves for 50 Years*. Translated by L. M. Ruitenberg-van Meersbergen. Heemstede, Netherlands: Gravé.

Jue, Melody. 2017a. "Intimate Objectivity: On Nnedi Okorafor's Oceanic Afrofuturism." *Women's Studies Quarterly* 45, nos. 1–2: 171–88.

Jue, Melody. 2017b. "From the Goddess Ganga to a Teacup: On Amitav Ghosh's Novel, *The Hungry Tide*." In *Scale in Literature and Culture*, edited by Michael Tavel Clarke and David Wittenberg, 203–24. New York: Routledge.

Jue, Melody. 2020. *Wild Blue Media: Thinking through Seawater*. Durham, NC: Duke University Press.

Julien, Claude. 2006. "The Enigma of Mayer Waves: Facts and Models." *Cardiovascular Research* 70, no. 1: 12–21.

Julien, Isaac. 2010. *Ten Thousand Waves*. London: Isaac Julien Studio and Victoria Miro Gallery.

Kahn, Jeffrey S. 2019. *Islands of Sovereignty: Haitian Migration and the Borders of Empire*. Chicago: University of Chicago Press.

Kajiura, Kinjiro. 1963. "The Leading Waves of a Tsunami." *Bulletin of the Earthquake Research Institute* 41: 535–71.

Kaplan, Caren. 2018. *Aerial Aftermaths: Wartime from Above*. Durham, NC: Duke University Press.

Karatzis, Emmanouil, Theodore G. Papaioannou, and Konstantinos Aznaouridis et al. 2005. "Acute Effects of Caffeine on Blood Pressure and Wave Reflections in Healthy Subjects: Should We Consider Monitoring Central Blood Pressure?" *International Journal of Cardiology* 98, no. 3: 425–30.

Keane, Stephen. 2006. *Disaster Movies: The Cinema of Catastrophe*, 2nd ed. London: Wallflower.

Keiner, Christine. 2004. "Modeling Neptune's Garden: The Chesapeake Bay Hydraulic Model, 1965–1984." In *The Machine in Neptune's Garden: Historical Perspectives on Technology and the Marine Environment*, edited by Helen M. Rozwadowski and David K. van Keuren, 273–314. Sagamore Beach, MA: Science History.

Keisling, Benjamin, Isabella La Bras, and Bonnie Ludka. 2021. "A Conversation on Building Safe Spaces for the LGBTQ+ Community in the Geosciences." *Nature Communications* 12. https://doi.org/10.1038/s41467-021-24020-z.

Keller, Catherine. 2003. *The Face of the Deep: A Theology of Becoming*. London: Routledge.

Keller, Sarah, and Jason N. Paul, eds. 2012. *Jean Epstein: Critical Essays and New Translations*. Amsterdam: Amsterdam University Press.

Kelley, Lindsay. 2014. "Tranimals." *TSQ: Transgender Studies Quarterly* 1, no. 1: 226–28.

Kelly, Shannon. 2012. "Dancing with the Water Wolf and Choreographing Urban Ecologies." In *Strootman: Strategies for the Sublime*, edited by Uje Lee, René van der Velde, and Véronique Hoedemakers, 154–71. Seoul: C3.

Kelty, Christopher, and Hannah Landecker. 2004. "A Theory of Animation: Cells, L-systems, and Film." *Grey Room* 17: 30–63.

Keown, Michelle. 2018. "Waves of Destruction: Nuclear Imperialism and Antinuclear Protest in the Indigenous Literatures of the Pacific." *Journal of Postcolonial Writing* 54, no. 5: 585–600.

Kervran, David Dumoulin, Mina Kleiche-Dray, and Mathieu Quet. 2018. "Going South: How STS Could Think Science in and with the South?" *Tapuya* 1, no. 1: 280–305.

Keulartz, Jozef. 1999. "Engineering the Environment: The Politics of 'Nature Development.'" In *Living with Nature: Environmental Politics as Cultural Discourse*, edited by Frank Fischer and Maarten Hajer, 83–102. Oxford: Oxford University Press.

Khan, Naveeda. 2019. "At Play with the Giants: Between the Patchy Anthropocene and Romantic Geology." *Current Anthropology* 60 (Supplement 20): S333–41.

Kilgour, Frederick G. 1963. "Vitruvius and the Early History of Wave Theory." *Technology and Culture* 4, no. 3: 282–86.

Kimura, Shuhei. 2016. "When a Seawall Is Visible: Infrastructure and Obstruction in Post-tsunami Reconstruction in Japan." *Science as Culture* 25, no. 1: 23–43.

King, Tiffany Lethabo. 2019. *The Black Shoals: Offshore Formations of Black and Native Studies*. Durham, NC: Duke University Press.

Kinsman, Blair. 1965. *Wind Waves: Their Generation and Propagation on the Ocean Surface*. Englewood Cliffs, NJ: Prentice-Hall.

Kipling, Rudyard. 1896. "The Bell Buoy." *Saturday Review*, Christmas supp., 1.

Kirksey, S. Eben, and Stefan Helmreich. 2010. "The Emergence of Multispecies Ethnography." *Cultural Anthropology* 25, no. 4: 545–76.

Kirsch, Stuart. 2020. "Why Pacific Islanders Stopped Worrying about the Apocalypse and Started Fighting Climate Change." *American Anthropologist* 122, no. 4: 827–39.

Kittler, Friedrich A. (1986) 1999. *Gramophone, Film, Typewriter*. Translated from the German by Geoffrey Winthrop-Young and Michael Wutz. Stanford, CA: Stanford University Press.

Klauder, J. R., A. C. Price, S. Darlington, and W. J. Albersheim. 1960. "Theory and Design of Chirp Radars." *Bell System Technical Journal* 39, no. 4: 745–808.

Klaver, Irene, Jozef Keulartz, Henk van den Belt, and Bart Gremmen. 2002. "Born to Be Wild: A Pluralistic Ethics Concerning Introduced Large Herbivores in the Netherlands." *Environmental Ethics* 24, no. 1: 3–21.

Kluzek, Zygmunt, and Alaiksandr Lisimenka. 2013. "Acoustic Noise Generation under Plunging Breaking Waves." *Oceanologia* 55, no. 4: 809–36.

Knight, Charles. 1855. *The English Cyclopedia: A New Dictionary of Universal Knowledge, Geography*, vol. 4. London: Bradbury and Evans.

Knight, David B. 2006. *Landscapes in Music: Space, Place, and Time in the World's Great Music*. Lanham, MD: Rowman and Littlefield.

Knorr Cetina, Karin. 1995. "Laboratory Studies: The Cultural Approach to the Study of Science." In *Handbook of Science and Technology Studies*, edited by Sheila Jasanoff, Gerald E. Markle, James C. Petersen, and Trevor Pinch, 140–66. Thousand Oaks, CA: Sage.

Knox, Hannah. 2018. "Not Flying: Steps toward a Post-carbon Anthropology." *Hannah Knox* (blog), March 24. https://hannahknox.wordpress.com/2018/03/24/not-flying-steps-towards-a-post-carbon-anthropology.

Kohler, Robert E. 2002. *Landscapes and Labscapes: Exploring the Lab-Field Border in Biology*. Chicago: University of Chicago Press.

Komel, Mirt. 2019. "The Wave of the Sign: Pyramidal Sign, Haptic Hieroglyphs, and the Touch of Language." In *The Language of Touch: Philosophical Examinations in Linguistics and Haptic Studies*, edited by Mirt Komel, 5–18. London: Bloomsbury Academic.

Komen, G. J., L. Cavaleri, M. Donelan, and K. Hasselmann et al. 1994. *Dynamics and Modeling of Ocean Waves*. Cambridge: Cambridge University Press.

Kondratieff, Nikolai. (1922) 2004. *The World Economy and Its Conjunctures during and after the War*. Translated from the Russian by V. Wolfson. Moscow: International Kondratieff Foundation.

Kraijo, Ineke. 2016. *Vechten tegen de golven*. Heerenveen, Netherlands: Royal Jongbloed.

Kraus, Werner. 2005. "Raden Saleh's Interpretation of the *Arrest of Diponegoro*: An Example of Indonesian 'Proto-nationalist' Modernism." *Archipel* 69: 259–94.

Kuhn, Thomas. 1962. *The Structure of Scientific Revolutions*. Chicago: University of Chicago Press.

LaDuke, Winona. 1983. "Native America: The Economics of Radioactive Colonialism." *Review of Radical Political Economics* (Fall): 9–19.

Lafrance, Adrienne. 2017. "The Algorithms behind *Moana*'s Gorgeously Animated Ocean." *The Atlantic*, May 31. https://www.theatlantic.com/technology/archive/2017/05/the-algorithms-behind-moanas-gorgeously-animated-pacific-ocean/528645.

Lamarre, Eric, and W. K. Melville. 1994. "Void-Fraction Measurements and Sound-Speed Fields in Bubble Plumes Generated by Breaking Waves." *Journal of the Acoustical Society of America* 95, no. 3: 1317–28.

Lamb, Horace. (1879) 1895. *Hydrodynamics*. Cambridge: Cambridge University Press.

Lanagan, David. 2003. "Dropping In: Surfing, Identity, Community and Commodity." In *Some Like It Hot: The Beach as a Cultural Dimension*, edited by James Skinner, Keith Gilbert, and Allan Edwards, 169–84. Oxford: Meyer and Meyer Sport.

Lane, Ferdinand C. 1947. *The Mysterious Sea*. New York: Doubleday.
Latour, Bruno. 1987. *Science in Action: How to Follow Engineers and Scientists through Society*. Cambridge, MA: Harvard University Press.
Latour, Bruno. 2004. "Why Has Critique Run Out of Steam? From Matters of Fact to Matters of Concern." *Critical Inquiry* 30, no 2: 225–48.
Latour, Bruno, and Steve Woolgar. 1986. *Laboratory Life: The Construction of Scientific Facts*. Princeton, NJ: Princeton University Press.
Lavery, Charne. 2020. "Thinking from the Southern Ocean." In *Sustaining Seas: Oceanic Space and the Politics of Care*, edited by Elspeth Probyn, Kate Johnston, and Nancy Lee, 307–18. London: Rowman and Littlefield.
Le Bon, Gustave. (1895) 1896. *The Crowd: A Study of the Popular Mind*. Translated from the French. New York: Macmillan.
Lee, Jia Hui. 2021. "Rat Tech: Transforming Rodents into Technology in Tanzania." Environment and Society Portal, *Arcadia* 5. http://doi:10.5282/rcc/9214.
Leenaers, Henk, comp. 2013. *Water Atlas of the Netherlands*. Groningen, Netherlands: Noordhoff Uitgevers B.V.
"Legacy of Genocide Resurfaces in Boston as Construction Is Planned on Burial Site." 2019. *Cultural Survival*, July 27. https://www.culturalsurvival.org/news/legacy-genocide-resurfaces-boston-construction-planned-burial-site-0.
Leheny, David, and Kay Warren. 2010. *Japanese Aid and the Construction of Global Development: Inescapable Solutions*. London: Routledge.
Lehman, Jessica. 2016. "A Sea of Potential: The Politics of Global Ocean Observations." *Political Geography* 55: 113–23.
Lehman, Jessica. 2021. "Sea Change: The World Ocean Circulation Experiment and the Productive Limits of Ocean Variability." *Science, Technology, & Human Values* 46, no. 4: 839–62.
Leibniz, Gottfried Wilhelm. (1765) 1982. *New Essays on Human Understanding*. Translated from the German by Peter Remnant and Jonathan Bennett. Cambridge: Cambridge University Press.
Leighton, Timothy G. 2014. "Bubble Acoustics." Paper presented at the Twelfth Particle Technology Forum 2014, Manchester, UK, September 16–17. https://eprints.soton.ac.uk/415945/1/2014_Leighton_2014_Particles_conference_Manchester_.pdf.
Lennert-Cody, Cleridy E., and Peter J. S. Franks. 2002. "Fluorescence Patches in High-frequency Internal Waves." *Marine Ecology-Progress Series* 235: 29–42.
Lentjes, Rebecca. 2017. "Surreal Conjunctions: An Interview with Annea Lockwood." *VAN Magazine*, July 26. https://van-magazine.com/mag/annea-lockwood/.
Leonelli, Sabina, Brian Rappert, and Gail Davies. 2017. "Data Shadows: Knowledge, Openness, and Absence." *Science, Technology, & Human Values* 42, no. 2: 191–202.
Lepore, Jill. 1998. *The Name of War: King Philip's War and the Origins of American Identity*. New York: Vintage.
Lewis, David. 1972. *We, the Navigators: The Ancient Art of Landfinding in the Pacific*. Canberra: Australian National University Press.

Lewis, Jason Edward. 2019 "An Orderly Assemblage of Biases: Troubling the Monocultural Stack." In *Afterlives of Indigenous Archives*, edited by Ivy Schweitzer and Gordon Henry Jr., 218–31. Hanover, NH: Dartmouth College Press.

Lewis, Martin W., and Kären Wigen. 1999. "A Maritime Response to the Crisis in Area Studies." *Geographical Review* 89, no. 2: 161–68.

Ley, Lukas. 2021. *Building on Borrowed Time: Rising Seas and Failing Infrastructure in Semarang*. Minneapolis: University of Minnesota Press.

Leya, Rabeya Sultana, Debanjali Saha, Sujit Kumar Bala, and Hamidul Huq. 2020. "Gender Vulnerability Assessment due to Flood in Northern Part of Bangladesh." In *Water, Flood Management and Water Security under a Changing Climate*, edited by Anisul Haque and Ahmed Ishtiaque Amin, 235–49. Cham, Switzerland: Springer International.

Liboiron, Max 2021. *Pollution Is Colonialism*. Durham, NC: Duke University Press.

Lien, Marianne. 2015. *Becoming Salmon: Aquaculture and the Domestication of a Fish*. Berkeley: University of California Press.

Lighthill, M. J. 1967. "Waves in Fluids." *Communications on Pure and Applied Mathematics* 20: 267–93.

Lindow, John. 2001. *Norse Mythology: A Guide to Gods, Heroes, Rituals, and Beliefs*. Oxford: Oxford University Press.

Lippit, Akira Mizuta. 2015. "Between Disaster, Medium 3.11." *Mechademia* 10: 3–15.

Littau, Karin. 2006. *Theories of Reading: Books, Bodies, and Bibliomania*. Cambridge: Polity.

Littlejohn, Andrew. 2021. "Ruins for the Future: Critical Allegory and Disaster Governance in Post-Tsunami Japan." *American Ethnologist* 48, no. 1: 7–21.

Longfellow, Henry Wadsworth. 1885. *The Poetical Works of Henry Wadsworth Longfellow*. Boston: Houghton Mifflin.

Longuet-Higgins, Michael, comp. 2010. "Group W at the Admiralty Research Laboratory." In *Of Seas and Ships and Scientists: The Remarkable Story of the UK's National Institute of Oceanography*, edited by Anthony Laughton, John Gould, 'Tom' Tucker, and Howard Roe, 41–66. Cambridge: Lutterworth.

Lorimer, Jamie. 2017. "The Anthropo-scene: A Guide for the Perplexed." *Social Studies of Science* 47, no. 1: 117–42.

Lorimer, Jamie, and Clemens Driessen. 2013. "Bovine Biopolitics and the Promise of Monsters in the Rewilding of Heck Cattle." *Geoforum* 48: 249–59.

Lotan, Gilad, Erhardt Graeff, Mike Ananny, and Devin Gaffney et al. 2011. "The Revolutions Were Tweeted: Information Flows During the 2011 Tunisian and Egyptian Revolutions." *International Journal of Communication* 5: 1375–405.

Lucente, Gregory. 1985. "An Interview with Italo Calvino." *Contemporary Literature* 26, no. 3: 245–53.

Luciano, Dana, and Mel Y. Chen. 2015. "Has the Queer Ever Been Human?" *GLQ* 21: 2–3.

Ludwin, Ruth S., Robert Dennis, Deborah Carver, and Alan D. McMillan et al. 2005. "Dating the 1700 Cascadia Earthquake: Great Coastal Earthquakes in Native Stories." *Seismological Research Letters* 76, no. 2: 140–48.

Luis-Brown, David. 2008. *Waves of Decolonization: Discourses of Race and Hemispheric Citizenship in Cuba, Mexico, and the United States*. Durham, NC: Duke University Press.

Lynteris, Christos. 2017. "Zoonotic Diagrams: Mastering and Unsettling Human-Animal Relations." *Journal of the Royal Anthropological Institute* 23, no. 3: 463–85.

Mack, John. 2011. *The Sea: A Cultural History*. London: Reaktion.

Mackenzie, Adrian. 2006. *Cutting Code: Software and Sociality*. New York: Peter Lang.

Mackenzie, Adrian. 2008a. "Codec." In *Software Studies: A Lexicon*, edited by Matthew Fuller, 48–55. Cambridge, MA: MIT Press.

Mackenzie, Adrian. 2008b. "Internationalization." In *Software Studies: A Lexicon*, edited by Matthew Fuller, 153–61. Cambridge, MA: MIT Press.

MacKinnon, Jennifer A. 2014. "Autobiographical Sketch. Women in Oceanography: A Decade Later." *Oceanography* 27, no. 4 (supp.): 164.

MacKinnon, Jennifer A., Zhongxiang Zhao, Caitlin B. Whalen, Amy F. Waterhouse, et al. 2017. "Climate Process Team on Internal Wave-Driven Ocean Mixing." *Bulletin of the American Meteorological Society* 98, no. 11: 2429–54.

Maleuvre, Didier. 2011. *The Horizon: A History of Our Infinite Longing*. Berkeley: University of California Press.

Mangolte, Babette. 2003. "Afterwards: A Matter of Time: Analog versus Digital, the Perennial Question of Shifting Technology and Its Implications for an Experimental Filmmaker's Odyssey." In *Camera Obscura, Camera Lucida: Essays in Honor of Annette Michelson*, edited by Richard Allen and Malcolm Turvey, 261–74. Amsterdam: Amsterdam University Press.

Manovich, Lev. 2001. *The Language of New Media*. Cambridge, MA: MIT Press.

Marcus, Adam. 2020. "Buoyant Ecologies: Interspecies Cooperation for Sea Level Rise Adaptation." In *Sustaining Seas: Oceanic Space and the Politics of Care*, edited by Elspeth Probyn, Kate Johnston, and Nancy Lee, 253–60. London: Rowman and Littlefield.

Markkula, Joanna. 2011. "'Any Port in a Storm': Responding to Crisis in the World of Shipping." *Social Anthropology* 19, no. 3: 297–304.

Marti, Don. 2003. "Letter to the US Department of Commerce on Exporting Linux to Iraq." *Linux Journal*, December 16. https://www.linuxjournal.com/article/7318.

Marx, Leo. 1964. *The Machine in the Garden: Technology and the Pastoral Ideal in America*. Oxford: Oxford University Press.

Masco, Joseph. 2014. *The Theater of Operations: National Security Affect from the Cold War to the War on Terror*. Durham, NC: Duke University Press.

Masco, Joseph. 2018. "The Six Extinctions: Visualizing Planetary Ecological Crisis Today." In *After Extinction*, edited by Richard Grusin, 71–106. Minneapolis: University of Minnesota Press.

Maury, Matthew Fontaine. 1860. *The Physical Geography of the Sea, Being a Reconstruction and Enlargement of the Eighth Edition of "The Physical Geography of the Sea."* London: Sampson Low, Son & Co.

Mauss, Marcel. (1934) 1973. "Techniques of the Body." *Economy and Society* 2, no. 1: 70–88.

McCormack, Derek P. 2018. *Atmospheric Things: The Allure of Elemental Envelopment*. Durham, NC: Duke University Press.

McCosker, Anthony. 2013. "De-framing Disaster: Affective Encounters with Raw and Autonomous Media." *Continuum* 27, no. 3: 382–96.

McKittrick, Katherine. 2013. "Plantation Futures." *Small Axe* 17, no. 3: 1–15.

McPhee-Shaw, Erika. 2006. "Boundary-Interior Exchange: Reviewing the Idea that Internal-wave Mixing Enhances Lateral Dispersal Near Continental Margins." *Deep Sea Research Part II: Topical Studies in Oceanography* 53, nos. 1–2: 42–59.

McTighe, Laura, and Megan Raschig. 2019. "Introduction: An Otherwise Anthropology." Theorizing the Contemporary, *Fieldsights*, July 31. https://culanth.org/fieldsights/introduction-an-otherwise-anthropology.

Meier, Allison. 2017. "The Octopus, a Motif of Evil in Historical Propaganda Maps." *Hyperallergic*, May 8. https://hyperallergic.com/375900/the-map-octopus-a-propaganda-motif-of-spreading-evil.

Melillo, John. 2016. "Breaking Waves and Aurality." Paper presented at Periods and Waves: A Conference on Sound and History, Stony Brook University, Stony Brook, NY, April 29–30.

Melville, W. K. 1983. "Wave Modulation and Breakdown." *Journal of Fluid Mechanics* 128: 489–506.

Melville, W. K., and R. J. Rapp. 1985. "Momentum Flux in Breaking Waves." *Nature* 317: 514–16.

Mentz, Steve. 2015. *Shipwreck Modernity: Ecologies of Globalization, 1550–1719*. Minneapolis: University of Minnesota Press.

Mentz, Steve. 2019a. *Break Up the Anthropocene*. Minneapolis: University of Minnesota Press.

Mentz, Steve. 2019b. "Experience Is Better than Knowledge: Premodern Ocean Science and the Blue Humanities." *Configurations* 27, no. 4: 433–42.

Metz, Tracy, and Maartje van den Heuvel. 2012. *Sweet and Salt: Water and the Dutch*. Rotterdam: NAi.

Mehtta, Megnaa. 2018. "Unembanking Habitations and Imaginations: The Politics of Life amidst the Ebbs and Flows of the Sundarbans Forests of West Bengal." In *Monsoon [+ other] Waters*, edited by Lindsay Bremner, 29–43. London: Monsoon Assemblages.

Mehtta, Megnaa. 2020. "The Sundarban after Cyclone Amphan." Interview by Raisa Wickrematunge. *Himal Southasian*, June 11. https://www.himalmag.com/what-the-sundarban-tells-us-podcast-2020.

Middleton, Fiona. 2020. "Brown Clouds." Ocean-Archive.org, n.d. https://ocean-archive.org/collection/119.

Mignolo, Walter D. 2011. *The Darker Side of Western Modernity: Global Futures, Decolonial Options*. Durham, NC: Duke University Press.

Mills, Eric. 1989. *Biological Oceanography: An Early History, 1870–1960*. Ithaca, NY: Cornell University Press.

Mills, Eric. 2009. *The Fluid Envelope of Our Planet: How the Study of Ocean Currents Became a Science*. Toronto: University of Toronto Press.

Mills, Mara. 2016. “Beyond Recognition: What Machines Don’t Read.” *American Foundation for the Blind* (blog), September 15. https://www.afb.org/blog/entry/beyond-recognition-what-machines-dont-read.

Minsberg, Talya. 2017. “On a California Ranch, Perfect Conditions for Pro Surfing.” *New York Times*, September 20, https://www.nytimes.com/2017/09/20/sports/world-surf-league-kelly-slater-test.html?_r=0.

Mitman, Gregg. 1999. *Reel Nature: America’s Romance with Wildlife on Film*. Cambridge, MA: Harvard University Press.

Mitsuyasu, Hisashi. 2009. *Looking Closely at Ocean Waves: From Their Birth to Their Death*. Tokyo: Terrapub.

Mirzoeff, Nicholas D. 2015. “The Drowned and the Sacred: To See the Unspeakable.” *How to See the World*, August 29. https://wp.nyu.edu/howtoseetheworld/2015/08/29/auto-draft-73.

Moore, Jason W., ed. 2016. *Anthropocene or Capitalocene? Nature, History, and the Crisis of Capitalism*. Oakland, CA: PM Press.

Moorthy, Shanti, and Ashraf Jamal. 2010. “Introduction: New Conjunctures in Maritime Imaginaries.” In *Indian Ocean Studies: Cultural, Social, and Political Perspectives*, edited by Shanti Moorthy and Ashraf Jamal, 1–31. London: Routledge.

Moretti, Franco. 2013. *The Bourgeois: Between History and Literature*. New York: Verso.

Morgan, Lewis Henry. 1868. *The American Beaver and His Works*. Philadelphia: J. B. Lippincott.

Morimoto, Ryo. 2012. “Shaking Grounds, Unearthing Palimpsests: Semiotic Anthropology of Disaster.” *Semiotica* 192: 263–74.

Morimoto, Ryo. 2014. “Message without a Coda: On the Rhetoric of Photographic Records.” *Signs and Society* 2, no. 2: 284–313.

Morimoto, Ryo. 2015. “Waves of Semiosis: Is It about Time? On the Semiotic Anthropology of Change.” In *International Handbook of Semiotics*, vol. 1, edited by Peter Pericles Trifonas, 547–64. Dordrecht: Springer.

Morita, Atsuro. 2013. “Traveling Engineers, Machines, and Comparisons: Intersecting Imaginations and Journeys in the Thai Local Engineering Industry.” *East Asian Science, Technology, and Society* 7, no. 2: 221–41.

Morita, Atsuro, and Wakana Suzuki. 2019. “Being Affected by Sinking Deltas: Changing Landscapes, Resilience, and Complex Adaptive Systems in the Scientific Story of the Anthropocene.” *Current Anthropology* 60 (Supplement 20): 286–95.

Morton, Jamie. 2001. *The Role of the Physical Environment in Ancient Greek Seafaring*. Leiden: E. J. Brill.

Morton, Timothy. 2013. *Hyperobjects: Philosophy and Ecology after the End of the World*. Minneapolis: University of Minnesota Press.

Moshe, Zach, Asher Metzger, Gal Elidan, Frederik Kratzert, et al. 2020. “HydroNets: Leveraging River Structure for Hydrological Modeling.” Paper presented at the International Conference on Learning Representations, April 26–May 1. https://ai4earthscience.github.io/iclr-2020-workshop/papers/ai4earth04.pdf.

Moten, Fred. 2003. *In the Break: The Aesthetics of the Black Radical Tradition*. Minneapolis: University of Minnesota Press.

Mouw, Colleen B., Sarah Clem, Sonya Legg, and Jean Stockard. 2018 "Meeting Mentoring Needs in Physical Oceanography: An Evaluation of the Impact of MPOWIR." *Oceanography* 31, no. 4: 171–79.

Mrázek, Rudolf. 2002. *Engineers of Happy Land: Technology and Nationalism in a Colony*. Princeton, NJ: Princeton University Press.

Mukerji, Chandra. 1989. *A Fragile Power: Scientists and the State*. Princeton, NJ: Princeton University Press.

Mukerji, Chandra. 2009. *Impossible Engineering: Technology and Territoriality on the Canal du Midi*. Princeton, NJ: Princeton University Press.

Mukharji, Projit Bihari. 2016. *Doctoring Traditions: Ayurveda, Small Technologies, and Braided Sciences*. Chicago: University of Chicago Press.

Mukherjee, Jenia. 2020. *Blue Infrastructures: Natural History, Political Ecology and Urban Development in Kolkata*. Singapore: Springer.

Munk, Judith, and Walter Munk. 1972. "Venice Hologram." *Proceedings of the American Philosophical Society* 116, no. 5: 415–42.

Munk, Walter. 1939. "Internal Waves in the Gulf of California." *Journal of Marine Research* 4, no. 1: 81–91.

Munk, Walter, and Carl Wunsch. 2019. "A Conversation with Walter Munk." *Annual Review of Marine Science* 11: 15–25.

Munn, Nancy D. 1992. "The Cultural Anthropology of Time: A Critical Essay." *Annual Review of Anthropology* 21: 93–123.

Musil, Robert. (1930–43) 1996. *The Man Without Qualities*, vol. 1. Translated from the German by Sophie Wilkins and Burton Pike. New York: Vintage.

Muto, Ichiyo. 2013. "The Buildup of a Nuclear Armament Capability and the Postwar Statehood of Japan: Fukushima and the Genealogy of Nuclear Bombs and Power Plants." *Inter-Asia Cultural Studies* 14, no. 2: 171–212.

Narayan, Siddharth, Michael W. Beck, Borja G. Reguero, and Iñigo J. Losada et al. 2016. "The Effectiveness, Costs and Coastal Protection Benefits of Natural and Nature-Based Defences." *PLOS ONE* 11, no. 5: 4.

National Resilience Program. 2019a. *Annual Programme Narrative Progress Report, Bangladesh, Reporting Period: 1 January–31 December 2019*, United Nations Development Programme, UN Women, and United Nations Office for Project Services. https://info.undp.org/docs/pdc/Documents/BGD/2.%20Annual%20Progress%20Report%202019-v4-30-04-2020.pdf.

National Resilience Program. 2019b. *Report on Gender Review of Development Project Proposal*. https://drive.google.com/file/d/17FUozWNdor5bks9BXNK8r1irjUdNejoU/view.

National Science Foundation (NSF). 1967. *International Indian Ocean Expedition* film. Washington DC: Screen Presentations, Inc.

National Science Foundation (NSF). 2016. "LIGO Detects Gravitational Waves." Press conference, National Press Club, Washington, DC, February 11.

Neale, Timothy, and Daniel May. 2020. "Fuzzy Boundaries: Simulation and Expertise in Bushfire Prediction." *Social Studies of Science* 50, no. 6: 837–59.

Neimanis, Astrida. 2012. "Hydrofeminism: Or, On Becoming a Body of Water." In *Undutiful Daughters: New Directions in Feminist Thought and Practice*, edited by Henriette Gunkel, Chrysanthi Nigianni, and Fanny Söderbäck, 85–99. New York: Palgrave Macmillan.

Neimanis, Astrida, Aleksija Neimanis, and Cecilia Åsberg. 2017. "Fathoming Chemical Weapons in the Gotland Deep." *Cultural Geographies* 24, no. 4: 631–38.

Nelson, Diane M. 2019. "Water as Genre." Roundtable on Form, Genre and Ethnographic Imaginations, 118th Annual Meeting of the American Anthropological Association, Vancouver, BC, November 20–23.

Nemani, Mihi. 2015. "Being a Brown Bodyboarder." In *Seascapes: Shaped by the Sea*, edited by Mike Brown and Barbara Humberstone, 83–100. Farnham, UK: Ashgate.

Newman, J. N. 1977. *Marine Hydrodynamics*. Cambridge, MA: MIT Press.

Newman, J. N. 2014. "Cloaking a Circular Cylinder in Water Waves." *European Journal of Mechanics-B/Fluids* 47: 145–50.

Nietzsche, Friedrich. (1882) 1974. *The Gay Science*. Translated by Walter Kaufmann. New York: Vintage.

Nixon, Rob. 2013. *Slow Violence and the Environmentalism of the Poor*. Cambridge, MA: Harvard University Press.

Nolan, Bruce. 2009. "Stacy Head Has Rubbed Some People the Wrong Way, but Supporters Say Her Brash Style Is Misunderstood." *Times-Picayune*, April 8. http://www.nola.com/news/index.ssf/2009/04/stacy_head_photo_for_terry.html.

Novak, David. 2013. *Japanoise: Music at the Edge of Circulation*. Durham, NC: Duke University Press.

Obama, Barack. 1995. *Dreams from My Father: A Story of Race and Inheritance*. New York: Times.

Oguma, Eiji. 2013. "Era." In *To See Once More the Stars: Living in a Post-Fukushima World*, edited by Daisuke Naito, Ryan Sayre, Heather Swanson, and Satsuki Takahashi, 240–43. Santa Cruz, CA: New Pacific.

Okorafor, Nnedi. 2014. *Lagoon*. London: Hodder and Stoughton.

Oldenziel, Ruth. 2011. "Islands: The United States as a Networked Empire." In *Entangled Geographies: Empire and Technopolitics in the Global Cold War*, edited by Gabrielle Hecht, 13–42. Cambridge, MA: MIT Press.

"On Creeds." 1809. Editorial. *Panoplist, and Missionary Magazine United*, 298–305.

O'Neill, Kevin Lewis. 2018. "On the Importance of Wolves." *Cultural Anthropology* 33, no. 3: 499–520.

Onneweer, Maarten. 2006. "An Ecology of Prospects: Time and Nature in Dutch Landscapes." *Etnofoor* 19, no. 2: 23–46.

Open University Course Team. 1994. *Waves, Tides, and Shallow-Water Processes*. Oxford: Butterworth-Heinemann.

Opperman, Serpil. 2019. "Storied Seas and Living Metaphors in the Blue Humanities." *Configurations* 27, no. 4: 443–61.

Oreskes, Naomi. 1996. "Objectivity or Heroism? On the Invisibility of Women in Science." *Osiris* 11: 87–113.

Oreskes, Naomi. 2000. "'*Laissez-tomber*': Military Patronage and Women's Work in Mid-20th-Century Oceanography." *Historical Studies in the Physical and Biological Sciences* 30, no. 2: 373–92.

Oreskes, Naomi. 2003. "A Context of Motivation: U.S. Navy Oceanographic Research and the Discovery of Sea-Floor Hydrothermal Vents." *Social Studies of Science* 33, no. 5: 697–742.

Oreskes, Naomi. 2020. *Science on a Mission: How Military Funding Shaped What We Do and Don't Know about the Ocean.* Chicago: University of Chicago Press.

Ortberg, Daniel Mallory. 2016. "Women Inexplicably Aroused by Waves in Art." *Toast*, January 25. http://the-toast.net/2016/01/25/women-inexplicably-aroused-by-waves-in-art.

Ortiz-Suslow, David G., Brian K. Haus, Sanchit Mehta, and Nathan J. M. Laxague. 2016. "Sea Spray Generation in Very High Winds." *Journal of Atmospheric Sciences* 73, no. 10: 3975–95.

Ozeki, Ruth. 2013. *A Tale for the Time Being*. New York: Penguin.

Palmer, Dexter. 2017. *Version Control*. New York: Vintage.

Pandian, Anand. 2015. *Reel World: An Anthropology of Creation*. Durham, NC: Duke University Press.

Pantti, Mervi. 2013. "Getting Closer? Encounters of the National Media with Global Images." *Journalism Studies* 14, no. 2: 201–18.

Parker, Bruce. 2012. *The Power of the Sea: Tsunamis, Storm Surges, Rogue Waves, and Our Quest to Predict Disasters*. New York: St. Martin's Griffin.

Parks, Lisa. 2014. "Drones, Infrared Imagery, and Body Heat." *International Journal of Communication* 8: 2518–21.

Parks, Lisa, and Nicole Starosielski, eds. 2015. *Signal Traffic: Critical Studies of Media Infrastructures*. Urbana: University of Illinois Press.

Parikka, Jussi. 2014. *The Anthrobscene*. Minneapolis: University of Minnesota Press.

Parry, Richard Lloyd. 2017. *Ghosts of the Tsunami: Death and Life in Japan's Disaster Zone*. New York: MCD.

Parsons, Keith, and Robert Zaballa. 2017. *Bombing the Marshall Islands: A Cold War Tragedy*. Cambridge: Cambridge University Press.

Parvin, Nassim, and Anne Pollock. 2020. "Unintended by Design: On the Political Uses of 'Unintended Consequences.'" *Engaging Science, Technology, and Society* 6: 320–27.

Patton, Kimberly C. 2007. *The Sea Can Wash Away All Evils: Modern Marine Pollution and the Ancient Cathartic Ocean*. New York: Columbia University Press.

Pattullo, June, Walter Munk, Roger Revelle, and Elizabeth Strong. 1955. "The Seasonal Oscillation in Sea Level." *Journal of Maritime Research* 14, no. 1: 88–155.

Pauly, Philip J. 2000. *Biologists and the Promise of American Life: From Meriwether Lewis to Alfred Kinsey*. Princeton, NJ: Princeton University Press.

Pauwelussen, Annet, and G. M. Verschoor. 2017. "Amphibious Encounters: Coral and People in Conservation Outreach in Indonesia." *Engaging Science, Technology, and Society* 3: 292–314.

Paz, Octavio. (1949) 1976. "Mi vida con la ola/My Life with the Wave" In *Eagle or Sun?* by Octavio Paz. Translated from the Spanish by Eliot Weinberger, 44–53. New York: J. Laughlin.

Pearson, Kent. 1979. *Surfing Subcultures of Australia and New Zealand*. St. Lucia, Australia: University of Queensland Press.

Pearson, Michael. 2003. *The Indian Ocean*. London: Routledge.

Pearson, Michael. 2006. "Littoral Society: The Concept and the Problems." *Journal of World History* 17, no. 4: 353–73.

Pearson, Michael, ed. 2015. *Trade, Circulation, and Flow in the Indian Ocean World*. New York: Palgrave.

Pereira, Clifford J. 2012. "Zheng He and the African Horizon: An Investigative Study into the Chinese Geography of Early Fifteenth-Century Eastern Africa." In *Zheng He and the Afro-Asian World*, edited by Chia Lin Sien and Sally K. Church, 248–79. Melaka, Malaysia: Perbadanan Muzium.

Peters, John Durham. 2015. *The Marvelous Clouds: Towards a Philosophy of Elemental Media*. Chicago: University of Chicago Press.

Petryna, Adrianna. 2022. *Horizon Work: At the Edges of Knowledge in an Age of Runaway Climate Change*. Princeton, NJ: Princeton University Press.

Pezman, Steve. 1978. "Interview: Dr. Timothy Leary: The Evolutionary Surfer." *Surfer*, no. 5, January, 100–103.

Philip, M. NourbeSe, as told to the author by Setaey Adamu Boateng. 2008. *Zong!* Middletown, CT: Wesleyan University Press.

Philippopoulos-Mihalopoulos, Andreas. 2022. "We Are All Complicit: Performing Law through Wavewriting." In *Laws of the Sea: Interdisciplinary Currents*, edited by Irus Braverman, 283–93. London:Routledge.

Pierson, Ryan. 2015. "Whole-Screen Metamorphosis and the Imagined Camera (Notes on Perspectival Movement in Animation)." *Animation* 10, no. 1: 6–21.

Pierson, Willard J., and Wilbur Marks. 1952. "The Power-Spectrum Analysis of Ocean-Wave Records." *Transactions of the American Geophysical Union* 33, no. 6: 834–44.

Pinet, Paul. 2009. *Invitation to Oceanography*. 5th ed. Sudbury, MA: Jones and Bartlett.

Pinkel, Robert. 1975. "Upper Ocean Internal Wave Observations from Flip." *Journal of Geophysical Research-Oceans and Atmospheres* 80, no. 27: 3892–3910.

Pinkel, Robert. 2016. "A Brief Review of Doppler Sonar Development at Scripps." *Methods in Oceanography* 17: 252–63.

Pinkel, Robert, Matthew Alford, Andrew J. Lucas, and Shaun Johnston et al. 2015. "Breaking Internal Tides Keep the Ocean in Balance." *Eos* (November 17). https://eos.org/project-updates/breaking-internal-tides-keep-the-ocean-in-balance.

Pinkel, Robert, Albert Plueddemann, and Robin Williams. 1987. "Internal Wave Observations from FLIP in MILDEX." *Journal of Physical Oceanography* 17, no. 10: 1737–57.

Pirie, David. 1996. "Wave Theory." *Sight and Sound* 6, no. 9: 26–27.

Pizzo, Nick. 2017. "Surfing Surface Gravity Waves." *Journal of Fluid Mechanics* 823: 316–28.

Pizzo, Nick, Luc Deike, and Alex Ayet. 2021. "How Does the Wind Generate Waves?" *Physics Today* 74, no. 11: 38–43.
Pollock, Anne. 2008. "The Internal Cardiac Defibrillator." In *The Inner History of Devices*, edited by Sherry Turkle, 98–111. Cambridge, MA: MIT Press.
Pombo, Pedro. 2018. "Soaking Cartographies: Of Waters, Landscapes and Materialities." In *Monsoon [+ other] Waters*, edited by Lindsay Bremner, 127–38. London: Monsoon Assemblages.
Porter, Roy. 1980. "The Terraqueous Globe." In *The Ferment of Knowledge: Studies in the Historiography of Eighteenth-Century Science*, edited by George Sebastian Rousseau and Roy Porter, 285–324. Cambridge: Cambridge University Press.
Povinelli, Elizabeth A. 2016. *Geontologies: A Requiem to Late Liberalism*. Durham, NC: Duke University Press.
Pretor-Pinney, Gavin. 2010. *The Wave Watcher's Companion: From Ocean Waves to Light Waves via Shock Waves, Stadium Waves, and All the Rest of Life's Undulations*. New York: Perigee.
Prince, Stephen. 1996. "True Lies: Perceptual Realism, Digital Images, and Film Theory." *Film Quarterly* 49, no. 3: 27–37.
Pritchard, Sara B. 2011. *Confluence: The Nature of Technology and the Remaking of the Rhône*. Cambridge, MA: Harvard University Press.
Probyn, Elspeth. 2021. "Doing Cultural Studies in Rough Seas: the COVID-19 Ocean Multiple." *Cultural Studies* 35, nos. 2–3: 557–71.
Prodanovich, Todd. 2020. "Whose Coast Are You Surfing in San Diego? A Look at the Indigenous History of One of California's Most Iconic Surf Zones." *Surfer Magazine*, August 17. https://www.surfer.com/features/whose-coast-are-you-surfing-in-san-diego/.
Quinn, Judy. 2014. "Mythologizing the Sea: The Nordic Sea-Deity Rán." In *Nordic Mythologies: Interpretations, Intersections, and Institutions*, edited by Timothy R. Tangherlini, 71–99. Berkeley, CA: North Pinehurst.
Raban, Jonathan. 2010. "The Waves." In *Driving Home: An American Journey*, by Jonathan Raban, 159–64. New York: Pantheon.
Raffles, Hugh. 2002. *In Amazonia: A Natural History*. Princeton, NJ: Princeton University Press.
Lord Raglan. 1960. "Canute and the Waves." *Man* 60, no. 4: 7–8.
Rahman, Ashiqur, Munsur Rahman, Shampa, and Anisul Haque et al. 2020. "Performance Evaluation of Bandal-like Structures for Sediment Management in Braided Jamuna River." *Proceedings of the 22nd IAHR-APD Congress 2020, Sapporo, Japan*, September 15–16. https://iahr-apd2020.eng.hokudai.ac.jp/htdocs/static/mirror/proceedings/pdf/1-1-7.pdf.
Rahman, Munsur, Tuhin Ghosh, Mashfiqus Salehin, and Amit Ghosh et al. 2020. "Ganges-Brahmaputra-Meghna Delta, Bangladesh and India: A Transnational Mega-Delta." In *Deltas in the Anthropocene*, edited by Robert J. Nicholls, W. Neil Adger, Craig W. Hutton, and Susan E. Hanson, 23–50. Cham, Switzerland: Springer Nature.
Rainger, Ronald. 2004. "'A Wonderful Oceanographic Tool': The Atomic Bomb, Radioactivity and the Development of American Oceanography." In *The*

Machine in Neptune's Garden: Historical Perspectives on Technology and the Marine Environment, edited by Helen Rozwadowski and David K. van Keuren, 96–132. Sagamore Beach, MA: Science History.
Raitt, Helen. 1956. *Exploring the Deep Pacific*. New York. W. W. Norton.
Raitt, Helen, and Beatrice Moulton. 1967. *Scripps Institution of Oceanography: First Fifty Years*. Los Angeles: Ward Ritchie Press.
Rankin, Bill. 2016. *After the Map: Cartography, Navigation, and the Transformation of Territory in the Twentieth Century*. Chicago: University of Chicago Press.
Rao, R. Prasada and Eugene C. LaFond. 1958. "The End of a Wave." *Current Science* 27, no. 11: 431–32.
Ray, T. K. 1986. "Ocean Waves and Their Response to Monsoon." *Mausum (मौसम)* 37, no. 2: 203–6.
Read, Richard. 2015. "*The New Yorker* Earthquake Article Unleashes Tsunami of Social Media: 8 Takeaways." *Oregon Live/The Oregonian*, July 15. https://www.oregonlive.com/pacific-northwest-news/2015/07/the_new_yorker_lets_cascadia_r.html.
Reguero, Borja G., Iñigo J. Losada, and Fernando J. Méndez. 2019. "A Recent Increase in Global Wave Power as a Consequence of Oceanic Warming." *Nature Communications* 10. https://doi.org/10.1038.
Rehbock, Philip F. 1979. "The Early Dredgers: 'Naturalizing' in British Seas, 1830–1850." *Journal of the History of Biology* 12, no. 2: 293–368.
Reidy, Michael S. 2008. *Tides of History: Ocean Science and Her Majesty's Navy*. Chicago: University of Chicago Press.
Reidy, Michael S., and Helen M. Rozwadowski. 2014. "The Spaces in Between: Science, Ocean, Empire." *Isis* 105, no. 2: 338–51.
"Remembering Deer Island: A Cause Worthy of Nipmuc Support." 1994. Editorial, *Nipmucspohke Newsletter*, vol. 1, no. 2, 3.
"The Revolutionary AuthaGraph Projection." 2017. *Decolonial Atlas* (blog), March 11. https://decolonialatlas.wordpress.com/2017/03/11/authagraph-world-maps.
Rezaie, Ali Mohammad, Celso Moller Ferreira, and Mohammad Rezaur Rahman. 2019. "Storm Surge and Sea Level Rise: Threat to the Coastal Areas of Bangladesh." In *Extreme Hydroclimatic Events and Multivariate Hazards in a Changing Environment*, edited by Viviana Maggioni and Christian Massari, 317–42. Amsterdam: Elsevier.
Rheinberger, Hans-Jörg. 1997. *Toward a History of Epistemic Things: Synthesizing Proteins in the Test Tube*. Stanford, CA: Stanford University Press.
Rhodan, Maya. 2016. "Donald Trump Calls Climate Change a Hoax, but Worries It Could Hurt His Golf Course." *Time*, May 23. http://time.com/4345367/donald-trump-climate-change-golf-course.
Rijser, Christy. 2016. "De Waterwolf is verslagen." *Die Uitkomst*, January 19.
Riles, Annelise. 2004. "Real Time: Unwinding Technocratic and Anthropological Knowledge." *American Ethnologist* 31, no. 3: 392–405.
Robinson, Kim Stanley. 2017. *New York 2140*. New York: Orbit.
Rodgers, Tara. 2016. "Toward a Feminist Epistemology of Sound: Refiguring Waves in Audio-technological Discourse." In *Engaging the World: Thinking after Iriga-*

ray, edited by Mary Rawlinson, 195–214. Albany: State University of New York Press.

Rodriguez, Stanley, Martha Rodriguez, and Elizabeth Newsome. 2021. "Kelp Road: Kumeyaay Oceanways." *Oceanographic Art and Science: Navigating the Pacific*. https://www.graphicocean.org/work-1/kelp-road.

Roelens, Nathalie. 2019. "The Wave as an Epistemological Challenge Today." Paper presented at Water Logics, Tulane University, New Orleans, April 11–12.

Rogers, W. Erick, and K. Todd Holland. 2009. "A Study of Dissipation of Wind-waves by Mud at Cassino Beach, Brazil: Prediction and Inversion." *Continental Shelf Research* 29, no. 3: 676–90.

Rooijendijk, Cordula. 2009. *Waterwolven: Een geschiedenis van stormvloeden, dijkenbouwers en droogmakers*. Amsterdam: L. J. Veen.

Roosth, Sophia. 2022. "The Sultan and the Golden Spike; or, What Stratigraphers Can Teach Us about Temporality." *Critical Inquiry* 48, no. 4: 697–720.

Rosenthal, Wolfgang, and Susan Lehner. 2008. "Rogue Waves: Results of the Max-Wave Project." *Journal of Offshore Mechanical and Architectural Engineering* 130, no. 2: 21006–13.

Ross, John. 2014. *The Forecast for D-Day and the Weatherman behind Ike's Greatest Gamble*. Guilford, CT: Lyons.

Ross, Nancy Wilson. 1943. *The WAVES: The Story of the Girls in Blue*. New York: Henry Holt.

Rotman, Brian. 2000. *Mathematics as Sign: Writing, Imagining, Counting*. Stanford, CA: Stanford University Press.

Rozwadowski, Helen. 2002. *The Sea Knows No Boundaries: A Century of Marine Science under ICES*. Seattle: University of Washington Press.

Rozwadowski, Helen. 2004. "Turning Heads: FLIP and the Technological Imagination in Postwar Oceanography." Paper presented at the annual meetings of the History of Science Society, Austin, November 18–21.

Rozwadowksi, Helen. 2005. *Fathoming the Ocean: The Discovery and Exploration of the Deep Sea*. Cambridge, MA: Harvard University Press.

Rozwadowski, Helen. 2010a. "Playing by—and on and under—the Sea: The Importance of Play for Knowing the Ocean." In *Knowing Global Environments: New Historical Perspectives on the Field Sciences*, edited by Jeremy Vetter, 162–89. New Brunswick, NJ: Rutgers University Press.

Rozwadowksi, Helen. 2010b. "Ocean's Depths." *Environmental History* 15: 520–525.

Rozwadowksi, Helen. 2019. *Vast Expanses: A History of the Oceans*. London: Reaktion.

Rozwadowski, Helen M., and David K. van Keuren, eds. 2004. *The Machine in Neptune's Garden: Historical Perspectives on Technology and the Marine Environment*. Sagamore Beach, MA: Science History.

Rueben, M., R. Holman, D. Cox, S. Shin, et al. 2011. "Optical Measurements of Tsunami Inundation through an Urban Waterfront Modeled in a Large-Scale Laboratory Basin." *Coastal Engineering* 58, no. 3: 229–38.

Ruskin, John. 1851. *The Stones of Venice. Volume the First*. London: Smith, Elder & Co.

Russi, Daniela, Patrick ten Brink, Andrew Farmer, Tomas Badura, et al. 2013. *The Economics of Ecosystems and Biodiversity for Water and Wetlands*. London: Institute for European Environmental Policy.

Rust, Stephen. 2013. "Hollywood and Climate Change." In *Ecocinema Theory and Practice*, edited by Stephen Rust, Salma Monani, and Sean Cubitt, 191–211. New York: Routledge.

Saket, Arvin. 2014. "Evaluation of ECMWF Wind Data for Wave Hindcast in Chabahar Zone." Presented at KOZWaves: Kiwi-Oz Waves Conference: First International Australasian Conference on Wave Science, Newcastle, Australia, February 17–19.

Saldanha, Arun. 2007. *Psychedelic White: Goa Trance and the Viscosity of Race*. Minneapolis: University of Minnesota Press.

Sammler, Katherine G. 2019. "The Rising Politics of Sea Level: Demarcating Territory in a Vertically Relative World." *Territory, Politics, Governance* 8, no. 5: 606–20.

Samuels, Richard J. 2013. *3.11: Disaster and Change in Japan*. Ithaca, NY: Cornell University Press.

Samuelson, Meg, and Charne Lavery. 2019. "The Oceanic South." *English Language Notes* 57, no. 1: 37–50.

Santiago-Fandiño, Vicente, Yevgeniy Kontar, and Yoshiyuki Kaneda, eds. 2015. *Post-Tsunami Hazard: Reconstruction and Restoration*. Dordrecht: Springer.

Santos, Boaventura de Sousa. 2018. *The End of the Cognitive Empire: The Coming of Age of Epistemologies of the South*. Durham, NC: Duke University Press.

Sanyal, Debarati. 2021. "Race, Migration, and Security at the Euro-African Border." *Theory & Event* 24, no. 1: 324–55.

Saussure, Ferdinand de. (1915) 1959. *Course in General Linguistics*. Translated from the French by Wade Baskin. New York: Philosophical Library.

Save the Waves. 2016. "#NatureTrumpsWalls: Stop Trump's Irish Wall." Save the Waves. https://www.savethewaves.org/nature-trumps-walls/.

Savelyev, Ivan, and Julian Fuchs. 2018. "Stereo Thermal Marking Velocimetry." *Frontiers in Mechanical Engineering* 4 (February). https://doi.org/10.3389/fmech.2018.00001.

Scammacca, Pietro, and Daniela Zyman. 2020. "The Indian Ocean Contact Zone." Ocean-Archive.org. https://ocean-archive.org/collection/119.

Scanlan, Emma, and Janet Wilson. 2018. "Pacific Waves: Reverberations from Oceania." *Journal of Postcolonial Writing* 54, no. 5: 577–84.

Scaramelli, Caterina. 2019. "The Delta Is Dead: Moral Ecologies of Infrastructure in Turkey." *Cultural Anthropology* 34, no. 3: 388–416.

Scaramelli, Caterina. 2023. "Swamps and Dragons." In *Swamps and the New Imagination: On the Future of Cohabitation in Art, Architecture, and Philosophy*, edited by Nomeda Urbonas, Gediminas Urbonas, and Kristupas Sabolius. Cambridge, MA: MIT Press.

Schama, Simon. 1987. *An Embarrassment of Riches: An Interpretation of Dutch Culture in the Golden Age*. New York: Alfred A. Knopf.

Schild, Jan. 2015. *Leven met het water*. Marken, Netherlands: Stichting Mooi Marken.
Schlee, Susan. 1973. *The Edge of an Unfamiliar World: A History of Oceanography*. New York: E. P. Dutton.
Schilling, Mark. 2015. "In the Cinematic Wake of the Fukushima Nuclear Disaster." *Japan Times*, March 4. https://www.japantimes.co.jp/culture/2015/03/04/films/cinematic-wake-fukushima-nuclear-disaster.
de Schipper, Matthieu A., Sierd de Vries, Gerben Ruessink, and Roeland C. de Zeeuw et al. 2016. "Initial Spreading of a Mega Feeder Nourishment: Observations of the Sand Engine Pilot Project." *Coastal Engineering* 111: 23–38.
Schlangher, Zoë. 2017. "Rogue Scientists Race to Save Climate Data from Trump." *Wired*, January 19. https://www.wired.com/2017/01/rogue-scientists-race-save-climate-data-trump.
Schmitt, Carl. (1950) 2003. *The* Nomos *of the Earth in the International Law of the* Jus Publicum Europaeum. Translated from the German by G. L. Ulmen. New York: Telos.
Schulz, Kathryn. 2015. "The Really Big One." *New Yorker*, July 20, http://www.newyorker.com/magazine/2015/07/20/the-really-big-one.
Schwartz, Hillel. 2011. *Making Noise: From Babel to the Big Bang & Beyond*. New York: Zone Books.
Schweitzer, Albert. 1911. *J. S. Bach*. Translated from the German by Ernest Newman. Leipzig: Breitkopf and Härtel.
Scott, Michael. 2017. "Getting More Real with Wonder: An Afterword." *Journal of Religious and Political Practice* 3, no. 3: 212–29.
Seaver, Nick. 2015. "Bastard Algebra." In *Data, Now Bigger and Better*, edited by Tom Boellstorff and Bill Maurer, 27–46. Chicago: Prickly Paradigm.
Sedley, David. 2005. "Plato's Tsunami." *Hyperboreus* 11: 205–14.
Semedo, Alvaro, Ralf Weisse, Arno Behrens, Andreas Sterl, et al. 2013. "Projection of Global Wave Climate Change toward the End of the Twenty-First Century." *Journal of Climate* 26, no. 21: 8269–88.
Shange, Savannah. 2020. "Comments at *The Case for Letting Anthropology Burn? Race, Racism, and Its Reckoning in American Anthropology*." Webinar sponsored by the Wenner-Gren Foundation for Anthropological Research and the Department of Anthropology, University of California, Los Angeles September 23. https://vimeo.com/461947530.
Sharpe, Christina. 2016. *In the Wake: On Blackness and Being*. Durham, NC: Duke University Press.
Shaw, Rosalind. 1992. "'Nature,' 'Culture' and Disasters: Floods and Gender in Bangladesh." In *Bush Base, Forest Farm: Culture, Environment and Development*, edited by Elisabeth Croll and David Parkin, 200–17. London: Routledge.
Sheen, Ivonne. 2020. "Visions du Réel 2020: Purple Sea de Amel Alzakout y Khaled Abdulwahed." *Desistfilm*, September 5. https://desistfilm.com/visions-du-reel-2020-purple-sea-de-amel-alzakout-y-khaled-abdulwahed.

Shepard, Jim. 2009. "The Netherlands Lives with Water." In *McSweeney's 32: 2024 AD*, edited by Dave Eggers, 189–212. San Francisco: McSweeney's Quarterly Concern.
Sherrif, Abdul. 2010. *Dhow Cultures of the Indian Ocean: Cosmopolitanism, Commerce and Islam*. Oxford: Oxford University Press.
Shewry, Teresa. 2015. *Hope at Sea: Possible Ecologies in Oceanic Literature*. Minneapolis: University of Minnesota Press.
Shiga, John. 2019. "The Nuclear Sensorium: Cold War Nuclear Imperialism and Sensory Violence." *Canadian Journal of Law and Society/Revue Canadienne Droit et Société* 34, no. 2: 281–306.
Shockley, Evie. 2006. *A Half-Red Sea*. Durham, NC: Carolina Wren.
Shockley, Evie. 2011. "African American Poetic Innovation and the Middle Passage." *Contemporary Literature* 52, no. 4: 791–817.
Shor, Elizabeth Noble. 1978. *Scripps Institution of Oceanography: Probing the Oceans, 1936–1976*. San Diego, CA: Tofua.
Shuto, Nobuo, and Koji Fujima. 2009. "A Short History of Tsunami Research and Countermeasures in Japan." *Proceedings of the Japanese Academy, Series B: Physical and Biological Sciences* 85, no. 8: 267–75.
Siegel, Deborah L. 1997. "The Legacy of the Personal: Generating Theory in Feminism's Third Wave." *Hypatia* 12, no. 3: 46–75.
Siegel, Greg. 2014. *Forensic Media: Reconstructing Accidents in Accelerated Modernity*. Durham, NC: Duke University Press.
Siegert, Bernhard. 2014. "The *Chorein* of the Pirate: On the Origin of the Dutch Seascape." *Grey Room* 57: 6–23.
Sinnett, Gregory, and Falk Feddersen. 2014. "The Surf Zone Heat Budget: The Effect of Wave Heating." *Geophysical Research Letters* 41, no. 20: 7217–26.
Sivasundaram, Sujit. 2020. *Waves across the South: A New History of Revolution and Empire*. London: William Collins.
Sloterdijk, Peter. (2004) 2016. *Foams: Spheres, Volume 3: Plural Spherology*. Translated from the German by Wieland Hoban. Cambridge, MA: MIT Press.
Slovic, Scott. 2016. "Seasick among the Waves of Ecocriticism: An Inquiry into Alternative Historiographic Metaphors." In *Environmental Humanities: Voices from the Anthropocene*, edited by Serpil Oppermann and Serenella Iovino, 99–111. Lanham, MD: Rowman and Littlefield.
Smit, Wim A. 1995. "Science, Technology, and the Military: Relations in Transition." In *Handbook of Science and Technology Studies*, edited by Sheila Jasanoff, Gerald E. Markle, James C. Peterson, and Trevor Pinch, 598–626. Thousand Oaks, CA: Sage.
Snider, Robert G. 1961. "The International Indian Ocean Expedition 1959–1964." *Discovery* (March): 114–17.
Snodgrass, F. E., G. W. Groves, K. F. Hasselmann, and G. R. Miller et al. 1966. "Propagation of Ocean Swell across the Pacific." *Philosophical Transactions of the Royal Society of London. Series A, Mathematical and Physical Sciences* 259, no. 1103: 431–97.
Sobchack, Vivian, ed. 2000. *Meta-morphing: Visual Transformation and the Culture of Quick-Change*. Minneapolis: University of Minnesota Press.

Sobczyk, Nick. 2020. "NOAA Watchdog Chides Agency for How It Handled Hurricane Dorian's 'Sharpiegate.'" *Science Insider*, July 10. https://www.science.org/content/article/noaa-watchdog-chides-agency-how-it-handled-hurricane-dorian-s-sharpiegate.

Sobel, Dava. 1995. *Longitude: The True Story of a Lone Genius Who Solved the Greatest Scientific Problem of His Time*. New York: Walker.

Søland, Birgitte. 2000. *Becoming Modern: Young Women and the Reconstruction of Womanhood in the 1920s*. Princeton, NJ: Princeton University Press.

Spencer, Jane. 2007. "Afterword: Feminist Waves." In *Third Wave Feminism: A Critical Exploration*, 2d ed., edited by Stacy Gillis, Gillian Howie, and Rebecca Munford, 298–303. New York: Palgrave Macmillan.

Spigel, Lynn. 2004. "Theorizing the Bachelorette: 'Waves' of Feminist Media Studies." *Signs* 30, no. 1: 1209–21.

Spillers, Hortense J. 1987. "Mama's Baby, Papa's Maybe: An American Grammar Book." *Diacritics* 17, no. 2: 64–81.

Spitz, Julia. 2010. "Nipmucs Add History to Memorial to Deer Island Internment." *Metro West Daily News*, October 24.

Spivak, Gayatri Chakravorty. 2003. *Death of a Discipline*. New York: Columbia University Press.

Srinivas, Tulasi. 2018. *The Cow in the Elevator: An Anthropology of Wonder*. Durham, NC: Duke University Press.

Starosielski, Nicole. 2013. "Beyond Fluidity: A Cultural History of Cinema under Water." In *Ecocinema Theory and Practice*, edited by Stephen Rust, Salma Monani, and Sean Cubitt, 149–68.

Starosielski, Nicole. 2015. *The Undersea Network*. Durham, NC: Duke University Press.

Starosielski, Nicole. 2019. "Infrared." *Environmental Planning D: Society and Space*. https://www.societyandspace.org/articles/infrared.

Steenhuis, Marinke, Lara Voerman, Marlies Noyes, and Joost Emmerik. 2015. *Waterloopkundig Laboratorium: Cultuurhistorische duiding, ruimtelijke analyse en essentiële principes*. Schiedam, Netherlands: SteenhuisMeurs.

Steinberg, Philip E. 2001. *The Social Construction of the Ocean*. Cambridge: Cambridge University Press.

Steinberg, Philip E. 2013. "Of Other Seas: Metaphors and Materialities in Maritime Regions." *Atlantic Studies* 10, no. 2: 156–69.

Steinberg, Philip E., and Kimberly Peters. 2015. "Wet Ontologies, Fluid Spaces: Giving Depth to Volume through Oceanic Thinking." *Environment and Planning D: Society and Space* 33, no. 2: 247–64.

Steinhardt, Stephanie. 2018. "The Instrumented Ocean: How Sensors, Satellites, and Seafloor-Walking Robots Changed What It Means to Study the Sea." PhD diss., Cornell University, Ithaca, NY.

Sterman, Nan, and Will Gullette. 2005. "Munks' Folly: Celebrated Couple's Garden and Sunken Treasure." *San Diego Home/Garden Lifestyles*, August, 62–65, 86–90.

Stewart, Kathleen. 2011. "Atmospheric Attunements." *Environment and Planning D: Society and Space* 29, no. 3: 445–53.

Stoler, Ann Laura. 2001. "Tense and Tender Ties: The Politics of Comparison in North American History and (Post) Colonial Studies." *Journal of American History* 88, no. 3: 829–65.
Stoker, Bram. 1897. *Dracula*. Westminster, UK: Archibald Constable.
Stokes, George. 1847. "On the Theory of Oscillatory Waves." *Transactions of the Cambridge Philosophical Society* 8, no. 287: 441–55.
Stokes, M. D., G. B. Deane, K. Prather, and T. H. Bertram, et al. 2013. "A Marine Aerosol Reference Tank System as a Breaking Wave Analogue for the Production of Foam and Sea-Spray Aerosols." *Atmospheric Measurement Techniques* 6: 1085–94.
Stranger, Mark. 1999. "The Aesthetics of Risk: A Study of Surfing." *International Review for the Sociology of Sport* 34, no. 3: 265–76.
Strathern, Marilyn. 1992. *After Nature: Kinship in the Late Twentieth Century*. Cambridge: Cambridge University Press.
Sturken, Marita, and Simon Leung. 2005. "Displaced Bodies in Residual Spaces." *Public Culture* 17, no. 1: 129–52.
Sturluson, Snorri. (ca. 1200) 1916. *The Prose Edda*. Translated from the Icelandic by Arthur Gilchrist Brodeur. New York: American-Scandinavian Foundation.
Subramaniam, Banu. 2001. "The Aliens Have Landed! Reflections on the Rhetoric of Biological Invasions." *Meridians: Feminism, Race, Transnationalism* 21, no. 1: 26–40.
Subramanian, Ajantha. 2009. *Shorelines: Space and Rights in South India*. Stanford, CA: Stanford University Press.
Subramanian, Lakshmi. 2010. "Commerce, Circulation, and Consumption: Indian Ocean Communities in Historical Perspective." In *Indian Ocean Studies: Cultural, Social, and Political Perspectives*, edited by Shanti Moorthy and Ashraf Jamal, 136–57. London: Routledge.
Sugden, Edward. 2018. *Emergent Worlds: Alternative States in Nineteenth-Century American Culture*. New York: New York University Press.
Supper, Alexandra. 2015. "Data Karaoke: Sensory and Bodily Skills in Conference Presentations." *Science as Culture* 24, no. 4: 436–57.
Sutherland, P., and W. K. Melville. 2013. "Field Measurements and Scaling of Ocean Surface Wave-Breaking Statistics." *Geophysical Research Letters* 40: 3074–79.
Sverdrup, Harald, Martin W. Johnson, and Richard H. Fleming. 1942. *The Oceans: Their Physics, Chemistry, and General Biology*. New York: Prentice-Hall.
Sverdrup, Harald, and Walter Munk. 1943. *Wind, Waves and Swell: A Basic Theory for Forecasting*. La Jolla, CA: Scripps Institution of Oceanography.
SWAMP Group. 1985. *Ocean Wave Modeling*. New York: Plenum.
Swapna, P., J. Jyoti, R. Krisgnan, N. Sandeep, and S. M. Griffies. 2017. "Multidecadal Weakening of Indian Summer Monsoon Circulation Induces an Increasing Northern Indian Ocean Sea Level." *Geophysical Research Letters* 44, no. 20: 10560–72.
TallBear, Kim. 2011. "Why Interspecies Thinking Needs Indigenous Standpoints." *Fieldsights*, November 18. https://culanth.org/fieldsights/why-interspecies-thinking-needs-indigenous-standpoints.
Taussig, Karen-Sue. 2009. *Ordinary Genomes: Science, Citizenship, and Genetic Identities*. Durham, NC: Duke University Press.

Taylor, Ula Y. 1998. "Making Waves: The Theory and Practice of Black Feminism." *Black Scholar* 28, no. 2: 18–28.

Taylor, G. I. 1938. "The Spectrum of Turbulence." *Proceedings of the Royal Society of London, Series A: Mathematical and Physical Sciences* 164: 476–90.

TeBrake, William H. 2002. "Taming the Waterwolf: Hydraulic Engineering and Water Management in the Netherlands during the Middle Ages." *Technology and Culture* 43, no. 3: 475–99.

Temmerman, Stijn, Patrick Meire, Tjeerd J. Bouma, Peter M. J. Herman, et al. 2013. "Ecosystem-based Coastal Defence in the Face of Global Change." *Nature* 504: 79–83.

ten Bos, René. 2009. "Towards an Amphibious Anthropology: Water and Peter Sloterdijk." *Environment and Planning D: Society and Space* 27, no. 9: 73–86.

Theweleit, Klaus. 1987. *Male Fantasies, Volume 1: Women, Floods, Bodies, History*. Minneapolis: University of Minnesota Press.

Thompson, Rachel. 2018. "A Dutch Garuda to Save Jakarta? Excavating the NCICD Master Plan's Socio-environmental Conditions of Possibility." In *Jakarta: Claiming Spaces and Rights in the City*, edited by Jörgen Hellman, Marie Thynell, and Roanne van Voorst, 138–56. London: Routledge.

Thorade, Henri. 1931. *Probleme der Wasserwellen*. Hamburg: Henri Grand.

Thornberry-Ehrlich, Trista L. 2017. *Boston Harbor Islands National Recreation Area: Geologic Resources Inventory Report*. Natural Resource Report NPS/NRSS/GRD/NRR—2017/1404. Fort Collins, CO: National Park Service, US Department of the Interior.

Thorpe, Steve. 2010. "Internal Waves and All That." In *Of Seas and Ships and Scientists: The Remarkable Story of the UK's National Institute of Oceanography*, edited by Anthony Laughton, John Gould, 'Tom' Tucker, and Howard Roe, 140–47. Cambridge: Lutterworth.

Tierney Kathleen, Christine Bevc, and Erica Kuligowski. 2006. "Metaphors Matter: Disaster Myths, Media Frames, and Their Consequences in Hurricane Katrina." *Annals of the American Academy of Political and Social Science* 604, no. 1: 57–81.

Tingley, Kim. 2016. "The Secrets of the Wave Pilots." *New York Times*, March 17. https://www.nytimes.com/2016/03/20/magazine/the-secrets-of-the-wave-pilots.html.

Todd, Zoe. 2015. "Indigenizing the Anthropocene." In *Art in the Anthropocene: Encounters Among Aesthetics, Politics, Environments and Epistemologies*, edited by Heather Davis and Etienne Turpin, 241–54. London: Open Humanities Press.

Toffler, Alvin. 1980. *The Third Wave*. New York: William Morrow.

Tolman, Hendrik L. 1989. "The Numerical Model WAVEWATCH: A Third Generation Model for Hindcasting of Wind Waves on Tides in Shelf Seas." Communications on Hydraulic and Geotechnical Engineering, report no. 89–2. Department of Civil Engineering Group, Delft University of Technology, Netherlands.

Tolman, Hendrik L., and Robert W. Grumbine. 2013. "Holistic Genetic Optimization of a Generalized Multiple Discrete Interaction Approximation for Wind Waves." *Ocean Modelling* 70: 25–37.

Tompkins, Dave. 2017. "*Moonlight*'s Forgotten Frequencies: How Miami Bass, Ocean Waves, and Pirate Radio Shaped the Film." *MTV News*, February 23. https://www.mtv.com/news/mbfohf/moonlight-forgotten-frequencies.

Toop, David. 1995. *Ocean of Sound: Aether Talk, Ambient Sound and Imaginary Worlds*. London: Serpent's Tail.

Traufetter, Gerald. 2008. "The Dutch Simulate Their Demise." Translated by Christopher Sultan. *Der Spiegel*, November 14. http://m.spiegel.de/international/europe/a-590489.html.

Trautmann, Thomas R. 1992. "The Revolution in Ethnological Time." *Man* 27, no. 2: 379–97.

Traweek, Sharon. 1988. *Beamtimes and Lifetimes: The World of High Energy Physicists*. Cambridge, MA: Harvard University Press.

Trower, Shelly. 2019. "Peripheral Vibrations." In *Unsound: Undead*, edited by Steve Goodman, Toby Heys, and Eleni Ikoniadou, 19–22. Falmouth, UK: Urbanomic.

Tsing, Anna. 2012. "On Nonscalability: The Living World Is Not Amenable to Precision-Nested Scales." *Common Knowledge* 18, no. 3: 505–24.

Tsing, Anna, Andrew Mathews, and Nils Bubandt. 2019. "Patchy Anthropocene: Landscape Structure, Multispecies History, and the Retooling of Anthropology." *Current Anthropology* 60 (Supplement 20): S186–97.

Tsutsui, William M. 2013. "The Pelagic Empire: Reconsidering Japanese Expansion." In *Japan at Nature's Edge: The Environmental Context of a Global Power*, edited by Ian Jared Miller, Julia Adeney Thomas, and Brett L. Walker, 21–38. Honolulu: University of Hawai'i Press.

Tucker, Tom. 2010. "Applied Wave Research." In *Of Seas and Ships and Scientists: The Remarkable Story of the UK's National Institute of Oceanography*, edited by Anthony Laughton, John Gould, 'Tom' Tucker, and Howard Roe, 182–90. Cambridge: Lutterworth.

Tukey, John W., and Richard W. Hamming. 1949. "Measuring Noise Color 1." *Memorandum MM-49-110-119*. Murray Hill, NJ: Bell Telephone Laboratory.

Turner, Sharon. 1820. *History of the Anglo-Saxons from the Earliest Period to the Norman Conquest*, 3d ed. London: Longman, Hurst, Rees, Orme, and Brown.

Turner, Victor. 1966. *The Ritual Process: Structure and Anti-Structure*. Ithaca, NY: Cornell University Press.

Valentine, David. 2017. "Gravity Fixes: Habituating to the Human on Mars and Island Three." *HAU: Journal of Ethnographic Theory* 7, no. 3: 185–209.

van Boxsel, Matthijs. 2004. "Birth of the Water Wolf." *De Gids* 167, nos. 5–6: 30–36.

van den Vondel, Joost. 1641. "Aan den Leeuw van Holland." In *Provisionneel Concept Ontwerp ende Voorslach dienende tot de bedyckinge van de groote Water Meeren*. Amsterdam.

van der Cammen, Hans, and Len de Klerk. 2012. *The Selfmade Land: Culture and Evolution of Urban and Regional Planning in the Netherlands*. Antwerp: Spectrum.

van der Heiden, Henri J. L. 2019. "Modelling Viscous Effects in Offshore Flow Problems: A Numerical Study." PhD diss., University of Groningen, Groningen, Netherlands.

van der Molen, Karin, ed. 2012. *Drift: Art in Nature Laboratory*. Blankenham, Netherlands: Stichting ReRun Producties.
van de Stadt, Leontine. 2013. *Nederland in 7 Overstromingen*. Zuphen, Netherlands: Walburg Pers.
van Dijk, Tomas. 2012. "Sand Engine Quells the Coast's Hunger for Sand." *Delft Outlook* 1: 6–9.
van Gent, Marcel R. A. 2015. "The New Delta Flume for Large-Scale Testing." *Proceedings of the 36th IAHR World Congress (The Hague)*: 2811–20.
van Keuren, David K. 2000. "Building a New Foundation for the Ocean Sciences: The National Science Foundation and Oceanography, 1951–1965." *Earth Sciences History* 19, no. 1: 90–109.
van Looy, Rein. 1955. *Sijtje en het Klompenland*. Gorssel, Netherlands: Terra Nostra.
van Veen, Johan. 1962. *Dredge Drain Reclaim! The Art of a Nation*. The Hague: Martinus Nijhoff.
van Vledder, Gerbrant. 2006. "The WRT Method for the Computation of Non-linear Four-Wave Interactions in Discrete Spectral Wave Models." *Coastal Engineering* 53, no. 2-3: 223–42.
van Vledder, Gerbrant. 2017. "The Book of Waves by Dolph Kessler." In *The Wave: Crossing the Atlantic*, by Dolph Kessler, 97–101. Oentsjerk, Netherlands: Mauritsheech.
Vardy, Mark. 2020. "Relational Agility: Visualizing Near-Real-Time Arctic Sea Ice Data as a Proxy for Climate Change." *Social Studies of Science* 50, no. 5: 802–20.
Vaughn, Sarah E. 2017. "Disappearing Mangroves: The Epistemic Politics of Climate Adaptation in Guyana." *Cultural Anthropology* 32, no. 2: 242–68.
Vierlingh, Andries. (1579) 1920. *Tractaet van Dyckagie*. Den Haag and Rotterdam: Martinus Nijhoff, Nederlandse Vereniging Kust en Oeverwerken.
Vigoureux, P., and J. B. Hersey. (1962) 2005. "Sound in the Sea." In *The Sea: Ideas and Observations on Progress in the Study of the Seas, Volume 1: Physical Oceanography*, edited by M. N. Hill, 476–97. Cambridge, MA: Harvard University Press.
Vink, Markus P. M. 2007. "Indian Ocean Studies and the 'New Thalassology.'" *Journal of Global History* 2: 41–62.
Vissering, G. 1916. *De watervloed van 13–14 Januari 1916*. Leiden: E. J. Brill.
von Storch, Hans, and Klaus Hasselmann. 2010. *Seventy Years of Exploration in Oceanography: A Prolonged Weekend Discussion with Walter Munk*. Berlin: Springer.
Wagner, Roy. 2001. *An Anthropology of the Subject: Holographic Worldview in New Guinea and Its Meaning and Significance for the World of Anthropology*. Berkeley: University of California Press.
Waitt, Gordon. 2008. "Killing Waves: Surfing, Space, and Gender." *Social and Cultural Geography* 9, no. 1: 75–94.
Walcott, Derek. 1978. "The Sea Is History." *Paris Review* 74. https://www.theparisreview.org/poetry/7020/the-sea-is-history-derek-walcott.
Walker, Isaiah Helelkunihi. 2011. *Waves of Resistance: Surfing and History in Twentieth-Century Hawai'i*. Honolulu: University of Hawai'i Press.
Wang, Chunzai, Liping Zhang, Sang-Ki Lee, Lixin Wu, and Carlos R. Mechoso. 2014. "A Global Perspective on CMIP5 Climate Model Biases." *Nature Climate Change* 4: 201–5.

Wang, Qing, Denny P. Alappattu, Stephanie Billingsley, Byron Blomquist, et al. 2018. "CASPER: Coupled Air-Sea Processes and Electromagnetic Ducting Research." *Bulletin of the American Meteorological Society* 99, no. 7 (July): 1449–71.

Wang, Qing, Robert J. Burkholder, Caglar Yardim, and Denny P. Alappattu, et al. 2019. "Sampling Spatial-Temporal Variability of Electromagnetic Propagation in CASPER-West." Keynote talk at the Thirteenth European Conference on Antennas and Propagation, March 31–April 5, Krakow, Poland.

Waterhouse, Amy F., Jennifer A. MacKinnon, Ruth C. Musgrave, and Samuel M. Kelly, et al. 2017. "Internal Tide Convergence and Mixing in a Submarine Canyon." *Journal of Physical Oceanography* 47, no. 2: 303–22.

Waters, Hannah. 2012. "FLIP: The FLoating Instrument Platform." *Smithsonian Ocean*, https://ocean.si.edu/human-connections/exploration/flip-floating-instrument-platform.

Watts, Laura. 2018. *Energy at the End of the World: An Orkney Islands Saga*. Cambridge, MA: MIT Press.

Watts, Laura, and Brit Ross Winthereik. 2018. "Ocean Energy at the Edge." In *Ocean Energy: Governance Challenges for Wave and Tidal Stream Technologies*, edited by Glen Wright, Sandy Kerr, and Kate Johnson, 229–46. New York: Routledge.

Weiner, Melissa F., and Antonio Carmona Báez, eds. 2018. *Smash the Pillars: Decoloniality and the Imaginary of Color in the Dutch Kingdom*. Lanham, MD: Lexington.

Wekker, Gloria. 2016. *White Innocence: Paradoxes of Colonialism and Race*. Durham, NC: Duke University Press.

Wenzel, Jennifer. 2014. "Planet vs Globe." *English Language Notes* 52, no. 1: 19–30.

Wesselink, Anna J., Wiebe E. Bijker, Huib J. de Vriend, and Maarten S. Krol. 2007. "Dutch Dealings with the Delta." *Nature and Culture* 2, no. 2: 188–209.

Westcott, Kathryn. 2011. "Is King Canute Misunderstood?" *BBC News Magazine*, 26 May, http://www.bbc.com/news/magazine-13524677.

Weston, Kath. 2002. *Gender in Real Time: Power and Transience in a Visual Age*. New York: Routledge.

Weston, Kath. 2017. *Animate Planet: Making Visceral Sense of Living in a High-Tech, Ecologically Damaged World*. Durham, NC: Duke University Press.

Whissel, Kristen. 2014. *Spectacular Digital Effects: CGI and Contemporary Cinema*. Durham, NC: Duke University Press.

White, Richard. 1995. *The Organic Machine: The Remaking of the Columbia River*. New York: Hill and Wang.

Wielenga, Friso. (2012) 2015. *A History of the Netherlands: From the Sixteenth Century to the Present Day*. Translated by Lynne Richards. London: Bloomsbury.

Wilensky, Hiroko. 2012. "Disaster Symbols and Cultural Responses." *International Reports on Socio-Informatics* 9, no. 2: 67–73.

Willet, Julie A. 2000. *Permanent Waves: The Making of the American Beauty Shop*. New York: New York University Press.

Wills, Clair. 2000. "Let Eriny Remember." *Times Literary Supplement*, no. 5070, June 2, 6.

Wilts, Alexandra. 2017. "Hurricane Irma Has Become So Strong It's Showing Up on Seismometers Used to Detect Earthquakes." *Independent*, September 5. https://www.independent.co.uk/news/world/americas/irma-hurricane-strength-category-earthquake-measurement-seismometer-a7931286.html.

Winkler, Raimund. 1901. "On Sea Charts Formerly Used in the Marshall Islands, with Notices on the Navigation of these Islanders in General." In *Annual Report of the Smithsonian Institution for the Year Ending June 30, 1899* 54: 487–508.

Winsberg, Eric. 2010. *Science in the Age of Computer Simulation*. Chicago: University of Chicago Press.

Winthereik, Brit Ross. 2019. "Is ANT's Radical Empiricism Ethnographic?" In *The Routledge Companion to Actor-Network Theory*, edited by Anders Blok, Ignacio Farias, and Celia Roberts, 24–33. London: Routledge.

Wittgenstein, Ludwig. 1953. *Philosophical Investigations*. Translated from the German by G. E. M. Anscombe. New York: Macmillan.

Wolfe, Cary. 2009. *What Is Posthumanism?* Minneapolis: University of Minnesota Press.

Wolfenstein, Eugene Victor. 2000. *Inside/Outside Nietzsche: Psychoanalytic Explorations*. Ithaca, NY: Cornell University Press.

Wood, Marcus. 2000. *Blind Memory: Visual Representations of Slavery in England and America 1780–1865*. Manchester: Manchester University Press.

Woods, Austin. 2006. *Medium-range Weather Prediction: The European Approach*. New York: Springer Science and Business Media.

Woolf, Virginia. 1931. *The Waves*. London: Hogarth.

Wooster, Warren S. 1965. "Indian Ocean Expedition." *Science* 150: 290–92.

Woznik, Bear, with Lou Aronica. 2012. *Deep in the Wave: A Surfing Guide to the Soul*. New York: Hachette.

Wylie, Alison. 2006. "Afterword: On Waves." In *Feminist Anthropology: Past, Present, and Future*, edited by Pamela L. Geller and Miranda K. Stockett, 167–76. Philadelphia: University of Pennsylvania Press.

Wyman, Kristen. 2010. "Remembering Deer Island." Interviewed by J. Kēhaulani Kauanui, *Indigenous Politics: From Native New England and Beyond* radio show, http://www.indigenouspolitics.com.

Yeh, Harry, Shinji Sato, and Yoshimitsu Tajima. 2011. "Waveform Evolution of the 2011 East Japan Tsunami." In *Proceedings of the Sixth International Conference on Asian and Pacific Coasts*, edited by Joseph Hun-Wei Lee and Chiu-On Ng, 131–138. Singapore: World Scientific.

Yuill, Simon. 2008. "Concurrent Version System." In *Software Studies: A Lexicon*, edited by Matthew Fuller, 64–69. Cambridge, MA: MIT Press.

Yusoff, Kathryn. 2018. *A Billion Black Anthropocenes or None*. Minneapolis: University of Minnesota Press.

Zalasiewicz, Jan, and Mark Williams. 2011. "The Anthropocene Ocean in Its Deep Time Context." In *The World Ocean in Globalisation: Climate Change, Sustainable Fisheries, Biodiversity, Shipping, Regional Issues*, edited by Davor Vidas and Peter Johan Schei, 19–35. Leiden: Martinus Nijhoff.

Zee, Jerry C. 2017. "Holding Patterns: Sand and Political Time at China's Desert Shores." *Cultural Anthropology* 32, no. 2: 215–41.

Zirker, J. B. 2013. *The Science of Ocean Waves: Ripples, Tsunamis, and Stormy Seas.* Baltimore, MD: Johns Hopkins University Press.

Zylinska, Joanna. 2014. *Minimal Ethics in the Anthropocene*. Ann Arbor, MI: Open Humanities Press.

INDEX

Page locators in italics refer to figures.